"If any disease needs to be rethought, it is surely diabetes, and that is the premise of Gary Taubes's latest book. . . . Taubes's message is important and should be heard."   —*The Wall Street Journal*

"Given the prevalence of type 2 diabetes, health and science journalist [Gary] Taubes . . . offers a crucial reassessment of the proper diet for controlling the disease. . . . When it comes to the connection between diet and chronic illness, Taubes has consistently proven himself to be a superb investigator."
—*Booklist* (starred review)

"A must-read book for diabetics. . . . Doctors will also learn a lot."
—*Kirkus Reviews* (starred review)

"[A] provocative study. . . . Exhaustively researched and providing cautionary insight into the fallibilities of medical advice, this intrigues."   —*Publishers Weekly*

"Decades of research have brought us shifting and even contradictory recommendations on managing diabetes. In lucid prose, Taubes sifts through and critically reviews what is known. *Rethinking Diabetes* convincingly makes the case for a new paradigm for the prevention and treatment of this common disorder."
—Abraham Verghese, M.D., author of *The Covenant of Water*

"With the rigor and clarity that mark all his work, Gary Taubes tackles the history of the science of diabetes, and he reveals why and how much of the conventional wisdom is misguided—and, even more important, how a better approach is possible. This book is indispensable reading for anyone who wants to understand this disease."   —Gretchen Rubin,
author of *The Happiness Project*

"Gary Taubes has once again done a masterful job exposing the bias of those who promoted and entrenched high-carbohydrate, low-fat diets as part of the management of people with type 1 and type 2 diabetes, and the lack of evidence supporting this dietary approach. *Rethinking Diabetes* is an outstanding book, a must-read for all people with diabetes and those who treat them."

—Martin Abrahamson, M.D., former chief medical officer, Joslin Diabetes Center, and coauthor of *Conquer Your Diabetes*

"Kudos to Gary Taubes for writing a remarkable tour de force that should be of considerable interest to all persons with diabetes and clinicians and researchers in the field. The historical narrative is both erudite and gripping. A must-read."

—Sanjiv Chopra, M.D., professor of medicine, Harvard Medical School, and coauthor of *Conquer Your Diabetes*

GARY TAUBES
Rethinking Diabetes

Gary Taubes is the author of six books, including the best sellers *Good Calories, Bad Calories* and *Why We Get Fat*. He is a former staff writer for *Discover* and correspondent for *Science*, and his writing has appeared on the cover of *The New York Times Magazine* and in *The Atlantic*, *Esquire*, and numerous "best of" anthologies, including *The Best of the Best American Science Writing*. He has received three Science in Society Journalism Awards from the National Association of Science Writers and the Robert Wood Johnson Foundation Investigator Award in Health Policy Research. He lives in Oakland, California.

Rethinking Diabetes

Rethinking
Diabetes

What Science Reveals About Diet, Insulin,
and Successful Treatments

GARY TAUBES

VINTAGE BOOKS
A DIVISION OF PENGUIN RANDOM HOUSE LLC
NEW YORK

Published in the United States by Vintage Books, a division of Penguin Random House LLC, New York, and distributed in Canada by Penguin Random House Canada Limited, Toronto. Originally published in hardcover in the United States by Alfred A. Knopf, a division of Penguin Random House LLC, New York, in 2024.

Vintage and colophon are registered trademarks of Penguin Random House LLC.

Grateful acknowledgment is made to the following for permission to reprint previously published material:

Edwin L. Bierman and Margaret Albrink: Excerpt from "Principles of Nutrition and Dietary Recommendations for Patients with Diabetes Mellitus" by Edwin L. Bierman and Margaret Albrink. Originally published in *Diabetes* by American Diabetes Association on September 1, 1971. Permission conveyed through Copyright Clearance Center, Inc.

David Eddy: Excerpt from "Variations in Physician Practice: The Role of Uncertainty" by David Eddy. Originally published in *Health Affairs* (June 21, 1984). Permission conveyed through Copyright Clearance Center, Inc.

The Library of Congress has cataloged the Knopf edition as follows:
Name: Taubes, Gary, author.
Title: Rethinking diabetes: what science reveals about diet, insulin, and successful treatments / Gary Taubes.
Description: First edition. | New York : Alfred A. Knopf, 2024.
Identifiers: LCCN 2022055539 (print)
Classification: LCC RC 660 .T387 2024 (print) | LCC RC660 (ebook) | DDC 616.462—dc23
LC record available at https://lccn.loc.gov/2022055539

Vintage Books Trade Paperback ISBN: 978-0-525-43575-4
eBook ISBN: 978-0-525-52009-2

Book design by Maggie Hinders
Author photograph © Michael Lionstar

vintagebooks.com

Printed in the United States of America
1st Printing

To Sarah Hallberg

For showing us how to live and helping change the world

The history of diabetes is marked by the recurrence of certain ideas which rise, decline and disappear, only to make a new appearance and go through a similar cycle again in an altered form, and a new generation. This is notably true of trends in diet.

—ROLLIN WOODYATT,
"Round Table Conference on Diabetes Mellitus. Dietary Trends," 1934

Diet has always generated passion, and passion in science is an infallible marker of lack of evidence.

—LYNN SAWYER AND EDWIN GALE,
"Diet, Delusion and Diabetes," 2009

Contents

Rethinking Diabetes

Introduction

The effect of any diet is to be judged by those who follow it, not by those who break it.

—JOSLIN'S DIABETES MELLITUS,
eighth edition, 1946

Of all the chronic diseases that are likely to end our lives prematurely, none has been so intimately and surely linked to our diets as diabetes. Since its first unambiguous appearance in the medical literature, clearly diagnosed by Hindu physicians in the sixth century BCE, physicians have speculated on what aspects of our diet might cause diabetes and how it could be treated or cured by dietary modifications.

By the mid-nineteenth century, when the disease was still so rare that even the large urban hospitals of the era could go years without seeing a case, those few physicians who studied the disorder were identifying two mostly distinct variations. One was the form that appeared in middle age and most often in those who also suffered from obesity. It appeared to be a problem of "excessive formation," as George Harley of University College London suggested in his 1866 book on the subject. Later physicians would come to say it was a disease of overnutrition because of its intimate association with obesity.

The far less common form struck acutely, most often in childhood or adolescence. These young patients died emaciated, as though starved for sustenance, often within weeks or months of their diagnosis. As Harley said, they appeared to suffer from "defective assimilation" of the food they ate. Twenty years after Harley's book, the French physician Étienne Lancereaux labeled these two diseases, clearly different in both their presentation and their expectation of future life, *le diabète gras* (fat) and *maigre* (thin, or emaciated). Today, the technical terminology for the former is "type 2," the chronic form that constitutes 90 to 95 percent of all diagnoses and that we're likely to get as we get older and fatter; "type 1" is the form that appears acutely and most often in childhood and adolescence.

Both diagnoses are characterized by elevated levels of blood sugar (blood glucose, technically). In the era before blood testing was commonly used, sugar in the urine was seen as the indictor of diabetes. Whatever the means of diagnosis, in both cases the body is failing to properly metabolize the carbohydrates in the diet—the *macronutrient* that makes up the great bulk of the calories in grains, starches, and sugars—to use them for fuel, as intended. For this reason, early diabetes specialists would often discuss the disorder as an intolerance for carbohydrates or, more simply, an intolerance for food itself. Until the discovery and purification of the hormone insulin in 1921, the first advice a doctor would likely give a newly diagnosed patient—occasionally the only advice that might confer any benefit—was what to eat and what not to eat to lessen the burden of the disease.

A few of the nineteenth-century physicians writing about the disease suggested that those with diabetes eat predominantly carbohydrate-rich foods—starchy vegetables and grains, sugar and potatoes—with the hope that their diabetic patients might compensate for their inability to metabolize these foods by consuming more of them. Most of the physicians of the era, however, took the opposite approach: since those with diabetes could not metabolize carbohydrate-rich foods but could metabolize fat, they told their patients that fat should make up the bulk of the diet. "Patients were always urged to take more fat," wrote Elliott Joslin, who, by the 1920s, was the most renowned and influential diabetes specialist in the United States, if not the world. "At one time my patients put fat in their soup, their coffee and matched their eggs with portions of fat of equal size. The carbohydrate was kept extraordinarily low."

For a century now, diabetes textbooks and chapters on diabetes therapy in medical texts have invariably included some variation on the statement that diet is the cornerstone or the foundation of diabetes treatment. As I write this, the most recent guidelines from the American Diabetes Association refer to dieting as "medical nutrition therapy" (MNT); the word "diet" is now seen as connoting only a temporary way to eat, something you go on and then off, not continue for a lifetime. MNT is seen as "integral" to diabetes therapy.

What constitutes MNT—the dietary recommendations themselves—has been driven over the decades not by any meaningful research comparing different dietary approaches but by advances in pharmacological therapy for diabetes and new methods to deliver insulin, lower blood sugar, and monitor blood sugar. Patients suffering from diabetes are no longer instructed to avoid specific foods or food groups. Rather, they are expected to follow the same "healthful eating pattern" that government agencies and health organizations recommend for all of us—"vegetables, fruits, legumes, dairy, lean sources of protein . . . nuts, seeds, and whole grains"—albeit with the expectation, if weight control is necessary, that they restrict their caloric intake.

Controlling the symptoms and complications of the diabetes is left to insulin injections and a pharmacopoeia of drugs that work, directly or indirectly, to maintain blood sugar levels near enough normal that the specter of diabetic complications may be reduced. High blood pressure and the other complications that accompany the diabetic condition can also be controlled by medications. It is assumed that this approach is easiest on the patients, allowing them to balance the burden of having to inject themselves with insulin regularly or wearing an insulin pump or taking the necessary *hypoglycemic* (blood-sugar-lowering) drugs against a greater enjoyment of the pleasures of the table and eating much as their healthy, nondiabetic friends and family do.

In 2009, the British diabetes specialist Edwin Gale, writing with his dietitian colleague Lynn Sawyer, described this situation aptly as paying "lip service to lifestyle change" as a means of controlling diabetes. "Life under communism," Sawyer and Gale wrote in an editorial in the journal *Diabetologia,* "was once summed up in the wry comment that 'we pretend to work, and they pretend to pay us.' Our patients might equally complain that we pretend to offer a diet, and they pretend to follow it." Sawyer and Gale, knowingly or unknowingly, were repeating what the New York City diabetes specialist Edward Tolstoi had written in 1950, when he explained his rationale for what were then called "liberal carbohydrate diets" or "free diets." New patients would come into his clinic at New York Hospital, Tolstoi explained, and they would leave with very carefully calculated diet plans and a program of minimal insulin doses that would keep their urine free of sugar, a sign that their disease was seemingly under control. "We learned quite frequently that a patient would leave the clinic, after having been complimented on his excellent co-operation, and would go at once to the hospital cafeteria for coffee and doughnuts or chocolate cake and sometimes we found him enjoying an ice cream soda. We then looked the other way while saying to ourselves, 'Oh well, [his urine] was sugar free.'"

Recent standard-of-care recommendations from the American Diabetes Association—the advice to physicians themselves, for instance, published in 2022—on what the organization calls lifestyle management is completely in line with this thinking, ironic as Sawyer and Gale (and even Tolstoi) may have meant it to be. The ADA suggests that because most individuals diagnosed with diabetes already eat a carbohydrate-rich diet little different from that of their healthy friends and relatives, then that's what physicians should recommend. The questionable logic being that if physicians tell their patients to keep eating what they've always eaten, the physicians can have confidence that their advice will be followed.*

The beginnings of this modern dietary thinking regarding diabetes therapy can be dated precisely, to March 1922. This was when

---

* For those who think that I am being unfair to the authors of the ADA guidelines, the precise wording of the statement is: "Most individuals with diabetes report a moderate intake of carbohydrate (44–46% of total calories). Efforts to modify habitual eating patterns are often unsuccessful in the long term; people generally go back to their usual macronutrient distribution. Thus, the recommended approach is to individualize meal plans with a macronutrient distribution that is more consistent with personal preference and usual intake to increase the likelihood for long-term maintenance."

University of Toronto researchers led by Frederick Banting and his student assistant, Charles Best, published their first report in the *Canadian Medical Association Journal*, describing not only the discovery, purification, and therapeutic use of the hormone insulin, but confirming that the hormone's absence or failure to control blood sugar was a fundamental defect in diabetes, and that even patients on the very brink of death could be restored to health with daily injections of the hormone. The discovery of insulin launched what medical historians would call a "therapeutic revolution." To the physicians of the era, it was quite a bit more than that. It was as close as medicine had ever come, and maybe ever has, to a miracle.

Emaciated patients, described as barely living skeletons, would recover, and some even would return to work within weeks. They would be resurrected, to use the biblical terminology, which physicians of the era often did. (Physicians today occasionally still do.) Diabetes specialists realized that the newly available insulin was not a cure for the disease, but it allowed their patients to metabolize carbohydrates and held the promise of allowing their patients to eat whatever and however they wanted. "Were I a diabetic patient," Fred Banting wrote in 1930 to a physician advocating for just such an unrestrictive diet, "I would go to the doctor and tell him what I was going to eat and relieve myself of the worry by demanding of him a proper dose of insulin." Insulin therapy could apparently keep their patients alive indefinitely, or so it seemed at the time. What was easier for the patients turned out to be easier for the physicians as well.

As the medical historian and physician Christopher Feudtner described it in *Bittersweet: Diabetes, Insulin, and the Transformation of Illness*, the discovery of insulin "set in motion a dialectic process—between novel treatments and the medical understanding of diabetes—that exemplifies the way in which much of modern clinical knowledge has emerged." Part of this shift in dialectic process was from diet versus disease to drug versus disease. In the decades that followed, diabetes organizations would advocate for dietary therapy for the disease—typically the careful counting and weighing of the carbohydrates consumed such that insulin or other medications could be properly dosed to "cover" them—while diabetes researchers would put virtually all of their efforts into drug therapy. The diets would then be prescribed to allow for the drugs to be used freely, not to minimize their use.

In his indispensable 2017 guide to living with diabetes, *Bright Spots & Landmines*, Adam Brown, diagnosed with type 1 when he was twelve years old, begins with a reminder of the reality of the disease.

> We have too much darkness in diabetes—negativity, confusion, frustration, exhaustion, blame, guilt, and fear. For those of us living with diabetes and the people we love, the cost of this darkness is high. We often don't know what to do, aren't doing what we "should" be doing, feel bad about what we are doing, or are told we're getting it wrong.

All of this is true far too often. One possible explanation for much of that confusion, though, is that the specialists treating diabetes in their clinics and studying it in their hospitals and laboratories, and their colleagues in the closely related disciplines of nutrition and obesity, have embraced some assumptions about the diabetes-diet relationship that may not, in fact, be true. If theoretically these errors in scientific judgment can be corrected, the burden of living with diabetes might lessen considerably. I am writing this book to address that possibility. Only by challenging accepted wisdom can we identify any errors it contains and correct them.

At the heart of a public health crisis that has been building for at least a century remain questions of what those with diabetes should eat: For those willing to make necessary dietary sacrifices to maximize their health and well-being, is there an ideal diet, a pattern of eating (or MNT, for those who prefer that term) that would maximize longevity and minimize the burden of the disease? If the answer is yes, is that diet as safe and effective for those with type 1 diabetes, who require insulin therapy, as it might be for those with type 2 diabetes, many of whom can live without insulin and even thrive without it?

These questions might not need to be asked if it was clearly the case that we truly understood the diabetes-diet relationship. Regrettably, as we'll discuss, the necessary studies have never been conducted to make such an understanding possible. This is one reason why Sawyer and Gale wrote, "From an historical perspective, it is easy to see that the investigators of diet have confused circular motion with progress." Diabetes specialists have often confused studies that compare the conventional thinking on an ideal diet for diabetes to no dietary therapy at all—giving some advice on healthy eating is clearly better than not doing so—with studies that would compare different approaches and philosophies of medical nutrition therapy to see whether one might indeed be superior, if not ideal.

The difficult, controversial questions about diabetes and diet might also be avoidable if fewer individuals were now suffering from diabetes than in the past. If the evidence suggested that diabetes was under control and that patients were thriving, ideally without the financial burden of expensive drug therapies, then we could assume the diabetes specialists and their dietitian colleagues had the necessary answers. But despite all the technological advances in diabetes management—both devices and ever more sophisticated drug therapies—diabetes remains a major problem, and the number of those afflicted is constantly rising.

Most obviously, the prevalence of diabetes exploded in the latter half of the twentieth century. The percentage of Americans diagnosed with diabetes increased 600 percent between the early 1960s, when the first concerted surveys were conducted by federal agencies to quantify this number, and 2015. One current estimate is that almost 30 million Americans have been diagnosed with diabetes, and another 8 to 9 million have diabetes but remain unaware of their condition. The very great majority of these cases are type 2 diabetes, and the remarkable increase in prevalence is blamed on the coincident epidemic rise in obesity. Most diabetes specialists assume that obesity or, at least, excess fat accumulation causes type 2 diabetes, although a significant percentage of patients with type 2 diabetes are lean—perhaps as many as one out of every five (depending on the population). The prevalence of type 1 diabetes has also steadily increased—by almost 50 percent between 2001 and 2017, as a recent nationwide survey published in *JAMA* (*Journal of the American Medical Association*) documented. Children are also being diagnosed now with type 2 diabetes at ever-increasing rates, despite its previously being considered a disorder exclusive to adulthood (and previously known as adult-onset diabetes). A reasonable estimate is that a century ago one in every three hundred to four hundred Americans had one of these two forms of diabetes. Today the number appears to be one in nine. Each year, physicians in the United States diagnose 1.4 million new cases.

Diabetes puts all of these individuals at increased risk of heart attack, stroke, cancer, blindness (retinopathy), kidney failure (nephropathy), nerve damage (neuropathy), gangrene, and lower limb amputation. Individuals living with type 2 diabetes also have increased rates of cognitive impairment and dementia as they get older. Their diabetes increases the likelihood that they will suffer from psychiatric disorders—most notably depression. Their life expectancy may be eight years shorter than those without diabetes. In this sense, diabetes can be thought of as accelerated aging, manifest most obviously in heart disease risk and the damage done to the arteries. As the problem was explained to me recently by James Foley, who founded the diabetes research program at the pharmaceutical giant Novartis, success in diabetes treatment often, if not typically, brings failure down the line. In 1960, he said, the average patient with type 2 diabetes would be diagnosed at sixty-five years old and the average lifespan was seventy. Now type 2 diabetes is often appearing in patients in their midforties and, thanks to the many advances in drug therapy and technologies, they're likely to live into their eighties. That's forty years of accumulating complications from the disease, rather than five. "Extending the average duration of type two diabetes eightfold," Foley said, "makes it likely that many patients will develop one or more of these microvascular complications in their lifetime, even though there are much better therapeutic options that enable physicians to achieve much tighter standards of care in their patients today than was possible in 1960."

Both type 1 and type 2 are still considered progressive chronic diseases, meaning the conditions of patients is expected to worsen, and more drugs and/or insulin will be required in response. A 2017 review, written by four leading diabetes specialists and based on a symposium on diabetes care held at the annual meeting of the American Diabetes Association, concluded that "first and foremost" it had to be recognized that type 2 diabetes would inevitably get worse with time. It is "characterized by a progressive deterioration of glycemic control," the authors wrote, "glycemic" referring to blood sugar levels.

Along with the physical and psychological burdens of diabetes, there is a staggering financial cost. Recent estimates suggest that the yearly cost of medical care in the United States as a result of diabetes has risen by over $100 billion in just a decade. "The remarkable magnitude" of the costs, as a 2018 editorial in the journal *Diabetes Care* said, reached $237 billion in 2017, the "elephant in the room," as the editorial's subtitle described the problem. "The cost of care for people with diabetes," the authors wrote, "now accounts for ∼1 in 4 health care dollars spent in the U.S. Care for a person with diabetes now costs an average of $16,752 per year." This is far more than twice what a healthy person of equivalent age might expect to spend.

Worldwide the situation is equally disheartening. The World Health Organization (WHO) estimates that the number of people suffering from diabetes around the globe increased fourfold between 1980 and 2014, from 108 million to over 400 million, with the greatest rise coming in the poorest countries. Some of this is because people are living longer; the longer we live, the more likely we are to be diagnosed with type 2 diabetes. But the risk of being diagnosed with diabetes *at any age* still more than doubled during those thirty-four years. According to the WHO, diabetes has become the ninth leading cause of death worldwide. In October 2016, Margaret Chan, then WHO director general, discussed this worldwide epidemic of diabetes in a keynote address at the annual meeting of the National Academy of Medicine in Bethesda, Maryland. She described the situation famously and quite simply: It is, she said, a "slow-motion disaster."

For those with type 1 diabetes, the control of blood sugar—the primary goal of therapy—seems to be getting worse, on average. As of 2018, fewer than one in five children and adults in the United States afflicted with type 1 diabetes were achieving even the relatively liberal blood sugar goals set by the American Diabetes Association—a smaller percentage than a decade earlier. In fact, children with this disorder—often, if not typically, diagnosed after a significant loss of weight—now have a prevalence of obesity "as high or higher" than the general population. Since obesity is so closely associated with type 2 diabetes—never before with type 1—specialists are now thinking of these children as suffering from "double diabetes," somehow a combination of the two major forms of the disease. More worrisome, even children with relatively well-controlled type 1 diabetes, meaning they keep their average blood sugar levels within the range recommended by diabetes associations, may experience stunting of growth; the higher the average blood sugar, the shorter their average adult height. They may also experience deleterious effects on brain growth and development. The diagnosis of type 1 diabetes in childhood, while not nearly as dire in its outcome as previously, still creates an expectation of a premature death, according to a 2018 analysis in *The Lancet*, by more than seventeen years in women and fourteen years in men.

As for the individuals most at risk for diabetes, a 2021 review in the ADA journal *Diabetes Care* by a team of researchers from institutions nationwide made clear "that diabetes affects racial and ethnic minority and low-income populations in the U.S. disproportionately, with relatively intractable patterns seen in these populations' higher risk of diabetes and rates of diabetes complications and mortality." The prevalence of diabetes, according to the report, increases the poorer you are, the less educated you are, and the poorer the neighborhood you live in. Having a family income below the poverty line is associated with a twofold higher risk of having diabetes. Those who have never graduated from high school have almost twice the risk of those with more than a high school education, twice the risk of dying prematurely than do college graduates. Adults with type 1 diabetes who do not have a college degree have three times the risk of dying prematurely from their disease than those who do. In all these categories, racial minorities are at greater risk than whites, and the situation is getting worse.

. . .

If we knew how to correctly treat and prevent this disease and this information was disseminated widely, would this situation and these numbers be as dire?

Since the 1970s, diabetologists (a word I will use for both the researchers who study the disease and for the physicians who specialize in its treatment) have been more than willing to test many of their fundamental assumptions about the benefits and risks of strict glycemic (blood sugar) control. They have done so in large clinical trials for both type 1 and type 2 diabetes. But they have done so, as I will discuss, almost exclusively in the context of diabetic patients taking medications and employing devices to control their blood sugar, while adhering to the conventional assumptions about diet and lifestyle.

A disconcerting proportion of those trials, aimed mostly at reducing heart disease risk in individuals with diabetes and/or prolonging life, have *failed* to confirm what the diabetologists had come to believe with almost unconditional faith. The $200 million Look AHEAD trial, for example, which tested the assumption that weight loss in those with type 2 diabetes would lengthen lives, was ended for "futility" in 2012. The ten-thousand-patient ACCORD trial was ended prematurely in 2008, "Halted After Deaths," in the words of a *New York Times* headline. "Medical experts were stunned," the piece said. Almost equally disappointing results were reported from the twenty-country ADVANCE trial (2008) and the 1,800-subject Veterans Affairs Diabetes Trial (2009). The latter three trials tested the assumption that intensive blood sugar control by medications would reduce the burden of type 2 diabetes— specifically heart disease—and the risk of premature death. Diabetes researchers have initiated major studies to test the benefits of various categories of drug therapy compared to one another, but effectively none that test the fundamental assumptions about the relationship between diet and diabetes—not only how the dreadful complications of diabetes can be mitigated, but how the disease itself might be prevented. "There is still a serious shortage of data on long-term effects of these preventive interventions," according to the third edition of *Diabetes in America,* published in 2018 by the National Institute of Diabetes and Digestive and Kidney Diseases.

From the 1920s through the 1970s, diabetes specialists argued over the best dietary prescriptions for diabetes, or even whether such prescriptions were necessary considering the ongoing innovations in insulin therapy and oral blood-sugar-lowering (hypoglycemic) medications. Nonetheless, the British Diabetes Foundation, established in 1934, and the American Diabetes Association, founded six years later, promulgated dietary guidelines based on a consensus of expert opinion that came to carry the weight of seemingly indisputable scientific fact.

These are the guidelines that are now promoted as "evidence based." This means they are no longer based on the clinical experience of physicians themselves, but rather on surveys of large patient populations and the results from clinical trials that have only been feasible and practical (and supported by government funding) since the 1960s. But clinical trials necessary to rigorously assess the risks and benefits of diets of significantly different proportions of protein, fats, and carbohydrates have never been conducted.

When researchers have surveyed whether patients with diabetes were actually following the diets their physicians prescribed, or were getting the outcomes that such diets are expected to produce—weight loss, in particular, considered the primary goal of dietary recommendations for those with type 2 diabetes—the results, once again, have been disappointing. "An Analysis of Failure" was the subtitle of a seminal 1973 article reporting on these survey results. Diabetologists, along with their physician colleagues who focus on treating obesity, have mostly drawn the conclusion that their patients are simply unwilling to follow (or incapable of following) the dietary advice they're given; that a diet restricting what they can eat is not worth whatever benefits they might experience. The immediate burden of the disease, by this thinking, is not equal to the burden of dietary restrictions. This is why, as Sawyer and Gale suggested in 2009, the physicians treating diabetic patients have also assumed that their patients would prefer to take whatever drugs are necessary to control their blood sugar and other complications of their disease, rather than adhere to a diet.

If the diet a patient is instructed to follow does not work, however, if it does not optimize the patient's health and significantly lessen the burden of the disease—if perhaps they only get fatter and more diabetic—then not following it seems a perfectly appropriate response. A strict dietary approach that theoretically minimizes your risk of disease decades in the future, but has no noticeable benefits in the present, may be hard to sustain.

Over the past two decades, the thinking on the diet-diabetes relationship has begun once again to shift. Physicians confronted with the failure of diet therapy among their patients, reflected either in poor glycemic control or the inability of the patient to achieve or maintain a healthy weight, have taken to experimenting with the kinds of diets that were standard for patients with diabetes in the decades before insulin was discovered. These diets restricted carbohydrates almost entirely, while replacing those calories primarily with dietary fat. In the nineteenth century, when this approach was widely prescribed by physicians in the United States and Europe for their diabetic patients, it was often known as the "animal diet." Today, the fashionable term for this way of eating would be "keto," although the state of ketosis (elevated levels of molecules called ketone bodies in the blood and urine) may not be necessary for the diet to be beneficial.

When diabetes specialists turned away from these diets because of the availability of insulin to control blood sugar, physicians continued to promote these diets in books, often best sellers. Their authors, most famously the New York cardiologist Robert Atkins, maintained that such carbohydrate-poor and fat-rich diets had remarkable efficacy for achieving weight loss. In the early 1980s, Richard Bernstein, an engineer-turned-physician who had been diagnosed with type 1 diabetes as an adolescent, pioneered the use of these diets for type 1 in the modern era.

These diets are based on the simple assumption that if people have a disease that prevents them from properly metabolizing carbohydrate-rich foods, then the fewer of those foods they consume, the healthier they'll be and feel. Since diabetes does not prevent the use of dietary fat for energy, fat takes over most of that role. Rather than using insulin and other medications to "cover the carbohydrates" they eat—as patients are typically taught to do by diabetes educators upon their diagnosis—the patient works to minimize both the carbohydrates consumed and the medications used. Bernstein called this the "laws of small numbers," proposing that the fewer carbohydrates consumed, the lower the insulin doses, the healthier the patient will be and the lesser the risk of dangerous complications. In 1928, when insulin therapy was still new and Elliott Joslin, the most influential diabetologist of the era, talked about the smallest-possible-dose philosophy, he said essentially the same thing: "Unquestionably large doses of insulin will allow patients to take more food and thus gain weight very rapidly, but with larger doses of insulin the patient is walking on longer insulin stilts and his equilibrium is therefore correspondingly endangered. With small doses, progress is more gradual, but quite as sure and certainly safer." Whether Bernstein's approach is superior to conventional therapy and whether it would be more acceptable to patients are questions that cannot be answered with certainty, because the trials necessary to do so have never been conducted.

In 2019, the American Diabetes Association released its latest assessment of nutrition therapy for adults with diabetes—both type 1 and type 2—and the related condition of prediabetes, a consensus report prepared by a fourteen-member committee of physicians, dietitians, and researchers.* It concluded that the evidence was insufficient to state which eating pattern, which of these nutrition therapies, was preferable for diabetic patients. Of all those assessed, though, the low-carbohydrate or very-low-carbohydrate, high-fat diets (aka keto) were the ones that had been tested most frequently in the past twenty years and the only ones that showed consistent benefits. All the others, including the Mediterranean-style diet, the DASH (Dietary Approaches to Stop Hypertension) diet, vegan and vegetarian diets, and the kind of low-fat diets that had been prescribed by governments and health organizations as generally healthy diets for decades and had become since the 1970s standard of care for diabetes, had produced at best mixed results

---

* This is not to be confused with the standard-of-care recommendations, authored only by physicians and discussed above.

in the scattering of short-term (typically a few months to a year) clinical trials that had been done to test them.

Meanwhile, physician researchers found cause to study what their colleagues had come to consider medical or dietary heresy—particularly the consumption of any diet high in fat and saturated fat—and were reporting remarkable results in those eating this way to control their diabetes. Researchers working with a San Francisco–based health-care start-up called Virta Health, for instance, that uses telemedicine (via online and smartphone technology) to teach patients with type 2 diabetes the ins and outs of eating very-low-carbohydrate, high-fat diets, reported in a series of articles, beginning in 2017, that more than half of their patients were able to put their diabetes into remission. They were able to get off insulin and all but the most benign of their diabetes medications (metformin) while maintaining apparently healthy levels of blood sugar. Almost half of the Virta patients kept up this state of remission for the five years (as of this writing) that their progress has been tracked.

In 2018, a collaboration led by the Harvard pediatric endocrinologists Belinda Lennerz and David Ludwig reported on their study of members of a Facebook group called TypeOneGrit dedicated to using the dietary therapy promoted by Richard Bernstein in his book *Dr. Bernstein's Diabetes Solution*. Both youths and adults surveyed maintained near-normal levels of blood sugar with surprisingly few signs of the kind of complications—including very low blood sugar (hypoglycemia)—that make the life of a patient with type 1 diabetes so burdensome, and that physicians have traditionally feared from this pattern of eating. Physicians who have prescribed either Bernstein's approach or the even more extreme higher-fat ketogenic diets have said they would not have believed such results were even possible in their patients had they not seen it for themselves.

This does not constitute sufficient evidence to claim that eating a diet that restricts sugars, starches, grains, and legumes is a risk-free proposition. As the authors of the TypeOneGrit survey put it, "These findings by themselves should not be interpreted as sufficient to justify a change in diabetes management. Additional research is needed." And such research is desperately needed—the sooner, the better.

I'm writing this book for physicians and dietitians who want to understand why the conventional dietary advice they are giving patients with diabetes isn't helping, and for the patients themselves. I want to explain why an alternative way of eating might be more effective. (As the TypeOneGrit paper noted, some of the participants in the study had chosen to eat this way *against* the advice of their health-care providers.)

This is my fifth book on the relationship between diet and chronic disease, each successive book further unpacking the implications of the decade of my journalistic research that began with a series of investigative articles for the journal *Science* and *The New York Times Magazine*. That research exposed what I contend are two significant errors that were made in the science of nutrition and obesity, both with profound implications for diabetes. The first is the belief that obesity is a disorder of energy balance—that we get fat because we consume more calories than we expend—rather than a hormonal dysregulation dominated by the hormone insulin. The second is the conviction that the primary problem with modern diets is the fat content, specifically the saturated fat.

One of my goals here is to establish what diabetologists might have concluded about dietary therapy for type 1 and type 2 diabetes had these misconceptions not been embraced as medical truths. I'm writing from a historical perspective because I believe understanding the history of how this troublesome situation developed is vitally important to thinking our way out of it. I believe journalists can be especially effective in such situations because we are expected to challenge authoritative positions to determine if they hold up to critical and skeptical scrutiny. Good science journalists do so independent of the groupthink that exists in professional societies and can benefit by interviewing all those who have done significant work in the field, whatever their point of view. My book *Good Calories, Bad Calories*, for instance, was based on interviews with more than six hundred such researchers, public health experts, and administrators. My most recent book, *The Case for Keto*, was based on interviews with more than 120 physicians (and another twenty dietitians and other medical practitioners) who have come to prescribe low-calorie, high-fat/ketogenic diets for obesity and type 2 diabetes, and who further informed my thinking on the subject. This book, while largely based on historical documents and the medical literature going back to the nineteenth century, is also informed by what I learned from interviews with another forty physicians, researchers, and individuals afflicted with diabetes, most often type 1.

While these books have had immediate implications about how to eat to maximize health, I have had another purpose in mind in writing them: I want to show that the practice of medicine often conflicts with the goal of medical science, by which I mean the intellectual endeavor that seeks to establish *reliable* knowledge about the subject of study. Physicians have to make decisions about diagnosis and treatment based on the fragmented, preliminary state of knowledge at any one time. Scientists are taught to withhold judgment and remain skeptical—"the first principle," as the Nobel laureate physicist Richard Feynman famously put it, "is you must not fool yourself—and you are the easiest person to fool."

Scientists have the luxury of time to work meticulously through the challenges and ambiguities of their research. Physicians do not: their patients are sick, and treatment cannot be postponed for the years or decades required for clinical trial results that *might* tell them the best course of treatment—if, indeed, such trials are even planned or underway.

Moreover, what is considered to be the right prescription or intervention is often the product of consensus: an implicit and occasionally explicit polling of those considered to be experts in the field. This is how conventional wisdom or dogma in any science takes shape, but it is particularly common in medical science because the rigorous experiments necessary to test hypotheses can be prohibitively difficult and expensive. This leaves medical researchers frequently drawing their conclusions based on lesser evidence, information that is sufficiently ambiguous that clinical researchers can and do interpret it to fit their preconceptions. (Francis Bacon made this point four hundred years ago, when he inaugurated the scientific or experimental method as a means of deriving reliable knowledge. He understood that as humans we are programmed to see what we want to see and believe what we want to believe; this is as true of scientists and physicians as anyone else.) The result can be a consensus that seems reliable—based, as it is, on the expertise of experts—but may not be.

The consensus can also be driven by one or just a few dominant and influential personalities. The narrower the discipline, the more influence a single individual will have. When medical disciplines are relatively new, meaning diseases and conditions are still relatively rare and funding difficult to come by, they will be underpopulated with researchers. The tools those researchers have to work with during these periods will be relatively primitive, often ill-suited to the tasks at hand. Unfortunately, this is when the thinking that emerges from the initial research and observations—the paradigm, as the philosopher of science Thomas Kuhn famously termed it—can easily be shaped, embraced as a consensus, and locked in, even if it's wrong. All of this depends on the scientific acumen and critical thinking of one or a few dominant personalities involved. The history of diabetes research and the thinking on dietary therapy is no exception.

Medical associations and government organizations have come to believe that they are obligated to produce guidelines for treatment, a standard of care. At the very least, they want to guide physicians on the appropriate therapeutic interventions based on the evidence as it currently exists. This position was argued most forcefully to me in my early research by Carol Foreman, an assistant secretary of agriculture at the U.S. Department of Agriculture in the late 1970s. Foreman was the driving force when the USDA began the process of creating and then publishing the first Dietary Guidelines for Americans. She was aware that an active controversy existed between well-respected researchers on all sides about what conclusions could be drawn reliably from the evidence—but she felt the obligation to present a best guess. As she would say to scientists, "I have to eat three times a day and feed my children three times a day and I want you to tell me what your best sense of the data is right now."

That's what the USDA has communicated with enormous influence ever since, and other organizations and medical associations around the world have followed its lead. That doesn't change the fact that its recommendations always depend on which experts or authorities are consulted. All too often, the beliefs that lead to these best guesses and then to the clinical decisions of the physicians become accepted as indisputable not because the evidence demands it, but rather because it justifies the decisions and the recommendations that have already been made. Science is supposed to protect against this dangerous tendency by requiring a rigorous gauntlet of hypotheses and tests. This has not been the case with regard to diet and chronic disease.

A common response to these issues is that science is self-correcting, that an accumulation of evidence will eventually lead to eliminating mistakes. But no one knows how long this corrective process will take—when reliable knowledge will be established—and the longer it does, the more evidence is required to overthrow ever-hardening dogma. The words of the philosopher of science Karl Popper, describing science as "one of the very few human activities—perhaps the only one—in which errors are systematically criticized and fairly often, in time, corrected," make clear that errors can linger indefinitely.

No one likes to accept that something they have faithfully believed can be wrong, and the longer they believe it, the more difficult it becomes to change their views. Physicians and medical organizations are no exception. Physicians will often tell the story of their medical school instructor who told them that half of what they will be taught is wrong, but not, of course, which half.

With diabetes, all of this is exacerbated, because the conventional thinking on the diet-disease relationship is firmly rooted in a period in the history of the disease, over a century ago, when the physicians were working blindly. Influential diabetologists realized, as Elliott Joslin wrote in the first edition of his textbook in 1916, that understanding the "pathological physiology" of the disease would make the correct treatment self-evident, but that physiology remained unknown.

What were thought to be the healthiest diets for the disease were based on the perceived needs of their most critically ill patients, typically young men and women, often children, afflicted with type 1 diabetes. These patients might appear in the clinic or the hospital on the brink of death: having lost startling amounts of weight or slipping into the condition known as diabetic ketoacidosis that, prior to the discovery of insulin, would invariably be followed by coma and death. The physician's concern was preventing that death and stabilizing the patient so that they could be discharged from the hospital and sent home to family. Once insulin therapy was incorporated into clinical practice in the 1920s, it turned diabetic ketoacidosis and coma from a death sentence into a preventable and curable disorder. And yet a dietary philosophy embraced in the half dozen years before insulin was discovered determined how the most influential diabetologists—Elliott Joslin, particularly—thought about dietary therapy afterward.

Diabetologists needed four more decades of clinical experience and research after insulin's discovery before realizing that they were dealing with, in effect, two entirely different diseases, two very different pathological physiologies. Both diseases—type 1 and type 2—involve a failure of the hormone insulin to regulate blood sugar, but through different mechanisms and, to some extent, even different organs. The acute form of the disease is defined by a deficiency of circulating insulin; hence, patients with type 1 diabetes require insulin to survive. Before the type 1/type 2 terminology was embraced in the 1980s, type 1 diabetes was known for twenty years as insulin-dependent diabetes. The chronic form of the disease, type 2, is characterized by a resistance of cells throughout the body to the action of circulating insulin (the condition of *insulin resistance*), which means that too much insulin circulates through the blood (*hyperinsulinemia*) as the pancreas compensates by secreting ever more of the hormone. This form was known as non-insulin-dependent diabetes before "type 2" became the accepted terminology. Until the latter stages of this disease, when the pancreas eventually exhausts itself, insulin therapy may be of little or no benefit to these patients. Yet the flawed thinking that emerged in the first decades of the twentieth century with regard to diet has largely determined the treatment for all of those with diabetes ever since. My aim here in part is to discuss the errors in clinical judgment that were made in diabetes research in pursuit of an ideal therapy.

Ultimately, the history that emerged in my research for *Rethinking Diabetes* serves as a cautionary tale, a case study in the medicalization of modern life and one that is particularly relevant today. As the pharmaceutical industry develops more effective treatments for chronic, progressive conditions—obesity, now, being among the most controversial—medical associations become ever more likely to consider them diseases, beyond the control of the patients themselves. Physicians, in turn, become increasingly comfortable prescribing lifelong drug therapy as treatment. While physicians will be aware of the potential for side effects and long-term complications, the unvoiced assumption is that, given time, new drugs or drug variants will be developed and the patients can always switch or add a new drug as necessary. The clear immediate benefits outweigh concerns about future risks.

As I write this, a new class of drugs has recently been shown to be remarkably effective at inducing weight loss in those who are overweight or obese. Based on a hormone secreted from the gut known as glucagon-like peptide 1 (GLP1) and originally developed and approved for type 2 diabetes, these drugs have been described as "game-changers" and "'the' transformative breakthrough" in obesity therapy. The scenario is not all that different from the introduction of insulin therapy a century earlier. Indeed, as the journal *Nature* reported, when the results of the first clinical trials testing these drugs for obesity were announced at a conference in November 2022, "sustained applause echoed through the room 'like . . . at a Broadway show.'"

The parallels with insulin therapy are clear: as with insulin, these new drugs have to be injected (although those days may already be over), albeit once a week rather than daily. As with insulin for type 1 diabetes, they almost assuredly will have to be taken for life. They don't cure the state of obesity, they only reduce the symptom of excess weight so long as the medication is taken regularly. Those who go off the drugs, for whatever reason—cost, intolerable side effects, health concerns for mother and fetus during pregnancy, lack of availability—will apparently gain back much to all of the weight they had lost.

Very similar decisions are being made now by physicians about their patients with obesity and by those individuals themselves as were made a century ago with insulin therapy for diabetes. The issue in this case, though, is not life and death, as it was then and is still with type 1 diabetes, but quality of life, which can seem of almost equal importance to many who are burdened with obesity. In January 2023, the American Academy of Pediatrics published guidelines suggesting that physicians should consider obese patients as young as twelve—children—as candidates for the use of these drugs, assuming, again, that they will be taken for a lifetime.

As with insulin and diabetes, the new drugs are already changing the discourse, the dialectic process, on obesity from diet vs. disease to drug vs. disease. The drugs, as *The Wall Street Journal* put it, are "ripping up long-held beliefs that diet, exercise and willpower are the way to weight loss." A likely scenario is that in just the next few years, millions to tens of millions of individuals struggling with obesity will be using a new class of pharmaceutical therapy, committed to it for life, as the medical community struggles to understand and manage whatever long-term risks accompany the very clear benefits of the present. These risks are, by definition, unknowable and may take years to decades to manifest themselves. As with insulin, diet, and diabetes, whatever mistaken assumptions are embraced as this process plays out may likely haunt medical practice, and so the health and well-being of patients, for generations.

Living with diabetes is often described as self-care, more so than medical care. The diabetic patient has to do the hard work of keeping the disease under control. To do so successfully, though, the patient must have the support and understanding of their physician along the way. My ultimate goal is to make the patient's work easier, to improve their ability to control their disease, while giving their physicians, diabetes educators, and dietitians the understanding necessary to offer informed encouragement and assistance. This comes, however, with a caveat: what I write here is supported by the work of many diabetes researchers and specialists, but their voices do not carry the authority of a consensus. Hence, for individuals struggling to achieve optimal control of their disease, I cannot overemphasize the value of working with a physician—and there are now many of them—who understands the science and the therapeutic implications of carbohydrate-restricted diets and can therefore help minimize any risks.

1

The Nature
of Medical Knowledge

In diabetes . . . the chief difficulty lies in the fact that the danger is one of the future. This is the insidious peculiarity of diabetes. We do not at all disturb for the present the general well-being of the diabetic if we treat him badly and overweight his weakened functions; yes, we may even improve by psychical influence his momentary well being if we permit a more liberal diet. We are, however, playing a dangerous game. We are thinking only of the present and forgetting the future, the fortunes of which depend upon the vigilance of the practitioner.

—CARL VON NOORDEN,
New Aspects of Diabetes: Pathology and Treatment, 1912

In the writer's experience, there is nothing more disturbing than the diabetic who acquires the disease in childhood; who apparently is a picture of robust health—who looks and feels perfectly well—but whose blood vessels have been degenerating insidiously for years; who, in the early 20's or 30's and probably married and with a family, is beginning to feel the effects of the degenerative changes, either because of a progressive hypertension, kidney failure, disturbance of sight due to retinitis, or a sudden attack of coronary thrombosis. . . . To prevent such cases, or at least reduce their occurrence, is the purpose of this report.

—ISRAEL RABINOWITCH,
Canadian Medical Association Journal, 1944

Let's begin with a case study: "A 32-year-old white male was seen at the [Mayo] Clinic in July, 1921, by Dr. [Russell] M. Wilder. His symptoms were polyuria, polyd[i]psia, polyphagia, weakness and loss of weight." The patient, a farmer from "the hinterland of Montana," had severe diabetes. The diagnosis was not difficult to make. He was urinating abundantly (polyuria), had an unquenchable thirst (polydipsia), and was constantly hungry (polyphagia). But no matter how much he ate, he was losing weight.

The remarkable aspect of the case, though, isn't what happened at the time of diagnosis, or even eighteen months later, when the patient was started on insulin therapy, but what happened when he reappeared at the Mayo Clinic in June 1950, twenty-nine years later. "Since 1921," wrote the two Mayo Clinic physicians who reported on the case at the staff meetings of the clinic, "he had faithfully and strictly followed a diabetic diet which by modern standards seems almost unbelievable." The situation, they said, was "unique" in their experience. That's why they were writing it up as a case study.

When the patient had initially been hospitalized, the Mayo staff, led by Wilder, began its procedures for treating the diabetes. They fasted the patient for "several days" until all signs of sugar disappeared from his urine. "Desugarizing the urine," as these physicians called the procedure, was the primary goal in therapy. Then the patient was served very small amounts of carbohydrate foods daily to establish how much he could metabolize without the sugar reappearing. Once his doctors had established that level, they added protein and fat to his diet. When the combination of protein, fat, and carbohydrates led to the appearance of ketones (technically ketone bodies) in his urine, it was seen by physicians at the time as a sign of imminent danger.

The Mayo doctors assumed that the patient had now reached the limit of how much fat he could eat safely. His protein consumption was kept at the minimum considered necessary for a healthy man of his size and weight. If sugar reappeared in his urine, he would be fasted again. If ketones appeared, "all fat was omitted from the diet," which meant most of the food he was allowed to eat, and he would restart the process. The Mayo physicians hoped by this approach to make the load on the patient's "weakened sugar-using function as light as possible in order to rest it and thus favor its restitution." Restitution, in fact, had rarely, if ever, been reliably documented, but that was their hope.

When the patient was released from the hospital after a month of these dietary manipulations, he was allowed to eat 15 grams of carbohydrates a day (the amount in a single thin slice of bread, although bread was not among the foods his diet allowed), 45 grams of protein (the protein in about half a pound of lean ground beef), and 150 grams of fat. This added up to a total of 1,590 calories a day, a meager ration for a hardworking farmer. He was also instructed to fast one day a week. Complicating matters further, the carbohydrate-containing foods he could eat—green vegetables— had to be boiled three times before serving to remove most of the digestible carbohydrates. He was allowed to eat bran muffins made from specially purchased bran, also boiled three times. "It was calculated," according to his doctors at the Mayo Clinic, that "there was no food value in the muffin prepared in this manner." This was life for a diabetic patient before the discovery of insulin.

The rigid diet did not restore the Montana farmer's health. When he returned to the Mayo Clinic a year and a half later, in December 1922, insulin had become available. He would be among the first patients at the Mayo Clinic to receive it. He would also be the beneficiary of the recent work of two physicians at the University of Michigan, Louis Newburgh and Phil Marsh, who had experimented with a high-calorie, high-fat diet for diabetic patients and reported that they fared remarkably well. They didn't have to live on a near-starvation regimen but rather could eat to satiety. This approach had been embraced by Russell Wilder at the Mayo Clinic. Now he and his colleagues taught their patient how to inject himself with 30 units of insulin before breakfast every morning. They also instructed him on the new diet, allowing him to eat more than 2,700 calories a day. He was still boiling the green vegetables three times and eating his food-value-free bran muffins, but over 80 percent of his calories came from fat. To eat that much fat, the Mayo physicians reported, it was "necessary for him to consume large amounts of rich cream and butter." He religiously followed that diet, along with his insulin, as though living in a time capsule, for very nearly three decades. "He had never tasted such common foods as potatoes or bread," they wrote. "He still fasted from a half to one day a week."

By 1950, when the Montana farmer returned to the Mayo Clinic, the context of how his physicians understood diabetes had changed dramatically. They had come to realize that they were in the midst of a new kind of diabetes epidemic, not just an "appalling increase" in diabetes prevalence, as Elliott Joslin had described it that year, but the diabetic patients confronting them were different. The discovery of insulin and the initiation of insulin therapy in the early 1920s had transformed the very nature of the patient experience with diabetes. On average, Joslin's patients in Boston were living three times longer. "With insulin," Joslin had written prophetically in 1923, in the third edition of his famous textbook (this edition dedicated to Banting and Best, the University of Toronto researchers who had discovered insulin just shortly before), "we shall learn the remote rather than the acute results of the disease." And they had.

Now diabetic patients were living long enough to die of the complications of the disease itself, and because it had struck many of them in their childhood or adolescence, they were still quite young when they died; the damage to their veins and arteries caused at least in part by their bodies' inability to control their blood sugar. By 1934, Wilder and his Mayo Clinic colleagues were reporting lesions of the retina, hemorrhages, obscuring vision in young diabetic patients who had been on insulin for ten years. By 1936, Harvard pathologists were reporting a new type of kidney disease in patients with diabetes; by 1950, Joslin's colleagues were reporting that 80 percent of their patients who had been using insulin for at least twenty years had hypertension and more than 20 percent had kidney disease. "Of those who died between 1944 and 1950," as the British diabetologist and medical historian Robert Tattersall reported it, "more than half had advanced kidney disease." In one 1949 study from the University of Minnesota, kidney lesions had been a hundred times more common in diabetics than nondiabetics.

Coronary artery disease in patients with diabetes had by then become so common that Joslin was suggesting to cardiologists that they should study these young diabetics on insulin therapy to understand the fundamental cause of the disease. Those patients who developed diabetes as adults, who had the less severe, chronic form of the disease, type 2, were living long enough to suffer heart attacks, kidney failure, blindness, strokes, nerve damage, and gangrene. Physicians were reporting the "extraordinary frequency" of lesions in the arteries of the heart, the legs, and the kidneys of those with diabetes, "especially when the diabetes was of long duration."

This was what the Mayo Clinic physicians expected to confront in the Montana farmer when he returned to the clinic in June 1950 for an examination. The fact that he was still alive after thirty years with the disease would have been reason enough for his doctors to want to know more. The most important question in the field of diabetes had become whether or not these tragic complications could be prevented or minimized if diabetic patients would make the necessary effort to control their blood sugar.

The results of the patient's examination provided the second remarkable aspect of his case. The results were normal. The patient appeared to be thriving. He had maintained a healthy weight (he was five feet eleven inches and 143 pounds). His blood pressure was normal. "Neurologic and ophthalmoscopic findings were normal," his physicians noted. His blood vessels were "open." Normal. The urine showed a little sugar but was otherwise normal. There was no sign of the protein albumin, which would be symptomatic of kidney lesions. His cholesterol level and blood fats were normal. The Mayo physicians x-rayed his pelvis and discovered some minor calcification in the femoral arteries, but "no more than could be expected without diabetes at age 61." The patient had no sign of vitamin deficiencies despite the fact that boiling his green vegetables three times before eating them would have effectively removed the vitamins they contained along with the carbohydrate calories.

So how to make sense of it? The Montana farmer had religiously complied with the advice he had received in 1922—but that was very much not the advice the Mayo Clinic physicians were giving in 1950. "A diet such as he maintained," they wrote, "containing very little carbohydrate and protein and large amounts of fat, is diametrically opposed to much of the current feeling which proposes a low fat diet for diabetes. . . . Whether his diet had any part to play in the freedom from complications is, of course, unknown."

Despite the fact that the patient was thriving, the Mayo physicians would now change his treatment. Once insulin therapy had become standard of practice, the high-fat, high-calorie diet for diabetics had been discarded. Patients were instead instructed to eat carbohydrate-rich foods, on a strict schedule, to prevent the insulin from causing low blood sugar—insulin shock—which also could be fatal.

As the epidemic of patients with crippling diabetic complications had started to fill waiting rooms, physicians and diabetologists argued for more and more carbohydrates in the diabetic diet and less and less fat. The patients wanted to eat the starchy and sugary foods they had always eaten, and their physicians, including Joslin and his colleagues at his clinic in Boston, worried that the fat their patients had consumed pre-insulin was making them fat and causing the heart disease they were seeing ten, twenty, and thirty years later. They worried that it might precipitate coma because their patients, *pre-insulin*, eating fat-rich diets, died in a coma more often than not. This is why the Mayo physicians told the farmer he could eat far more carbohydrates, for which he could take higher doses of insulin. And he'd have to eat far less fat to balance out the calories. "For the first time in twenty-nine years he ate a normal meal," his physicians reported in the Mayo Clinic case study. Had they lengthened his life by changing what he ate, when his previous diet may have been responsible for keeping him so remarkably healthy? Had they done no harm? They had no idea. They had followed the prevailing fashion.

At the heart of any medical progress is a question that is simple only in the asking: How does the physician know enough to decide which therapy to use or prescribe? If the patient's ailment is acute or the patient is on the brink of death, then anything that can restore relative good health would seem to be a good thing. In the years immediately before and after the discovery of insulin and the introduction of insulin therapy, anything that could prevent diabetic coma for an indefinite period could seem worth the risk. "When I was a student and young doctor" in the 1920s, as the University of Edinburgh diabetologist Derrick Dunlop (later Sir Derrick) described thirty years later, "we were entirely occupied, so far as diabetes was concerned, with endeavoring to keep the patient alive for a while . . . we were taught little of its ultimate complications, for relatively few patients had by then lived long enough to develop them."

If an intervention restores the patient to health, but is associated with premature death or disease months or years later from side effects or complications, the treatment might still be readily justified. This was the case with insulin therapy for type 1 diabetes, as it is with many cancer therapies today. But the "ultimate complications" cannot be ignored. If the physician has a choice of two treatments that will restore the patient to health in the short run, then the long-term effects must be weighed before a choice is made.

For a chronic disease or disorder, one like heart disease or type 2 diabetes that disables and kills prematurely but does so only years or decades in the future, physicians had been forced to speculate on whether what they were doing would cause more good than harm. But with the invention of the modern clinical trial in the late 1940s—technically known as a randomized controlled trial— these doctors benefited from one of the great advances in medical science. For the first time, they had a means to assess the long-term risks and benefits of medical interventions and to compare interventions to establish—for an idealized, average patient—what would most likely be the safest, most effective one. The proliferation of clinical trials by the 1980s had launched the era of evidence-based medicine.

With its reliance on clinical trials to dictate accepted practice, medicine left behind the notion of basing therapeutic decisions on clinical experience and observations. But by the time diabetologists started to embrace the notion of an "evidence base" for their beliefs on the nature of a healthy diet, they had already succumbed to the biases formed decades earlier.

2

The Early History

In the matter of diet, it must be borne in mind, that while the exclusion of starch and sugar is the measure which gives us the most control over the disease, yet it is by no means paramount to every other consideration. . . . The object of treatment is to promote the general health and well-being of the patient, and the reduction of the out-put of sugar is to be regarded only as a means to that end, and never as an end in itself. It is a much more serious thing to the patient to have his digestion thoroughly disordered; his nutrition and strength impaired; and his morale broken down as a result of too strict a diet, than to pass an ounce or two more of sugar per diem.

—ANDREW H. SMITH,
Diabetes. Mellitus and Insipidus, 1889

There is no cure for diabetes. Only dieting relieves the sufferer.

—GRAHAM LUSK,
"Metabolism in Diabetes," 1909

The history of diabetes therapy, or at least successful therapy, begins with case reports. In an era when a physician might diagnose a case of diabetes only once in his lifetime, John Rollo, a Scottish doctor, saw it three times and published a pamphlet documenting his success in the second of his three cases. "The ingenious author of the work now before us," as a 1797 review in the Edinburgh journal *Annals of Medicine* put it, "recommends a mode of treatment, which, in some instances, has been decidedly productive of remarkable benefit. It may justly, therefore, be considered as well meriting a fair trial in future cases." Rollo had been trained in Edinburgh, joined the Royal Artillery in 1776, and eventually rose to the rank of surgeon general. He may have been the first physician to successfully bring a case of diabetes under control. Variations on his approach would become the standard of care for the next 125 years.

Until Rollo came along, physicians considered diabetes inevitably a progressive and quickly fatal disease. Their attempts to treat it were scattershot and ineffective; "whatever was available in medicine seems to have been employed against diabetes for centuries," as one medical historian has put it: "massage in the sun, hot and cold baths, steam baths, wine, whey, milk diet, various nostrums, bleeding, emetics, narcotics, and astringents."

Rollo had seen his first case of diabetes in 1777, but the patient had been discharged shortly afterward and Rollo learned little from the experience. On October 16, 1796, he diagnosed his second diabetic patient: a Captain Meredith of the Royal Artillery, formerly corpulent, now much diminished in size—"fallen away in fat and flesh considerably," as Rollo described him. Meredith had experienced symptoms for seven months by the time he saw Rollo, complaining of "great thirst and a keenness of appetite." He was also urinating copiously, but neither Meredith nor his regular physician had paid attention because he was drinking so much to slake his thirst, "the quantity of urine had appeared to him a necessary consequence." When Rollo tasted the urine, a common diagnostic method in that era, it was noticeably sweet, confirming that his patient had diabetes.

Rollo theorized that the cause of the disease was the formation of carbohydrates in the stomach—an excess of "saccharine matter." Assuming that the substance could only come from vegetable foods, he concluded that they should be restricted. He therefore prescribed an "animal diet" (and various salves and concoctions, including opium) as a treatment for Meredith. His patient was expected to eat puddings "made of blood and suet only" for lunch, and old meats and fat "as rancid as can be eaten" for dinner. He was allowed milk for breakfast and lunch, and bread and butter, so his diet was not free of vegetable foods, though nearly so. The diet rendered Meredith's urine sugar-free and returned him to health. Within a month, Rollo had prohibited Meredith from drinking the milk (which contains carbohydrates in the form of lactose, although Rollo would not have known that) and replaced it with what he called beef tea—we would call it broth or stock—made from boiling fat beef (or mutton) with water, and then straining the result to produce a clear liquid.

By the end of the year, Meredith seemed "free of disease," wrote Rollo, "rapidly gaining flesh," and was allowed to eat more bread and to exercise. By the following March, Meredith *seemed* cured. He "might, we apprehend, now eat and drink any thing with impunity," Rollo noted. On May 10 of the following year, Meredith wrote to Rollo that he continued "in perfect health." (Historians later established that Meredith remained, as his wife would write to a relative in 1805, "in tolerable health but quite thin." He died in March 1809, twelve years after last consulting with Rollo.)

Rollo's third case didn't go as well, but confirmed, Rollo thought, the principles of the animal diet. The patient was a fifty-seven-year-old general in the British army who had been ill for at least three years. Rollo first saw him in January 1797 and was not optimistic that he could return him to health. "We are satisfied the saccharine matter and morbid action of the stomach may be removed, yet the sequela of the disease may be such as to prevent the return of perfect health." He reported, once again, that so long as his patient adhered strictly to the animal diet, his condition improved. By now Rollo's dietary prescription was less reliant on rancid old meat and fat, and allowed meat of any kind.

Whenever the general's health deteriorated, Rollo would interrogate his patient and conclude that he had been eating vegetable matter or fruit or drinking beer. "We have on the whole to lament our patient's inclination to variety," Rollo wrote, "and his extreme impatience under restrictions, as otherwise we have no doubt he would have returned in a much better state to his family." When the general returned home, already cheating on the animal diet, a local physician further encouraged him "to eat what he pleased, and to drink wine," Rollo reported, and the general did. He was soon dead.

Rollo then compiled the two case studies and his speculations into a pamphlet and posted it: he wrote, "to every person in England or Scotland, who I thought were likely to meet with the disease; and I solicited a trial of the mode of cure, with an account of the results." Physicians were encouraged to write back to him with their experience, and some two dozen did. He compiled those into a book with his two cases and published it in multiple editions. The conclusion, again, was that the animal diet worked. Removing the vegetable matter from the diet resulted in mostly sugar-free urine, a resolution of the thirst, and normalizing of the appetite and urination. The patients felt healthier.

The letters to Rollo also confirmed that the animal diet was seen by patients, and often their physicians, as only a short-term necessity. The patients would ease off the dietary restrictions as soon as they started experiencing beneficial results. Rollo, for his part, was confident the diet would work if the patient would follow it. "We have to lament, that our mode of cure is so contrary to the inclinations of the sick," Rollo wrote. "Though perfectly aware of the efficacy of the regimen, and the impropriety of deviations, yet they commonly trespass, concealing what they feel as a transgression on themselves. They express a regret, that a medicine could not be discovered, however nauseous, or distasteful, which would supersede the necessity for any restriction in diet."

Rollo hoped, as did the other physicians trying his approach, that once the disease was in abeyance, it was cured, or maybe it could be cured. Building up the body's ability to tolerate carbohydrates—vegetable matter—seemed to be the obvious therapeutic goal: establish a maximum amount of carbohydrates the patient could tolerate without the symptoms reappearing. Rollo recommended that when the patient seemed to be well, the physician should suggest "a gradual return to the use of bread, and those vegetables and drinks which are the least likely to furnish saccharine matter, or to become acid in the stomach." He also worried about causing "scurvy,* or something akin to it" without any vegetables in the diet.

---

* A disease that nutritionists in the twentieth century would come to understand as caused by a deficiency of vitamin C.

It all came down to a balancing act between physician and patients, between the strict animal diet that might keep the disease at bay and what the patients preferred to eat and the physicians thought they should. This conflict was most apparent with children, which is still the case today. One physician in the London area wrote to Rollo in February 1798 describing his trial of Rollo's animal diet with a twelve-year-old girl. She was "of a thin habit of body, tall of her age," he wrote, accustomed "to eat much fruit, sweetmeats and pickles," and now afflicted by diabetes. As her health would seemingly improve on the animal diet, the physician would either give her a "small quantity of bread" or a few biscuits to see if the sweet urine returned. Occasionally the physician would learn that his young patient had been indulging herself in forbidden foods. Thirst, headache, and sweet urine betrayed the deviation. In his letter to Rollo, the physician notes repeated transgressions followed, invariably, by assurances from the young patient of "more steadiness in future." If nothing else, the physician concludes, Rollo has gifted medicine with a way to control diabetes. After that, it was up to the patient.

By the mid-nineteenth century, Rollo's animal diet had become accepted practice for treating diabetes. Prominent diabetologists often found reason to dismiss Rollo's specific instructions—rancid meat and old fat were beyond the pale—but they embraced the animal diet with fresh meat and fat and extreme carbohydrate restriction as the only therapy that could keep their patients alive. "The great point, then, in the treatment of diabetes, is to accustom the patient gradually to live entirely on meat," Thomas King Chambers, physician to the prince of Wales and perhaps the leading British medical nutritionist of the Victorian era, said in his 1862 lectures. Chambers did not see a "patient to be made carnivorous" as a deprivation or unnatural. Other cultures ate that way and thrived. "This need not seem a mighty hardship," he said; "the iron-framed Esquimaux do it, and the wiry, tough, half-breeds of the Pampas, with a bill of fare certainly less varied in flesh-meat than our European meadows afford." Allowing a diabetic patient to eat starches and grains, Chambers believed, was actively harmful and could not be justified.

Frederick Pavy, the leading British diabetologist of the era, also invoked the Inuit and the "Gauchos of South America" as evidence that humans can thrive on carnivorous diets. "There can be no reason, therefore, against a strictly animal regimen being adequate to all the physiological requirements of the Diabetic," he wrote in 1862 in the first English-language diabetes textbook; "and, I believe, if persevered in for a short time, it would come to be taken without giving rise to any feeling of material privation. Indeed, I have been told by patients, whom I have restricted to animal food; that, after a little while, they have experienced no desire to take any other."

Because uncontrolled diabetes can so easily prove fatal—a cold, an infection, an injury that a healthy person can easily withstand might lead to death in a patient with diabetes—Pavy was among the first to argue for strict control of the blood sugar, which at the time meant a strict animal diet. Those with diabetes have "a very insecure hold upon life . . . ," he wrote. "I do not go so far as to say, that the restriction in diet will in every case prevent Diabetes from proving fatal; . . . but, I do think, that the secondary ailments may be, in a great measure, if not entirely, staved off by it."

In France, Apollinaire Bouchardat, considered among the pioneers of modern diabetes therapy—"easily the most brilliant clinician in the history of diabetes," as Frederick Allen of the Rockefeller Institute for Medical Research put it*—took up Rollo's treatment and pleaded with his colleagues in the 1840s to use it as the only way to keep their diabetic patients alive. Bouchardat explicitly removed the rancid meats from the diet and suggested green vegetables as a side dish. He apparently also inaugurated the practice of boiling the vegetables three times before serving. Bouchardat recognized that milk contained significant carbohydrates—lactose—and prohibited that. Bouchardat was also the first to advocate for regular

---

* Allen may have been so excessively complimentary because Bouchardat's thinking so clearly aligned with his own. Both, in effect, counseled their patients with Bouchardat's dictum: "*Manger le moins possible*" (Eat as little as possible).

exercise as a way of keeping the disease under control, and daily testing of the urine to assure that the program of diet and exercise was doing the job.

In Italy, the animal diet was taken up by Arnoldo Cantani of the University of Pavia and then Naples. Cantani was certain that the principal cause of diabetes was the excessive consumption of starch and sugar, and that the remedy "is not in the drugstore but in the kitchen," a statement with which many diabetologists would still agree. Cantani prescribed only meats and fat, "followed with absolute rigor and complete perseverance"—"the simplest [treatment] in the world," he called it—and kept his patients "under lock and key," quite literally, to assure they didn't cheat. If they went two months without sugar in their urine tests, he would allow them green vegetables "and later, wine, cheese, nuts, sugar-poor fruits, and finally small quantities of farinaceous [starchy] foods."

Cantani's approach was then taken up in Germany by Bernhard Naunyn—"the foremost diabetic authority of the time," as Frederick Allen described him. Naunyn conclusively demonstrated that a significant percentage of the protein in the diet could also excite what these physicians described as the sugar-forming apparatus. Almost 60 percent of the protein consumed by a diabetic on an animal diet (as first reported by the German physician Wilhelm Griesinger after doing multiple experiments with a single patient) would apparently be converted to glucose, the form of carbohydrate that dominates in the circulation and constitutes the blood sugar itself, and appear in the urine in severe cases. Now the focus was not just on eating animal diets and the avoidance of carbohydrates or vegetable matter, but the importance of fat in the diabetic diet to make up for the carbohydrate calories that were missing and the protein that might be converted to carbohydrate and have to be restricted as well. As Naunyn's thinking was later described by Frederick Allen, "The introduction of fat is the most important art in diabetic cookery."

By the end of the nineteenth century this was the prescription in medical textbooks on both sides of the Atlantic, with lists of allowed foods (all meat, fish, and fowl and green vegetables) and those disallowed (all other vegetable or plant-based foods, as we would say today), little different from that of Pavy's or Bouchardat's. "The carbohydrates in the food should be reduced to a minimum," wrote William Osler, considered the father of modern medicine in the United States and Canada, in the 1893 edition of his seminal textbook *The Principles and Practice of Medicine*. Under such a regimen, "all cases are benefited and some are cured."

In the 1970s, the American Diabetes Association would flip this prescription on its head and other diabetes associations worldwide would follow suit. Not only would the ADA take to recommending high-carbohydrate diets for diabetes with *more than half* the calories coming from those foods specifically prohibited in the pre-insulin era, but they would insist that anyone taking insulin should be eating considerable amounts of carbohydrate-rich foods throughout the day—at mealtimes, but also during preplanned snacks. It's not unusual for children and adults diagnosed with type 1 diabetes today to find it difficult to eat as many carbohydrates as their diabetes educators insist that they should, as one Texas mother told me of her daughter, diagnosed at age eleven. The introduction of insulin in 1921 would be a primary reason why this shift occurred, but not, by any means, the only one.

When Osler was writing in *The Principles and Practice of Medicine* that "all cases are benefited and some are cured" by a diet with minimal carbohydrates, he was overstating the case. It's unlikely *all* cases benefited, or that anyone was being cured. Older patients with type 2 diabetes could control their disease if they stayed on their fat-rich animal diets. But the young patients, those with type 1, were continuing to die within weeks or a few years after diagnosis, either from infections or disease—tuberculosis, most commonly—or in diabetic comas or some combination of the three. Some died simply because they were diagnosed with the disease at such a late stage that they succumbed before the physician had any meaningful chance to intervene. As for the rest, no one knows whether they died because they didn't follow the diets prescribed, or because the diets didn't help or help enough—as Frederick Allen and Elliott Joslin came to assume—or because the physicians continued to believe that the diets were temporary, the goal being to build up the carbohydrate tolerance of the patient.*

Given the choice between a drug and a restricted diet, physicians assumed that most patients would prefer the drug, no matter the side effects. One French physician is said to have remarked in defense of this thinking that "patients in France were less willing than those in other countries to adhere to restricted diet, and demanded a cure which would enable them to eat freely." The fact that the available drugs didn't help was another explanation for the continued deaths. By the late 1890s, the list of compounds being marketed in the United States as antidiabetic, as cataloged in one government document, ran to over forty, and included uranium nitrate and arsenic. The only drug that appeared to help—and was prescribed widely—was opium, not because it influenced the progression of the disease in any meaningful way, but because it eased the mental strains of the patient. Meanwhile, physicians continued to experiment with dietary modifications.

For those patients in dire straits, still on the verge of coma despite eating some variation of the animal diet, physicians would try radical variations. If the patients appeared to get healthier, even if just over the course of the few days or weeks they remained in a hospital or clinic, the physicians would write it up in a journal or a pamphlet, just as Rollo had, and it would enter into the medical armamentarium. Since animal diets and abstinence from carbohydrates was the first line of dietary therapy on diabetes patients, these alternative diets were almost invariably carbohydrate rich.

These would later be invoked as a reason to embrace the relatively carbohydrate-rich diets that diabetes associations in the post–World War II years began to promote, but the evidence supporting their use was always slim at best. There was the sugar diet, for instance, promoted by Pierre Piorry, one of the most prolific and controversial French physicians of the nineteenth century,

---

* "It is desirable to continue the use of strict diet for some days or weeks in order that the tissues may regain part of their lost sugar-consuming power, and the duration of this regimen must depend upon the condition of the patient," as the British diabetologist Robert Saundby explained in 1908 in *A System of Medicine*. "Whenever a change becomes desirable we should increase the carbohydrate food."

once described as "a man who loved to turn everything on its head." Piorry theorized that diabetics were losing sugar in their urine, so it was sugar they should eat to excess. A British diabetologist working at the Leicester Infirmary reported in 1858 that he had tried it on three patients—they were given half a pound each day of treacle (a syrup made from molasses). It didn't help. One well-known physiologist of the era took it up after becoming diabetic himself and his disease "ran a quickly fatal course, apparently because of the treatment." The sugar cure faded into history. The London physician Arthur Donkin claimed in 1871 that he could "produce complete relief from suffering" in his diabetic patients with "almost magical *rapidity*" by feeding them seven to eight pints a day of skim milk. Donkin argued that the problem with milk was not the carbohydrate content (the lactose) but the dairy fats. "It has not come into favor," wrote Elliott Joslin in the first edition (1916) of his textbook. By the late 1890s, diabetologists had tried diets of rice and cereals, and of mostly to only potatoes, none of which panned out.

The one dietary therapy that had some staying power was the oatmeal cure promoted by Carl von Noorden, who succeeded Naunyn in the early twentieth century as the leading German authority on diabetes. The treatment was hailed by those physicians who were looking for evidence that diets rich with starches and grains could be beneficial for diabetic patients. But von Noorden's cure required serving the oatmeal with an "equal amount of butter," which meant that as much as 75 percent of the calories would have come from fat. Von Noorden was insistent that patients start the cure with a day or more of fasting, followed by several days of a "vegetable diet," "when only bacon, butter and a few eggs are taken besides green vegetables." That was followed by, at most, three days of oatmeal and butter. In maybe one in ten cases, Viennese authority Wilhelm Falta wrote, "the success can be called really remarkable." By the 1900s, Osler was prescribing it for severe cases—a "most excellent" diet, he called it—serving the butter-oatmeal mixture, 250 grams of each, four times a day to his patients along with the whites of half a dozen eggs.

But the oatmeal cure was no cure, just a short-term treatment with some poorly understood benefits regarding urinary sugar excretion. Even von Noorden felt compelled to disavow it eventually. The oatmeal had only temporary value for severely ill patients, rendering their urine sugar-free, he said, which was always the first goal of treatment. "We do not yet know why the oatmeal cure sometimes does good and sometimes does harm," he explained in 1912 in a series of lectures he gave to physicians in New York City. "Only temporary use must be made of such methods of cure." He then told his audience that he had spent five years striving to convince his peers that the oatmeal cure was not quackery, and then had to spend the following five arguing that it was being overused. "I have myself," von Noorden said, "warned people most emphatically that the oatmeal cure was not to be regarded as a panacea for diabetes; this warning has not been sufficiently taken to heart."

Von Noorden's preferred therapy for his diabetic patients remained the animal diet with green leafy vegetables, and for severe cases, limiting the consumption of large amounts of protein—beefsteak, most noticeably. "The general direction that the intake of carbohydrates should be restricted or excluded," he told the physicians in New York, "stands today in the foreground of diabetic therapy, just as it did in the previous century."

In 1905, seven years after Elliott Joslin opened his clinic in Boston, the first in North America to specialize in diabetes, Joslin reviewed what he had learned from both his studies of diabetes patients at Boston's Massachusetts General Hospital and his clinical experience. "The diabetic's chance for life depends upon his ability to eat fat," he wrote, "and consequently great care must be exercised not to prejudice him or even ourselves against its digestibility."

3

Diabetes in Retrospect

Each science confines itself to a fragment of the evidence and weaves its theories in terms of notions suggested by that fragment. Such a procedure is necessary by reason of the limitations of human ability. But its dangers should always be kept in mind.

—ALFRED NORTH WHITEHEAD,
Modes of Thought, 1938

Diabetes, recognizable clinically, is rather the end-result of a series of catastrophes than a primary condition. As in any endocrine disturbance, the original error in diabetes may be overlaid by symptoms caused by the failure of other systems which have been thrown out of gear by the initial defect. The hyperglycaemic state, the result of the original error in diabetes, is in itself symptomless. It is not until this state produces secondary troubles such as polyuria and thus thirst, or arterial degeneration, or wasting, that the condition becomes clinically recognizable as diabetes.

—DENNIS EMBLETON,
"Dietetic Treatment of Diabetes Mellitus
with Special Reference to High Blood-Pressure," 1938

Until 1889, when physicians speculated on the organ responsible for diabetes, they typically assumed it was either the brain or the liver, largely because of work done by the legendary French physiologist Claude Bernard at his Paris laboratory at Collège de France. In 1849, Bernard had famously demonstrated that puncturing the floor of the fourth ventricle of the brain causes the animal to pass sugar in its urine. So maybe diabetes was a disorder of the brain. Physicians speculated that animals (then) did not seem to get diabetes naturally and humans did because we had larger brains and more brain power to go awry, and that is why diabetes in the nineteenth century appeared more commonly among the educated classes. Their nervous systems supposedly took on "more wear and tear" than those of the uneducated.

Bernard's reputation, though, was ultimately founded on his studies of liver function and his demonstration that the liver did indeed produce and secrete glucose, even when no vegetable matter, and so no carbohydrates, were consumed. Bernard is credited with the discovery of *glycogen* (he coined the term), which is the chemical form in which the liver and muscle cells store glucose to be used as fuel when needed.* Bernard postulated that if the liver could influence blood sugar levels by producing and secreting its own glucose, independent of any carbohydrates in the diet, that could be the cause of diabetes. The British diabetologist George Harley, who had spent time studying with Bernard in Paris, considered it possible. In Harley's 1866 book on diabetes, *Diabetes: Its Various Forms and Different Treatments,* he suggested that overproduction of sugar by the liver was why his "fat and ruddy" patients became diabetic. Bernard and Harley were more right than wrong in their thinking, but oversimplification of a complex physiology

---

* Bernard's demonstration that a puncture of the fourth ventricle of the brain caused blood sugar to rise was later understood as the brain stimulating the liver through the vagus nerve to secrete the glucose it had stored up as glycogen. For this reason it was a temporary effect, as the blood sugar would return to normal when the liver ran out of glycogen.

would lead future generations of diabetologists away from the right answer rather than toward it.

All of this changed in April 1889, with the first great discovery in diabetes research. Two versions exist of this event and they're both worth recounting. Oskar Minkowski, the scientist who played the critical role, described the discovery as a "lucky accident." Minkowski was working in Bernhard Naunyn's institute at the University of Strasbourg in what was then Germany and would become France after the First World War. Minkowski was having what we might nowadays describe as a watercooler conversation with one of his associates, Joseph von Mering, discussing the role of the pancreas and whether or not it could be removed from a dog without killing it. (Dogs were the animals used most commonly by laboratory researchers of the era for their investigations. Their lives were not enviable.)

The pancreas is a large gland that sits behind the stomach, its head nestled in a bend of the small intestine called the duodenum. Von Mering had broached the idea of removing the gland, not because he was thinking about diabetes, but because he was studying secretions that flowed from the pancreas into the intestine and might play a role in the digestion of fats. Removing the pancreas would have been an obvious way to test if that was the case, assuming the dog could survive the surgery. Most researchers at the time would have guessed not—that was the conventional wisdom largely because Claude Bernard had tried and failed—but von Mering had a dog available and Minkowski, proud of his surgical skills, was willing to give it a try.* (This was before the days of ethics committees, as the British diabetologist Robert Tattersall later wrote.) With von Mering assisting, Minkowski removed the dog's pancreas and the animal survived. Von Mering had to leave Strasbourg the next day, and the dog, now *pancreatectomized,* was left in a cage in Minkowski's lab.

In a letter written in 1926 explaining why he had deservedly

---

* Minkowski later said that if he had known that Bernard had "stated that it was impossible for dogs to survive the total surgical removal of the pancreas," he and von Mering would never have tried it themselves.

received credit for the discovery and not von Mering, Minkowski said he had noticed that the dog, previously house-trained, had urinated on the laboratory floor. When he scolded an assistant for not taking the dog outside to do its business, the assistant replied that he had. But "as soon as it comes back," the assistant said, "it passes water again even if it has just done so outside." Minkowski tested the dog's urine for sugar and found plenty. Minkowski removed the pancreas from three more dogs. Two died shortly after the surgery, but both had sugary urine before they did, and the third "had a persistent diabetes just like the first animal's." When von Mering returned to Strasbourg after a week's absence, Minkowski said to him, "Do you know, von Mering, that all pancreatectomized dogs become diabetic?" The pancreas was playing a critical role in diabetes. That much seemed certain.

After publication of Minkowski's revelation, researchers set out to discover what this organ might be secreting that could have such a profound effect on carbohydrate metabolism. By the 1880s, they had identified the cells responsible for the secretion—known as the islets of Langerhans, after the German medical student Paul Langerhans, who had described them first in his 1869 dissertation.

In 1921, Frederick Banting, Charles Best, and their colleagues at the University of Toronto purified insulin from pancreatic tissue, and diabetologists had their central dogma: diabetes was a pancreatic disorder of carbohydrate metabolism and insulin deficiency. Most thinking about therapy would follow from that. To Minkowski's credit, he believed only that they had discovered the "existence of *a* pancreatic diabetes" (my italics), not that they had established the cause of all diabetes. Evidence that the liver—as Bernard had suggested—and perhaps other organs were playing a critical role in the disease state would be "virtually ignored," in the words of the University of Chicago endocrinologist Samuel Soskin in 1941, and this continued to be true for decades after.

In the second version of the discovery, which dates back at least to 1919, Minkowski noticed that the urine from the pancreatectomized dog had attracted a swarm of flies, and that in turn prompted him to taste the urine and find it sweet. This was the version preferred by the University of Texas Southwestern physiologist

J. Denis McGarry in a 1992 article in the journal *Science* that would become something of a legend in the diabetes literature. McGarry was a biochemist, among the most celebrated of his era. He would die in 2002 of a brain tumor at age sixty-one, ending prematurely his brilliant career and immediate influence on the thinking of diabetologists. But his research, as a remembrance in the journal *Diabetes Care* described it, constituted "a foundational pillar" for understanding human metabolism and nutrition. It is critical to understanding the pathology of diabetes as well.

In his article, McGarry was asking why a century of diabetologists had made such little progress in understanding a disease that had become so tragically common in their lifetimes. "Although much has been learned," McGarry wrote, "current knowledge remains largely descriptive, consisting mainly of an ever expanding list of the metabolic, vascular, and neurological abnormalities that accompany the active disease process. What has not yet emerged, despite immense investment of resources, is a clear understanding of the basic pathophysiological mechanisms of diabetes and their temporal relations to each other." In simpler language, despite all that immense investment in time, effort, and money, diabetologists had failed to discover what causes diabetes and in what order bad things happen.

McGarry's 1992 article had a memorable title, "What If Minkowski Had Been Ageusic?"—"*ageusic*" meaning the absence of a sense of taste. McGarry suggested that had Minkowski failed to taste the sweetness of the dog's urine, he might have noticed instead the pungent sweet-rotten aroma of acetone, from the ketone bodies that would also have been passed in the dog's urine. (Acetone is best known as the pungent solvent in nail polish removers.) Ketone bodies are a product of fat metabolism. In the absence of insulin, fat metabolism malfunctions, not just carbohydrate metabolism, and ketone bodies are produced and excreted in excess. Having smelled the acetone, McGarry wrote, Minkowski "would surely have concluded that removal of the pancreas causes fatty acid metabolism to go awry." Then McGarry suggested that the major conclusion of the work of Banting and Best and their Toronto colleagues, the discovery of insulin and its use to treat diabetes, might have been that insulin was necessary to keep fat metabolism under control and that's what insulin therapy accomplished. Diabetologists would then come to think of diabetes as a dysregulation of fat metabolism, not carbohydrate metabolism, and they would have crafted their theories and their therapies from that perspective, understanding why the eating of carbohydrates induced a disorder of fat metabolism, an entirely different perspective on the problem.

Among the lessons of McGarry's article was a simple one: the order in which scientists learn something is likely to determine how they think about it, right or wrong, ever after. The way diabetologists had come to think about the disease—focusing on glucose in the urine and blood and on carbohydrate metabolism rather than fat metabolism or, ideally, both—hampered their future efforts to understand diabetes properly and so to treat it appropriately.

In 2001, McGarry was asked to give the prestigious Banting Memorial Lecture at the annual meeting of the American Diabetes Association, and he made this point once again. As McGarry explained it then, the summer before his death, diabetologists had insisted on thinking about insulin and diabetes from a simplistic perspective—insulin as a hormone that merely regulates blood sugar and diabetes as a pancreatic disorder of carbohydrate metabolism. By doing so, they had derailed their progress; they were still "grappling with the enormous complexity of a disease process in which almost every aspect of the body's metabolism goes awry."

The problem of oversimplifying a very complex system also dictated the way the disease would come to be treated. Diabetologists set out to desugarize the urine in their patients and, they hoped, prevent them from dying in a diabetic coma. Once Minkowski's work established that the pancreas plays a critical role and Banting and Best and their University of Toronto colleagues isolated insulin from pancreatic tissue, that job could be handled by insulin therapy. That then left them with the job of treating the longer-term complications of the disease, and they have been struggling with that ever since. In doing so, however, these diabetologists rarely thought of the complications as something that might have been caused by *how* they were treating diabetes, by the insulin therapy or the dietary approach. They didn't grasp the "enormous complexity of a disease process."

Instead, they did what innovative physicians do; they developed and employed new therapies, second and then third and fourth lines of defense. They produced drugs and devices that could treat the complications individually: blood pressure medications for the hypertension associated with diabetes, laser surgeries and medications for the retinal hemorrhages, statins and other drugs for the heart disease and atherosclerosis, dialysis machines and drugs for the kidney failure, even drugs now for the weight gain that seems part and parcel of diabetes therapies. Oral medications, better insulins (first longer acting, then faster acting), better devices for injecting or infusing the insulin (insulin pumps) were created and put to use. All the complications of diabetes would be treated as distinct ailments, requiring their own unique therapies. They still are.

All of this is a consequence of the way in which the field of endocrinology itself evolved, and the disconnect in medicine between theory and practice.

Medical historians date the birth of endocrinology to the 1850s, and the first time a physiologist—Claude Bernard, once again—used the term "internal secretion" to describe an organ or a gland discharging a substance into the circulatory system that could have effects throughout the entire body. When Bernard used the phrase, he was referring specifically to the glucose secreted by the liver, but it would come to be used for secretions from the adrenal glands and thyroid glands, the gonads and uterus, until eventually these internal secretions—thanks to the suggestion of the British biochemist Ernest Starling in 1905—came to be called "hormones" instead. "Hormone" derives from a Greek term meaning "excite" or "arouse," and the idea was that these chemicals are secreted by one organ or tissue and travel through the blood to arouse or excite an action in another. The term "endocrine," literally meaning "secreting internally," and the field of endocrinology itself date to a few years later still. Endocrine glands are those that lack an obvious pathway, a duct, into, say, the gut or a neighboring organ for their secretions to flow. They were known as "ductless" glands well into the twentieth century, until "endocrine" eventually became the preferred term.

It was another French physician, the flamboyant Charles-Édouard Brown-Séquard, who pioneered the thinking on internal secretions and provided his fellow doctors both the motivation to study them and the plan of attack by which they would do so. In 1869, while doing research on the adrenal glands, Brown-Séquard suggested that all ductless glands "give to the blood, by an internal secretion, principles which are of great importance if not necessary." Hence, if physicians "could safely introduce the principle of the internal secretion of a gland taken from a living animal into the blood of men suffering from the lack of that secretion, important therapeutic effects would thereby be obtained."

Finding what was missing or malfunctioning and replacing it became the strategy for essentially all that has come after. Physicians followed Brown-Séquard's lead, attempted to link defective glands to particular diseases and then isolate the same secretions from the glands of animals, with the hope of injecting them into sick patients to cure their ills or even into not-so-sick patients to restore their vigor. This gave the pursuit an enduring taint of quackery. In 1889, Brown-Séquard gave a lecture in which he claimed that he had proved the efficacy and safety of this kind of medical therapy by using himself as an experimental subject. He had injected himself with the sterilized fluid collected from dogs' testicles, and, so he said, at seventy-two years of age had been physically rejuvenated; he could lift heavier weights than previously, run up stairs, and urinate 25 percent farther (considered a measure of bladder strength). "Though many jeered at him as the discoverer of the secret of perpetual youth," the *British Medical Journal* editorialized four years later, "the notion [of internal secretions] has steadily gained ground that there is, after all, something in it."

The *BMJ* editors were worried that Brown-Séquard's work would launch an era of quack hormonal cures, and it did. But it also prompted serious research and set a context for understanding how human diseases might be caused by either an excess of these internal secretions or a deficiency. Minkowski's discovery that the pancreas produced an internal secretion, and that it played a critical role in carbohydrate metabolism, was quickly embraced as compelling evidence that Brown-Séquard was on the right track. Further evidence was generated in the 1890s by the discovery that goiter, cretinism, and myxedema (severe hypothyroidism) could be treated by extracts of the thyroid gland. "Evidently the gland produced a secretion whose deficiency could be supplied artificially," as the medical historian Michael Bliss wrote.

For all this progress, though, what the research nonetheless left out was the discipline's fundamental organizing principle. Only in the mid-twentieth century did the notion of *homeostasis* begin to appear, in passing, in the textbooks. And yet it may have been, may still be, after Darwin's theory of evolution, the single most profound theoretical concept in all of medicine and biology. For diabetic patients, it is the restoration of homeostasis that any therapy is ultimately trying to achieve. Diabetologists would acknowledge the truth of this and expand their understanding in the post–World War II decades but they would never come to terms with its implications.

The term *homeostasis*, derived from the Greek words for *similar* and *standing still*, was coined in 1926 by the Harvard physiologist Walter Cannon to describe a concept that, once again, originated with Claude Bernard. The prerequisite for the life of any multicellular organism, Bernard observed in 1849, is that it maintains the internal environment of its cells—the *milieu intérieur*, in Bernard's famous phrase—within a narrow range of physiological values conducive for healthy functioning.

Warm-blooded animals—humans, for instance—require a body temperature that remains within a few degrees of 99 degrees Fahrenheit. (The body temperature of birds, on average, runs at about 104 degrees.) Should it stray higher or lower for whatever reason, regardless of external conditions, cells will begin to malfunction. The chemical reactions that make up the multitude of cellular metabolic processes are temperature dependent: they will run at different rates at different temperatures, and the demands and products of these metabolic processes will no longer be balanced. As the cells malfunction, tissues, organs, and the body itself will follow suit. "All the vital mechanisms" of a living multicellular organism, wrote Bernard, "however varied they may be, have only one object, that of preserving constant the conditions of life in the internal environment."*

This concept of homeostasis is very much a wholistic one, a point that has traditionally been lost in the limited discussion of it in medical textbooks. At every moment of the day, the body is working as a "harmonious ensemble," in Bernard's words, to maintain the necessary stability of the internal environment. No physiological system works in isolation. Perturb one system and the entire harmonious ensemble of the organism reverberates with corrections and countercorrections. In this sense, we can think of the human body as a fantastically complex, interrelated web of signals and countersignals, actions and reactions, in which the autonomic nervous system and the endocrine system of hormones play critical roles. The brain, and particularly the hypothalamus, sitting just above the brain stem, monitors the universe around us and sends signals through the nervous system to organs and glands prompting them to act in response to however the environment is changing or even is likely to change, whether it's the time of day or season of the year, changes in temperature or weather, the presence of predators or potential mates, or the opportunity of prey.

Among the key actions of those nervous system signals is the stimulation of endocrine secretions—hormones—which travel from the organs that produce them through the circulation, signaling cells throughout the body to respond as necessary. And the response, in turn, signals the nervous system and other organs, endocrine and otherwise, to respond, and on it goes. When Bernard demonstrated that puncturing the brain at the juncture of the hypothalamus and the vagus nerve stimulated the liver to secrete glucose into the circulation, he was interrupting a system by which the brain could stimulate the autonomic nervous system to signal the liver to make fuel available for whatever energetic need had to be either anticipated or immediately satisfied. This system is initi-

---

* "No more pregnant sentence was ever framed by a physiologist," the British biologist J. B. S. Haldane said a half century later of Bernard's observation.

ated automatically, for instance, when the brain detects that blood sugar is dropping dangerously low and must be raised.

In 1923, Ernest Starling described Bernard's conception as each organ and each cell working for all others, and the circulatory system as the "common medium from which each cell can pick up whatever it requires for its needs, while giving off in return the products of its activity." The hormones are the most conspicuous chemical messengers of this system, sending messages in response to the changes in the internal environment and the stimulation of the nervous system. No signal is generated without a countersignal, *counterregulatory hormones*, working to keep the system under control (known as a negative feedback loop). Countersignals in turn generate counterregulatory effects of their own, and, as I said, on it goes.

With this wholistic conception in mind, biologists and physiologists since Bernard have cautioned against the dangers of reductionism in understanding the human body in health and disease states because nothing the body does works in isolation. "We really must learn," as Bernard wrote, "that if we break up a living organism by isolating its different parts, it is only for the sake of ease in experimental analysis, and by no means in order to conceive them separately. Indeed when we wish to ascribe to a physiological quality its value and true significance, we must always refer it to this whole, and draw our final conclusion only in relation to its effects in the whole." A century later, the Nobel laureate biologist Hans Krebs paraphrased this lesson: If we neglect "the wholeness of the organism," he said, "we may be led, even if we experimented skillfully, to very false ideas and very erroneous deductions."

But this thinking was (and mostly still is) the contextual, intellectual stuff of research scientists, not practicing physicians. Physiologists like Bernard, Cannon, and Krebs could think like this; physicians didn't and can't, forced as they are to make immediate decisions to address symptoms or disease. Those making discoveries in the field of endocrinology leading up to insulin and beyond, and translating those discoveries into practical advances in medicine, have done so precisely in the manner that Bernard warned against: disease by disease, gland by gland, hormone by hormone.

Throughout the twentieth century, chapters in the textbooks that focused on the growing body of endocrine research would be organized by the endocrine glands themselves—a chapter for the adrenals, another for the thyroid, another for the pituitary gland, and so on—with little attention paid to the interaction between the organs or the tissues that were on the receiving end of the internal secretions themselves. The textbooks might open with a discussion of homeostasis as the organizing principle but then quickly leave Bernard's conception behind. They would acknowledge that all hormones have multiple effects, probably even multiple effects within the same cells and certainly different and often contrary effects in different tissues, and that any effect observed would be the result not just of one hormone, but the counterregulatory hormones and any others that might be triggered in endless downstream cascades. Then the textbook authors would warn against reductionist thinking, just as Bernard had, and default to the needs of everyday medicine.

This problem was compounded by the absence of a critical experimental tool necessary for any scientific endeavor: the ability to accurately measure the quantity or concentration of the substance of study. Endocrinologists and physicians could tell whether a hormone was dysfunctional or missing entirely by the effect it had on the human body (or their laboratory animals)—dwarfism or gigantism (acromegaly), for instance, and pituitary secretions, diabetes and the pancreatic secretion, insulin. But they could not measure the amount of a hormone present in the circulation; they could not determine whether the problem was with the hormone itself—too much or too little, or perhaps a biochemical defect—or with the cells and tissues that were on the receiving end of that hormonal signal.

Only in 1960 did two researchers—Solomon Berson, a physician, and Rosalyn Yalow, a physicist—working at a New York City Veterans Administration hospital, publish the details of a laboratory assay, a test, that could accurately measure hormone levels in the circulation. They called it the radioimmunoassay, and they first demonstrated its efficacy with insulin. The revelations that emerged from Yalow and Berson's research—their earliest papers—

would force a revolution in the understanding of type 2 diabetes. It was not a disorder of insulin deficiency, hence of the pancreas, as physicians had assumed since Minkowski; rather, individuals with this chronic form of diabetes tend to have too much insulin in their circulation—hyperinsulinemia. The fact that their blood sugar is also high at the same time means their cells are not paying sufficient attention to the insulin secreted, hence insulin resistance.

Eventually, researchers would demonstrate that the primary defect in type 2 diabetes is probably in the cells of the liver—as Bernard, Pavy, and Hardy had speculated—not the pancreas, but this came only after diabetologists had made up their minds about how to treat the disease with both drugs and diet. This breakthrough began to explain the connection between obesity and type 2 diabetes, since individuals with obesity, as Yalow and Berson also reported, had excessive insulin in their circulation and a tendency toward elevated blood sugar, suggesting that they, too, were insulin resistant. In 1977, Yalow would win a Nobel Prize for the discovery of the radioimmunoassay, five years after Berson had died. The Nobel Committee would say that their discovery had "brought about a revolution in biological and medical research." It completely transformed the understanding of type 2 diabetes as a disease entity, yet this new understanding had remarkably little effect on therapeutic principles.

Physiologists would come to accept that insulin has what McGarry called "a bewildering array of metabolic responses in target cells," and so the human body. This is the nature of homeostasis, but that's not how physicians had come to think of it. They had been taught that insulin controls blood sugar, and that's what concerned them therapeutically. When Banting and Best and their University of Toronto colleagues suggested the name "insulin" for the hormone they had purified from pancreatic tissue in 1921, they described it as a hormone that controls blood sugar—because that's what they were trying to find. Extracts of their hormone, they wrote, "when injected subcutaneously into normal rabbits cause the percentage of sugar in the blood to fall within a few hours." A century later, Google will tell you much the same thing: insulin is "a hormone produced in the pancreas . . . which regulates the amount of glucose in the blood." After 1960, when researchers using Yalow and Berson's radioimmunoassay started quantifying the high levels of insulin in the circulation of patients with type 2 diabetes and thinking that their cells must somehow be resistant to the insulin they were secreting, they measured and defined insulin resistance only in terms of the hormone's effect on blood sugar and what is known technically as *glucose uptake*. The rest of the bewildering array of effects that are dysregulated in diabetes, including all the effects on fat metabolism that McGarry had noted, were rarely considered.

Perhaps the best way to think about insulin and its function in the human body is as the dominant hormone orchestrating a process that is technically known as fuel partitioning. All the various processes under the influence of the endocrine system—reproduction, growth and development, protecting ourselves from illness and healing after injury or illness, maintaining homeostasis—require energy; all depend on the fuel available and the ability and readiness of the relevant cells to use that fuel to generate energy. That fuel is derived either from the meals we eat or from what we've stored from past meals. Those meals are composed of fats, carbohydrates, and protein, the three macronutrients in our diets. The mitochondria in our cells can use any of these three to generate energy (metabolize them or oxidize them), but the body has to prioritize how best to do so to maximize both our short- and long-term health.

These three different fuel sources have different utility for our bodies. Fats are ideal for fuel storage because they're the most energy dense of the three macronutrients—more energy can be stored for future use in less space than protein or carbohydrates. Fat, and hence the fat we eat, is also used for thermal regulation—keeping us warm—for cushioning our bones and organs, as structural components of cells and tissues. Fat is the primary component of cell membranes. Our brains are 60 percent fat, chiefly serving as insulation. Proteins are required for effectively all cellular processes and for growth and repair of cells and tissues. We can also store protein in muscles for future use as energy, if fat and glycogen supplies are depleted. Carbohydrates are unique in that they are used exclusively for generating energy and for limited storage of that energy—perhaps 2,000 calories in the form of glycogen, or a day's worth of (sedentary) energy.

The foods we eat are digested in the stomach and gut and the macronutrients they contain are released into the circulation to be put to work or stored for future work. The endocrine response to these macronutrients and the needs of the body will largely determine how this is accomplished. Every hormone has an influence on fuel metabolism because hormones are prompting functions that require energy to perform. In his book *Bittersweet*, a history of insulin therapy and the transformation of our understanding of type 1 diabetes from an acute to a chronic disease, Chris Feudtner aptly described this hormonal or endocrine regulation of energy metabolism and fuel partitioning as that of a "Council of Food Utilization." Organs communicate with one another "via the language of hormones," and the continuous dialogue that emerges "reflects the amount and type of food the body has eaten recently as well as the body's requirements for food energy. The rest of the body's tissues listen to this ongoing discussion and react to the overall pattern of hormonal messages."

Insulin dominates this process.* Its functions have to do with this fuel partitioning, affecting different organs and tissues in different ways. It facilitates the uptake of glucose (blood sugar) into cells, which is the function most obviously disturbed in diabetes and the one on which the medical community has focused most of its attention. But it also stimulates fat synthesis and storage in fat (adipose) tissue. It stimulates the uptake of fat by the cells, the conversion of carbohydrates into fat, as necessary, particularly in the liver, and then inhibits the release of fat by the fat cells. It stimulates the synthesis of glycogen in the liver and muscle cells, and so the storage of carbohydrates. It inhibits the liver's synthesis of ketone bodies, which is one reason why those who suffer from type 1 diabetes, who

---

* The process requires a counterregulatory hormone, as we'll discuss, called glucagon, also secreted by the pancreas.

lack insulin, were so likely to die of diabetic ketoacidosis, characterized by a toxic buildup of ketone bodies in the circulation and their
appearance in the urine at high concentrations. Insulin also stimulates the synthesis of proteins and the function, repair, and growth
of cells; it is an *anabolic* hormone in that sense, working to build up
the body. It stimulates the synthesis of RNA and DNA molecules in
the nucleus of cells. It does all this while assuring that fuel is available for all these functions. It inhibits the excretion of salt (sodium)
in the urine, which works to retain water as well. It's tempting to
assume that this function is accomplished because insulin will
stimulate the storage of glucose as glycogen—in response to rising
blood sugar—and each glycogen molecule requires three molecules
of water for every one of glucose.

All of these processes will be disturbed in disorders of insulin
signaling, in type 1 and type 2 diabetes. All of them are also what
endocrinologists and metabolism researchers, working too often in
their isolated academic silos, learned about insulin in the century
after its discovery. But as diabetologists developed their thinking
on how best to treat the disease in the early years of the twentieth
century and with the discovery of insulin itself, all they knew was
that those they diagnosed with diabetes had elevated blood sugar,
sugar in their urine, and all too often died of diabetic ketoacidosis.

Because these early diabetologists only knew how to measure the
amount of glucose in the blood or urine, they made lowering those
numbers the focus of their therapies. Largely, they still do. By the
1970s, endocrinologists had extended the use of Yalow and Berson's
radioimmunoassay to hormones beyond the pancreatic hormones,
and they took to diagnosing other endocrine diseases—hypo- or
hyperthyroidism, for instance—by the relative levels of the hormones involved. If the level of the hormone was too high, they tried
to lower it; if too low, they tried to add it back.

"The striking exception," as the British diabetologist Robert Tattersall has written, "is diabetes, which is still defined by blood glucose levels rather than by the amount of the hormone in the blood."
If type 2 diabetes was diagnosed by its insulin levels—invariably
elevated until the stage at which the pancreas exhausts its ability to
keep up with the demand of the insulin resistance—then treatments

might have been aimed at lowering insulin and judged by how well they accomplished that task. Rather, the focus has remained on blood sugar.

Today we understand the many and varied complications of diabetes. The physicians treating the disease a century ago rarely did. If we're to understand where they might have erred, we have to understand and compare those two perspectives: what we know now, and what they knew then.

4

The Fear of Fat

The history of diabetic therapy, so far as it is significant or valuable, consists merely in an interweaving or alternation of two principles. One is restriction of the sugar-yielding elements of the diet, namely carbohydrate and protein. The other is diminution of the total caloric value of the diet.

—FREDERICK ALLEN,
"Investigative and Scientific Phases of the Diabetic Question," 1916

When nutrition is the paramount consideration we must administer every possible calorie.

—ERICH GRAFE,
Metabolic Diseases and Their Treatment, 1933

Before diabetologists could embrace the idea of prescribing carbohydrate-rich diets to patients suffering from a disorder defined first and foremost by an inability to properly metabolize the carbohydrates in the diet, the physicians had to believe that carbohydrates are benign to those with diabetes so long as they either take their insulin or maintain a healthy weight. And they had to believe that dietary fat was not.

Until 1913, specialists generally believed that copious dietary fat was a necessity in the diabetic diet—"the staple food in diabetes," as Erich Grafe of the University of Würzburg described how the German authorities perceived fat in his seminal 1933 textbook, *Metabolic Diseases and Their Treatment*. It is the only major source of energy in our diet, other than alcohol,* that can be metabolized and generate energy for the human body without requiring a functional pancreas to control blood sugar while it happens. As long as that was seen as true, it seemed more prudent to have those with diabetes get their necessary calories from fat, which they could seemingly metabolize without consequence, rather than carbohydrates, which they could not.

By 1921, that thinking had been confirmed by diabetologists from the University of Michigan, Northwestern University Medical School in Chicago, and the University of Lund in Sweden. It had been embraced by leading clinicians and nutritionists in the United States and Europe and applied to practical benefit in clinics throughout both continents. The seminal paper on the relevant biochemistry had been referred to as "a masterpiece" by Graham Lusk, the most respected nutrition authority of that era in the United States, in his classic textbook, *The Elements of the Science of Nutrition*.

That is not, however, the scientific legacy of this era. Instead, in the United States, fat became a thing to be feared in diabetic diets—possibly "poison," Elliott Joslin would say, believing it was the fat in the diet of diabetic patients that precipitated "the dreaded acidosis" that caused heart disease, that led to obesity. That fear of fat has haunted the field of diabetes since, its pernicious influence ever present. Once insulin was discovered in 1921 and put to use in therapy, ever more carbohydrate-rich diabetic diets were employed. The fact that diabetic diets could be both high in fat and high in calories, and that those with diabetes had reportedly thrived on these diets, would simply be disregarded. Elliott Joslin, who would become the most influential diabetes specialist in the world in this era, would play the critical role.

---

* And fructose, which is the sweetest of the carbohydrates and is one-half of a sugar (sucrose) molecule.

Joslin grew up in the late nineteenth century in Oxford, Massachusetts, then a small town just fifty miles west of Boston. His introduction to diabetes had come when he was a child, living next door to his aunt Ellen, who had been diagnosed with the illness. She'd been "a jolly, plump, middle-aged woman" until the disease manifested itself, Joslin would reminisce many decades later: "Remedy after remedy was tried in accordance with the custom of the day . . . until at length she died, wan and emaciated, in diabetic coma." Then two more neighbors came down with the disease, and then his mother, in the spring of 1899, when she was sixty years old. Joslin's mother, like his aunt, had been obese by today's classification, but she had been losing weight steadily before her diagnosis. Joslin's mother was "Case No. 8" in his clinical records (though he may never have made that public), which would eventually detail tens of thousands of cases, all meticulously documented.

By the time of his mother's diagnosis, Joslin was a working physician, a graduate of Yale College and Harvard Medical School, beginning his lifetime of obsessive focus on diabetes and its treatment. He'd seen his first case professionally while a medical student in 1893. That patient had been a young woman from Ireland working as a domestic in the Boston suburbs. Joslin described her in his notes as having "a severe form" of the disease. Shortly after his mother's diagnosis and on her account, Joslin traveled to Europe to learn the latest in diabetic therapy from the great German diabetologist Bernhard Naunyn. This was "the heyday of German Medical science," as Joslin would describe it. It was at Naunyn's institute in Strasbourg that Joslin met Oskar Minkowski and many of the other great European diabetologists of that era and embraced Naunyn's thinking. When he returned to Massachusetts, he convinced his mother to adhere strictly to Naunyn's dietary principles. "A diabetic diet is really an ordinary diet in which the carbohydrates are replaced by fat," as Joslin described this thinking a few years later. "The diabetic's chance for life depends upon his ability to eat fat." His mother's diet then consisted almost exclusively of "vitamin-rich, low-carbohydrate vegetables and an abundance of meat and fat."

Joslin discussed Case No. 8 in his textbook because his mother had lived to be seventy-three years old, outliving many of her nondiabetic relatives, and had done so eating the fat-rich diet of the era. She had survived both a carbuncle and pneumonia, both of which were often fatal in that era to those with diabetes. She died of a cerebral hemorrhage. When she would occasionally have sugar in her urine at her regular checkups, she would be told to restrict her carbohydrate consumption to a bare minimum, under 30 grams a day—the carbohydrate content of a slice or two of bread—and that would solve the problem. Case No. 8 had responded so well to treatment that she had remained, Joslin wrote, "unusually strong and vigorous" until the very end of her life.

Joslin attributed his mother's success in holding the seemingly inexorable complications of the disease at bay for so many years in part to "absolute adherence" to her carbohydrate-restricted, high-fat diet. He seemed to believe that all diabetics who contracted the disease in middle age or older—type 2 diabetes—could do the same.

But Joslin, in fact, had come to believe that fat-rich diets were dangerous for diabetic patients, even if his mother had thrived on just such a diet. Joslin was being intellectually honest in discussing Case No. 8, but it also represented a remarkable contradiction as he challenged the conventional thinking he had once embraced.

This same contradiction appears again in the 1950s. By then insulin therapy had virtually abolished diabetic coma as a cause of death, and Joslin's Boston clinic had patients who had been diagnosed with grave cases of diabetes early in the insulin era, if not a year or two before, and yet were now still alive, surviving more than a quarter century with the disease. Some of these patients seemed to be in almost perfect physical condition. Who better to learn from about maximizing health and minimizing complications in the context of diabetes than those who did it best?

"On meticulous examination," Joslin wrote in 1950, a very small subset of all the patients they had treated at his clinic had managed to remain "physically sound, free from eye complications as certified by ophthalmologists and from degenerative changes in the arteries as certified by radiologists." In recognition of this remarkable accomplishment, Joslin and his colleagues took to offering "Quarter Century Victory Medals" to those individuals with diabetes anywhere in the world who satisfied the criteria of having had the severe form of the disease for twenty-five years without manifesting *any* detectable complications. When Joslin lectured in 1955 to the International Diabetes Federation, he predicted that the patients who received the medals would be "the ones who have controlled their disease most meticulously."

In 1970, eight years after Joslin's death, his colleagues at the clinic published an analysis of what they had learned from their Quarter Century Victory Medal winners.* They had awarded medals by then to 124 recipients. All of them had tended to maintain their weight at a healthy level. Most, as far as the Joslin physicians could tell, had worked hard to maintain control of their blood sugar. What Joslin earlier had described as "especially vigorous control of the disease in its earliest months" had been an essential requirement.

Other data, though, confounded their expectations. The Joslin physicians had awarded seventy-seven of the medals in the first ten years that patients were eligible. These were patients who began receiving insulin when it was first available in 1923 or in the decade that followed. They awarded only forty-seven medals in the next ten years, to those individuals who would have been diagnosed after 1933, when insulin therapy was well established and diet had evolved into an adjunct of insulin therapy. As the incidence of diabetes had steadily climbed and Joslin's clinic was admitting ever more patients each year and doing so ever earlier in the diagnosis, as insulin therapy became increasingly sophisticated and the physicians themselves accumulated more and more clinical experience, the number of Victory Medals awarded had fallen away. Fewer and fewer patients were living twenty-five years past their diabetes diagnosis with no complications to show for it.

In 1946, Joslin had speculated that the diabetic complications they were seeing in their aging diabetics had been caused by diets that were "too low in carbohydrates and calories" and, implicitly, too rich in fat. But this trend among the recipients of the Quarter

---

* Joslin's son, Allen P. Joslin, who had followed his father into medicine and diabetes research at his clinic, was a prominent author on the paper.

Century Victory Medal suggested the opposite. Diabetics who had done the very best tended to be the ones who had been diagnosed in an era when the diets had allowed the greatest proportion of calories from fat.

The Joslin physicians offered no explanation when they first published their analysis of these medal winners in the peer-reviewed medical literature. They did, though, in the 1971 edition of Joslin's textbook, the first published after his death. They were still debating "the value of fat" to the diabetic patient, as we still are today:

> In the first 20 years of this century, fat provided far more calories than 40% of the diabetic patient's calories. It is surprising how readily individuals would eat two or three times this amount.* Far more significant is the fact that our Quarter Century Victory Medal Diabetics, with their sound bodies and superb eyes and arteries, were exposed in the first few years following onset of their diabetes to diets proportionately much higher in fat than those of today.

Joslin's colleagues considered the possibility that it was the relatively fat-rich diets of the era that had made the difference. But that idea would rarely be taken seriously, despite considerable evidence to the contrary.

That the period in which fat first became feared in the diabetic diet coincided with the First World War and the years of the tragic flu epidemic that followed may not be a coincidence. Until 1914, European physician researchers had dominated thinking on diabetes and metabolic diseases. This is why American physicians like Joslin would travel to Europe to learn from these authorities, particularly the Germans and Austrians. Prior to the Second World War, as the Stanford University historian of science Robert Proctor has phrased it, Germany was the "world's most powerful scientific culture. . . . German science and medicine were the envy of the

---

* It is impossible to eat three times 40 percent fat in the diet, as that would be over 100 percent fat. I assume Allen Joslin and his colleagues meant they would eat three times the quantity of fat that constituted 40 percent of the calories in a typical diet.

world."* These researchers had the benefit of working together in institutes like Naunyn's and critically assessed one another's work and interpretations, minimizing errors and maximizing the reliability of the knowledge that emerged. They had a history and a culture of rigorous science that had evolved over several hundred years.

World War I took those European researchers and institutes and even the journals in which they published offline, to use modern terminology. Many of the major researchers—Minkowski, for instance—were engaged in the war effort of their respective nations, no longer doing basic research. Scientific communication between the European continent and the United States and Britain—the journals passing in both directions, lecture tours, working visits— came to a halt. The research void was filled primarily by Americans, an ocean away from the battlefields. Joslin and his friend Frederick Allen, both affiliated with Harvard Medical School at the time, emerged as the most influential thinkers on diabetes. Without the steadying influence of the European authorities who had dominated thinking on diabetes for half a century, Allen and Joslin were able to shift the paradigm of diabetes therapy for long enough to do what may have been irreparable damage.

Frederick Allen published the work that launched his career and his public acclaim as a diabetologist in 1913, a 1,200-page monograph called *Studies Concerning Glycosuria and Diabetes*. His influence rose and then peaked almost precisely in sync with the war and the flu epidemic that followed. Joslin disseminated Allen's thinking widely—his conclusions and speculations about the diet-diabetes relationship—beginning with the very first edition of Joslin's textbook, *The Treatment of Diabetes Mellitus,* published in 1916. By the time the Europeans returned to influence a decade later, insulin had been discovered and what Allen and Joslin believed about diabetic diets had become dogma in the United States. They defined what became a consensus of opinion on scientific truth not

---

* By 1940, German scientists alone had won or shared seventeen Nobel Prizes in chemistry and nine in physiology and medicine. Americans had won or shared three in chemistry and four in physiology and medicine.

because of its proved veracity but because they had the podium, in effect, during these critical years.

Joslin would continue to hold that position in the decades that followed. He had been the first physician in the Americas to open a clinic dedicated solely to the treatment of diabetes. By 1916, he'd already seen a thousand diabetic patients, giving him far more clinical experience than any other American physician. Between that ever-accumulating experience and the numerous editions of *The Treatment of Diabetes Mellitus*, considered the diabetologist's bible as it still is,* and of his lay guide to therapy, *A Diabetic Manual for the Mutual Use of Doctor and Patient,*† Joslin would become what later diabetologists would call "the god of diabetes" to multiple generations of physicians and medical researchers. Even across the Atlantic, according to R. D. Lawrence, a founder in 1934, with the author H. G. Wells, of the British Diabetic Association, Joslin's textbook "was the English book which everybody read in those days." He called it "the great book of Joslin."

To say Joslin's word was gospel on diabetic therapy is not strictly true, because clinicians would argue and debate with him throughout his life. But Joslin's opinions certainly carried far more weight in the United States and perhaps even worldwide from the 1920s onward than those of any other diabetologist. What Joslin believed would be the default thinking in the field, what physicians and researchers accepted as true until remarkable evidence—often their own clinical experience—convinced them otherwise. The longer they believed in Joslin's opinions, though, the more they assumed that these opinions were based on sound scientific evidence. Joslin was a tireless and prolific communicator of his thinking on diabetes and its treatment; he played an invaluable role in promoting awareness of the disease and disseminating the knowledge necessary to live with the disease as a patient and treat it as a physician. But on critical questions of science, he was often more wrong than right.

Joslin came to assume that dietary fat was dangerous to diabetics largely because of Frederick Allen. By 1915, Joslin had come

---

* Now known as *Joslin's Diabetes Mellitus* and in its fourteenth edition.

† First published in 1918 and with nine more editions by the time of Joslin's death in 1962.

to believe that Allen was responsible for what he considered the greatest advance in diabetic therapy "since Rollo's time." The key to keeping diabetic patients alive was to underfeed them, Allen argued, to starve them, if necessary. That meant restricting carbohydrates, protein, *and* fat, such that the body weight remained low and the strain on the pancreas was minimized. And if that was true, then dietary fat could be as harmful to the diabetic patient as carbohydrates. Joslin would never leave Allen's thinking behind, even as a series of studies suggested that Allen's interpretation of the science was incorrect and that very-high-fat diets could be particularly effective in *all* cases of diabetes.

The rationale behind Allen's work emerged from the struggle to treat those patients with the most severe form of the disease, typically children and adolescents diagnosed with type 1 diabetes and near death by the time they arrived at the hospital or clinic. Insulin would be a lifesaver for these patients. In the pre-insulin era, they often died within weeks or months of their diagnosis, doing so more often than not in a diabetic coma, what physicians came to think of as the endstage of diabetic ketoacidosis—with ketones in their urine and ketones accumulating in their blood.*

By Joslin's estimate, two-thirds of his patients in the pre-insulin era died of coma. It was not uncommon for these children and adolescents to be admitted to a clinic or hospital close to death, if not already comatose. These patients might die within hours. Any advance was welcome. Since diet was the only meaningful factor that these diabetologists could manipulate before the discovery of insulin, it was from dietary manipulations that these advances would have to come. Dietary manipulations, though, can be a tricky business for physicians and researchers.

People and animals tend to eat until their metabolisms have a sufficient supply of food—of calories—to run at a healthy level, providing enough energy for basic needs (technically known as "basal metabolism") and for all necessary physical activity that daily life requires. This means that any decrease in the proportion of one

---

* The blood and urine become more acidic and less alkaline when this happens; physicians often referred to the condition as acidosis or acid intoxication, assuming that the intoxication led to the coma.

macronutrient in the diet—fat, for instance, or carbohydrate—requires an increase in another to compensate. For any macronutrient that is increased, another macronutrient is decreased.

A diet for diabetes that restricted carbohydrates but kept patients sufficiently well nourished—with enough calories providing the energy for health and healthy functioning in work and life—would have to replace the absent carbohydrate calories with either fat or protein. (Alcohol was also a possibility, and patients with diabetes could metabolize it, which is why diabetic diets, even for children, often allowed alcohol in this early era.) Once it became clear that protein can be a source of glucose for the body—converted from the amino acids that are its building blocks—diabetologists began to suspect that those with severe diabetes respond to the glucose converted from protein no better than they do the glucose from carbohydrates. But protein is essential in a healthy diet. The body needs at least some protein to sustain and repair cells, tissues, and organs.

So diabetic patients had to eat some protein, but how much? And if a diabetic diet replaced the absent carbohydrate calories with enough fat to run the metabolism at a healthy rate, as physicians were advising until Allen suggested otherwise, was that really harmless to those patients whose diabetes was severe? The fact that the Inuit could eat such a diet, or the "Gauchos of South America," might be irrelevant. If diabetic patients could not, if that was the cause of diabetic ketoacidosis and coma as many diabetologists feared, what was the alternative?

These were the questions that Frederick Allen claimed to have answered. Allen was an enigmatic figure in the diabetes world in the pre-insulin era. He seems to have been the very first clinician in the United States to approach the challenge of understanding diabetes purely from a scientific perspective, rather than as a clinician treating patients. When he published his first book on the subject, the one that made his name in 1913—*Studies Concerning Glycosuria and Diabetes*—it included no discussion of his clinical experience because he had yet to treat human patients.

Allen had come late to medicine and medical science. When he received his medical degree from the University of California, Berkeley, in 1907, he was in his late twenties. In 1909, he took a poorly paid position at Harvard Medical School and then conducted an unprecedented series of animal experiments—several hundred—on the nature of diabetes and dietary therapy. He may have been the first diabetologist ever to base a clinical therapy purely on animal models of the disease rather than human patients. This was one of the consequences of Oskar Minkowski's work twenty years earlier. By establishing that dogs become diabetic upon removal of the pancreas, Minkowski had done more for diabetes than give researchers a reason to search for the pancreatic secretion responsible for the disease. He had turned diabetes into a subject amenable to experimental research. Once researchers like Allen could create the symptoms of the disease in laboratory animals, they could study not only how the symptoms progressed in these animals, but how these diabetic animals responded to therapies, dietary or otherwise. They couldn't know if what they observed in their animals held for humans as well, but it would create the possibility.

This led to an explosion in the medical literature on diabetes, and Allen's *Studies Concerning Glycosuria and Diabetes* was the first book in the English language attempting to make sense of it all. By this time, diabetologists were already referring to this literature as "voluminous" and throwing up their metaphorical hands in despair at their inability to assimilate it. "The subject of diabetes heretofore has been what William James might have called 'a big, blooming, buzzing confusion,'" Allen wrote in the preface. "It is believed that the cure of diabetes is now a feasible experimental problem." Allen's lengthy monograph included well over 1,500 references, and he discussed with a relentlessly critical eye what might be construed from these reports. He also discussed at length what he had learned from his innumerable animal experiments, from the dogs, cats, rabbits, rats, and guinea pigs in which he had tried to establish and study diabetic conditions. It was the animal experiments that impressed Joslin so much. That Allen's thinking on diabetes and how to treat it by diet was based on "patient scientific experimentation," Joslin wrote in 1915, gave it all the more reason to be taken seriously.

Allen's revelation was that diabetes in the dog mimicked better the disease in humans if only most of the pancreas was removed, keeping the blood supply intact. (Other European researchers had come to the same conclusion independently, as Allen himself would scrupulously note.) By doing so, he wrote, "diabetes varying in intensity from the mildest to the very severe, can thus be produced at will." Some of these dogs, the more mild cases, would thrive indefinitely on all-meat diets. Those that didn't, he discovered, the more severe cases, did better when they were fasted and then kept undernourished and underweight. Restricting carbohydrates and protein wasn't enough to keep these dogs alive, though the restriction of all food seemed effective. It desugarized the dog's urine quickly and kept them from succumbing to acidosis and coma.

The initial fast might have to be "measured in weeks rather than in days," Allen wrote, but if the animals could be kept underweight and their urine sugar-free, "the animals may remain lively and strong though thin." He even held out hope that pancreatic function could be restored to some extent as time went on. Add fat back to the diets of these dogs, however, and it "produces . . . an appearance of spontaneous aggravation of condition as striking as anything witnessed in human patients."

That observation, along with the very questionable assumption that the physiology of his depancreatized dogs mirrored that of diabetic humans, drove Allen to rethink the very nature of the disease. It led Joslin to consider dietary fat to be the likely reason that comas were killing his patients—they were being fed or told to eat fat-rich diets, as part of the therapy—and generations of diabetologists afterward came to fear high-fat diets as a result.

The problem, Allen speculated, was the body's inability to metabolize *all* food. Whatever was wrong with the pancreas, it could be negatively affected by carbohydrates, fat, and protein, and the heavier the patient, the harder the pancreas had to work and the greater the strain. He supported this thinking by reassessing the carbohydrate-containing therapies used on diabetes in the past—the milk cure, sugar cure, potato cure, oatmeal cure, and others—and concluding that maybe they all worked, if they ever did, because they underfed the patient. "Most of the diets represented some degree of undernutrition," Allen wrote; hence, the patient lost weight and the strain on the pancreas was lessened.

Allen's thinking was a radical revision on what physicians at the time had come to consider the most rational approach to treating diabetes, particularly their severe cases, which was to put weight back on them, overfeeding them as necessary with fat to do so. If emaciation was a symptom of uncontrolled diabetes, if patients typically appeared first at the doctor's clinic or in the hospital only after losing significant weight, then restoring the patient to a healthier weight, even if requiring copious dietary fat to do it, seemed all to the good.

In 1913, based on the reception to *Studies Concerning Glycosuria and Diabetes*, Allen was offered a junior position at the newly opened hospital of the Rockefeller Institute for Medical Research in New York City. There he would have the opportunity to test his program of prolonged fasting and undernutrition on diabetic patients. Allen's first opportunity came the following February, when a twenty-eight-year-old woman arrived at Rockefeller weighing less than ninety pounds, thirty pounds underweight, and close enough to coma that Allen and the Rockefeller staff were anxious about embarking upon Allen's experimental therapy. Still, on the third day after she was admitted, her situation turned sufficiently dire that they did. They had, Allen wrote, "no choice but to take the chance of beginning the proposed treatment."

Allen discussed her treatment five years later, when he and two colleagues published another book, *Total Dietary Regulation in the Treatment of Diabetes*, of nearly a thousand pages.* The book covered all that Allen had learned about dietary therapy for diabetes with detailed reports on all his cases. What his first patient was fed and how much, if anything, seemed based on day-to-day decisions regarding her status and what might be necessary to either build up her strength or lessen the strain on her pancreas. She would be fasted and fed only vegetables on some days and only oatmeal on others (following von Noorden's program) to desugarize her urine or lower the level of urinary ketones. When she seemed safely free of sugar and acidosis, Allen and his colleagues would feed her a more traditional protein-and-fat diet, even extra fat, to build up

* Allen says in his unpublished memoirs that he wrote every word of *Total Dietary Regulation in the Treatment of Diabetes*, although it was coauthored with two colleagues.

her strength. Allen believed, though, that his patient's recalcitrance stymied his best attempts at treatment. (He thought that of many patients.) When she was discharged the following December, she had lost eleven more pounds. But she was still alive—that was "the actual accomplishment," Allen wrote—despite their "bungling and inadequate treatment." She died a year after her discharge, having returned, Allen wrote, to "eating everything at will, including much candy."

This was a common pattern in response to Allen's program of fasting and undernutrition. Patients would follow along when they were in the hospital, under the watchful eye of the Rockefeller staff, then they would revert to their old eating habits upon returning to their normal life.

By May 1915, Allen had treated forty-four patients at Rockefeller, all severe cases, and was calling the results "uniformly beneficial." By February 1916, Allen and the Rockefeller staff had used his approach on sixty patients, and *The New York Times* reported it as "radical and revolutionary in method." In a symposium at the New York Academy of Sciences, Allen claimed that despite first fasting and then continued undernutrition of emaciated patients, many had managed to return to a level of health that is almost unimaginable. "Many of our patients run up the eight flights of our stairs at the hospital of the institute twenty times a day," he said. "We are making athletes of them, thin as they are."

Using Allen's program at his clinic, Joslin calculated that deaths in the first year after diagnosis had dropped from almost 70 percent of cases to under 20 percent. Moreover it was simple for both patients and physicians. "Fasting is never so rigorous as doctors or patients expect," Joslin wrote. "Patients are more ready to undergo it than physicians to prescribe it. Quite as often it is as much a relief to the patient as it is discomfort."

Other physicians waxed enthusiastic about Allen's approach, too. John Williams, a diabetes specialist at Highland Hospital in Rochester, New York, reported in 1921 that he had used Allen's protocol on over three hundred patients: "I have witnessed many patients, practically moribund, temporarily saved from impending coma and death by its use," he wrote. "I have had bad cases live for more than two years on diets averaging below 700 calories," a level of nutrition he aptly described as "exceedingly low." He also acknowledged that "a certain number of cases will succumb in spite of all fasting and underfeeding." Williams believed the problem wasn't the starvation therapy itself, but the patient's "unfaithfulness" or "lack of courage." Williams, like Allen and Joslin, thought the benefits clearly worth the costs, but they weren't the ones who had to starve themselves while already struggling with the ravenous hunger, the polyphagia, of uncontrolled diabetes.

Other physicians found the trade-off unacceptable. "The ironies, the Hobson's choices, the catch-22's of the treatment were staggering," as the University of Toronto medical historian Michael Bliss described it in his 1982 history, *The Discovery of Insulin*:

> An adult diabetic, weak, emaciated, wasted to perhaps ninety pounds, would be brought into hospital and ordered to fast. If the patient or the patient's family complained that he or she was too weak to fast, Dr. Allen replied that fasting would help the patient build up strength. If the patient complained about being hungry, Allen said that the fasting would help ease the hunger. Suppose the method didn't seem to work and the symptoms seemed to get worse. The answer, Allen insisted, was more rigorous under-nourishment: longer fasting, a maintenance diet even lower in calories. . . . Where was the limit to the dieting? Where would you stop? In fact there was no limit. In the most severe cases the choice came to this: death by diabetes or death by what was often called "inanition."

If the patients died of inanition—of starvation—Allen considered it proof that they had an "inability to acquire tolerance for any living diet." And even death by inanition, Allen believed, entailed distinctly less suffering than "dying with active diabetic symptoms produced by lax diets or by violations of diets."

By 1917, with the United States having entered World War I, Allen had left Rockefeller and been assigned to head up a diabetic service for the military in New Jersey. He had apparently alienated enough of the staff at Rockefeller, or enough of them had bridled at his starvation therapy, that the hospital did not invite him back after the war. The medical staff had continued to give most of the patients—"contrary to my belief and wish," Allen would say—high-calorie diets, and Allen seemed to believe that the "severe cases" had died because of it.

In 1920, Allen opened his own clinic in New Jersey. Among his most famous patients was Elizabeth Hughes, daughter of Charles Evans Hughes, who had served as secretary of state in the Harding administration and would later become chief justice of the Supreme Court. Elizabeth "disliked her diet, and found the fast days a special nightmare" and apparently disliked Allen. The therapy, nonetheless, would contribute to keeping her alive long enough to receive insulin from Banting in Toronto and become a poster child for its success. Her recovery symbolized the rapidity with which insulin therapy could return an emaciated patient at the brink of death not just to life but to good health. Allen, though, eventually categorized two-thirds of his patients as treatment failures, fatalities. "It can only be said," he wrote, "that the treatment was ineffectual."

While these physicians, none more so than Joslin, were dedicated to saving the lives of their patients with diabetes, the clinical decisions they were making about how best to do that had profound scientific implications. They were a response to fundamental questions about human metabolism and how it goes awry in diabetes. This was a conflict between the art of medical practice and medical science itself.

Allen's starvation therapy spoke directly to the question of the cause of diabetic ketoacidosis, and why healthy people and those with diabetes respond so differently in this respect to dietary manipulations. Fast a healthy person, as diabetologists knew, and their liver synthesizes ketone bodies and the ketone concentration in their urine goes up. "If a healthy individual lives for three successive days upon a carbohydrate-free diet," as Joslin wrote in his textbook, "the urine voided upon the subsequent morning will show the presence of diacetic acid and acetone [i.e., ketone bodies]."*

---

* When the Viennese diabetologist Wilhelm Falta, one of the pioneers in the science of endocrinology, lectured in New York City in 1909, he explained that it was vitally important for physicians to know the difference between what he called "ketonuria"—or ketones in the urine—and the acidosis from which diabetics often died. "The formation

Feed a healthy individual a fat-rich diet, absent carbohydrates, and the same thing happens: their livers synthesize ketones from the fat and they enter ketosis, a benign state that became well known in the early 1970s with the Atkins diet for weight loss. Fast patients with diabetes, though, and their livers synthesize fewer ketones. Hence, the one question at the heart of these dietary manipulations was the effect of a fat-rich diet, absent carbohydrates, on a patient with diabetes. If the absence of carbohydrates would induce the synthesis of ketones in a healthy patient (as fasting did or the kind of high-fat diet Naunyn had been recommending and Joslin and others had been using until Allen convinced them otherwise), could it cause excessive ketone synthesis, diabetic ketoacidosis, and so death in a patient with diabetes? It was certainly a possibility. Put simply, what aspect of the diet—which macronutrient—was causing ketone synthesis and the accumulation of ketones in the circulation to become a pathological problem in diabetes?

Allen's animal experiments, as we have seen, had suggested that dietary fat was the trigger of ketoacidosis in pancreatectomized or partially pancreatectomized dogs, and that's what Allen and Joslin had come to believe was true for humans also. If "the incautious use of fat," as the Rochester diabetologist John Williams described it, "was responsible for many of the failures" in treatment, which was still a critical question, then a diet had to restrict fat, as well as carbohydrates and protein. If so, Allen's undernutrition and fasting were necessary to keep patients desugarized and safe from ketoacidosis and coma. This chain of linked assumptions is why Joslin and other physicians in the United States embraced it. But what they had demonstrated was that fasting and undernutrition could prevent their patients from succumbing to diabetic ketoacidosis and coma and keep them alive longer than otherwise, not that it was the best or only way to do it.

If another method could be found that would prevent the ketoacidosis and coma, and that did not require that the patients embrace

---

of ketone bodies depends, as you know," he said, "on the lack of combustion of carbohydrates. Ketonuria may be produced even in the normal person by means of inanition or an exclusive protein and fat diet. The fact that such higher grades of ketonuria appear in diabetes is easily understood, since here there is also failure to consume the sugar arising from the [c]atabolism of protein."

the Hobson's choice of Allen's starvation therapy—and all the complications that came with inanition—wouldn't that be worth considering? And if it was, would that change the interpretation of the science, the understanding of the mechanisms involved in triggering the ketoacidosis in the first place?

Between 1920 and 1923, two physicians at the University of Michigan, Louis Newburgh and Phil Marsh, published a series of four articles in *Archives of Internal Medicine* on the treatment of diabetes with a fat-rich diet, discussing the theory behind it as well as their ever-growing clinical experience using the approach. Newburgh and Marsh believed that any diet that threatened a patient with starvation was unacceptable, perhaps because their work suggested that it wasn't necessary. If the University of Michigan physicians were right about the benefits of a fat-rich diet, then Allen and, more important, Joslin were wrong about its dangers.

Louis Newburgh is a problematic figure in this history. His legacy in the dietary therapy of diabetes is for the most part beneficial, while his work on obesity, the disorder most intimately associated with diabetes, had far more influence and did harm. Born in 1883, four years after Allen, Newburgh attended Harvard as an undergraduate and then Harvard Medical School. He trained for sixteen months at Massachusetts General Hospital in Boston and spent a year working in Carl von Noorden's clinic in Vienna. Returning to Boston, Newburgh continued his research and clinical work, treating local immigrants for constipation by feeding them fiber-rich diets, publishing studies on kidney disease and pneumonia, and not pulling his rhetorical punches when he believed, as he often did, that he had established where he was right and others wrong. "These theories," as he wrote about one such conclusion, "to anyone acquainted with the facts, are clearly without foundation in quantitative experiment." He was at Harvard Medical School coincident with Allen, but whether they knew each other or Newburgh knew Joslin then is unknown. By the time Newburgh joined the medical faculty at the University of Michigan in 1916, he clearly considered himself capable of accomplishing great things in medicine. In 1919, after he had already begun treating his diabetic patients with a high-fat diet, he was joined in the work by Phil Marsh, a young internist who had just recently received his medical degree after what Newburgh's University of Michigan biographer later called "a stormy time with the Promotions Committee."

Newburgh and Marsh accepted that Allen's approach could protect diabetic patients from ketoacidosis, but they were troubled by the trade-off. If the diet replaced the fat with any amount of carbohydrates or protein, the result would be sugar in the urine. Allen's treatment addressed that by restricting the calories sufficiently that no extra fat was consumed (no butter, no cream, etc.). The urine would be sugar-free, but now the patient was, at best, constantly hungry and undernourished: "Because of the low energy intake, [the physician] renders him unfit for the ordinary activities of life," wrote Newburgh and Marsh:

> It is evident that the two horns of the dilemma can be avoided if the diabetic can safely be given enough calories to maintain metabolic equilibrium, without producing hyperglycemia [high blood sugar] or acidosis. Since carbohydrate cannot be used, and since protein is . . . unsatisfactory, we have dared to ignore the belief concerning the danger of fat in the diet of diabetics, and have investigated in the clinic the effect of a diet whose energy comes largely from fat, to which is added sufficient protein . . . and the minimal carbohydrate necessitated in making up a diet that a human being can eat over a long period of time.

To test their idea, Newburgh and Marsh started off diabetic patients who entered the clinic on a fat-rich diet—80 to 90 percent of all calories—that would still keep them underfed, supplying only 900 to 1,000 calories. If that succeeded in desugarizing the patient's urine and keeping ketosis in check, they let their patients eat more fat-rich foods—working up to as much as 2,500 calories each day. The fat content of the diet, though, always constituted at least 80 percent of all calories. By starting the patients off undernourished, Newburgh and Marsh assumed they could prevent what Joslin feared, which was overwhelming the patient with fat and triggering an irreversible ketoacidosis. By keeping the protein content of the diet low, they would also minimize the amount of glucose to which the patient's pancreas had to respond.

In their very first paper, in 1920, Newburgh and Marsh stated clearly that their program negated the need for fasting and under-nutrition. They discussed what they had learned from using their treatment on seventy-three patients. They assumed that all these patients were suffering from serious diabetes because otherwise physicians in the state would not have transferred them to University Hospital—Michigan's "court of last appeal," as Newburgh and Marsh called it. Patients were often admitted on the verge of coma, they reported, and then they would be desugarized on the fat-rich diet, regain their strength, and walk out "sugar free."

Fear of fat in the diabetic diet "is entirely ungrounded," Newburgh and Marsh wrote. Of those seventy-three patients, four had apparently died in the hospital, three within a day of entering (from pneumonia, sepsis, and coma) and "the fourth patient refused to limit herself to the diet, and went into coma after eating a bag of oranges brought by a relative." All the others had been discharged with their diabetes seemingly under control.

By 1923, Newburgh and Marsh were documenting their experience with 190 cases, all treated with their fat-rich diet, still claiming (as had Allen, too, initially about semi-starvation) that it was almost uniformly beneficial.* For the great proportion of patients treated, the fat-rich diet avoided "the evils of fasting and under-nutrition," kept the urine sugar-free and blood sugar apparently controlled, avoided acidosis, "cause[d] its disappearance when present (short of coma)," and supplied sufficient energy such that the patients could be active and even earn a living. Newburgh and Marsh published a comparison suggesting that their patients did at least as well as those in Allen's clinic or Joslin's, and Joslin would later acknowledge this seemed to be true. But it did so without

---

* As said earlier, a common problem with these kinds of case studies is that the physicians invariably bias their interpretation of the patient histories in favor of their therapies. Newburgh and Marsh did much the same as Allen: when patients died of coma after having been discharged from the hospital without sugar in their urine—as five did in the Michigan case history—Newburgh and Marsh naturally assumed that they had "discarded the diet."

the dangers of undernutrition or the threat of starvation, without requiring courage to stick with it.*

A year after Newburgh and Marsh published their first paper on their high-fat-diet therapy, Rollin Woodyatt of Rush Medical College in Chicago proposed a theory to explain what the Michigan physicians had observed: why their high-fat diet might save patients from ketoacidosis and coma, why the fat-rich diet might be so successful. Woodyatt was a diabetologist and world-renowned authority on carbohydrate metabolism. He was also, in the words of Walter Campbell, the University of Toronto diabetologist who would oversee the first clinical use of insulin, "the respected intellectual giant among diabetic clinicians of his day." Woodyatt considered Newburgh and Marsh's success "striking," and his attempt to communicate how he had come to understand that success was the article that Graham Lusk, the leading American nutrition researcher of the era, had referred to as a "masterpiece" in his textbook.

The critical fact that has to be kept in mind—one too often ignored today—Woodyatt explained, is that our cells do not necessarily metabolize the food we eat in the form that we've eaten it. They respond to what happens to that food after it has been digested and absorbed into the circulation, even after it has been first stored in various depots—fat tissue, muscle, the liver—and perhaps even after it has been run through various metabolic processes. That's why protein can be a problem in diabetic diets: it is broken down into amino acids and some of those amino acids may be converted to glucose. Then the pancreas has to respond to the glucose, just as if it came originally from a slice of bread or a potato. If not enough energy is available in the diet for metabolism to function at a healthy level—which would be the case with any patients who are undernourished or fasted on Allen's program—then the body will find the fuel it needs by metabolizing its own stores of fat and protein. This use of *endogenous* fuel (the fuel already stored

---

* The Montana farmer we met earlier who was treated at the Mayo Clinic in 1921 received something very close to the Allen therapy the first time he was admitted to the clinic, and then, when he returned in December 1922, the Newburgh-Marsh high-fat approach was applied. It's the latter that he ate for the next three decades.

in the body) rather than *exogenous* (the food we eat), Woodyatt explained, is what has to be taken into account whenever diabetologists, physiologists, biochemists, or nutritionists try to understand the effect of any diet on metabolism and diabetes.

If the patient with diabetes is overweight to begin with, then the body will prioritize using the fat it has stored; when the overweight patient is fasted or undernourished, as in Allen's program, the patient's body will still be metabolizing and responding to a mostly fat fuel supply, just as if the patient was eating a fat-rich diet. In that context, in overweight or obese patients with diabetes—i.e., type 2 diabetes—Allen's undernutrition and Newburgh and Marsh's fat-rich diet could be expected to have similar, if not identical, beneficial results. In both cases, fat will be the body's primary fuel.

If the patient with diabetes is emaciated already, as would be the case with more severe cases and with type 1 diabetes, the body will have no choice but to use the protein stored in the muscles (lean tissue) for fuel; some of the amino acids from that protein will still be converted to glucose, and that glucose might still be too much for the patient's dysfunctional pancreas to handle. In that case, the patient might fail to be desugarized, even with long fasts. These would be patients for whom Allen's undernutrition program might fail. Feeding those patients fat, Woodyatt proposed, would relieve the strain on the pancreas by preventing the patient's body from having to cannibalize its protein and cope with the glucose from the amino acids. A fat-rich diet would provide all the necessary fuel for the body while minimizing the work of the pancreas and keeping protein where it belongs and the body needs it to be, in the lean tissues and organs.

"For diabetes itself," Woodyatt wrote, "and particularly for diabetes associated with undernutrition, why for the purpose of desugarization should the patient be compelled to draw from his tissues the fat that he might draw from a diet, especially if in drawing from his tissues he lowers his fat reserves to the extent that he increases his protein losses?" Woodyatt's summation of the science was a reiteration of what Newburgh and Marsh had argued: "When nutrition is the paramount consideration we must administer every possible calorie, and this method may help us do so. . . . A case that can be desugarized by fasting can be desugarized with a fat replacement if [glucose] is made low enough."

As with Allen's starvation therapy, diabetologists having read Newburgh and Marsh's papers and Woodyatt's rationale tried this new method on their patients. At Columbia University's College of Physicians and Surgeons in New York and the Medical Clinic at Johns Hopkins University and Hospital in Baltimore, William Ladd and Walter Palmer concluded that they could desugarize their patients more quickly with a fast and undernutrition, but the fat-rich diet also worked. So long as the protein content of the diet was kept low, and the fat content and total calories were ramped up slowly, there was no danger of ketoacidosis. The patients tolerated the high-fat diet, they wrote, "with an increase in the[ir] strength and subjective well-being."

Russell Wilder, who had been a student of Woodyatt's, concluded much the same with his experience at the Mayo Clinic. Wilder came to believe that diabetic coma was probably caused not by the fat in the diet, as Allen and Joslin were claiming, but by the glucose from the protein and the "extreme limitation of the ability of the organism to utilize glucose." In Wilder's experience, his patients succumbed to ketoacidosis and coma only if they had an infection or from what he called a "dietary indiscretion, usually in the consumption of carbohydrate (candy) or protein, rarely of fat."

European diabetologists would also experiment with and then embrace the fat-rich, low-protein diet for their patients, prompted less by the work of Newburgh and Marsh than of a Swedish diabetologist, Karl Petrén, at the University of Lund. Petrén had been treating his patients with fat-rich diets since 1914. He had been a student of Naunyn's in Strasbourg, as had a German diabetologist named Wilhelm Weintraud, who had argued as early as 1893 that diabetic ketoacidosis and comas could be avoided by limiting both protein and carbohydrates in the diabetic diet. This implied that fat could be fed to diabetic patients with impunity, which is what Petrén did, claiming to get results even better than Marsh and Newburgh's. His 1923 monograph on the subject, though, had been published only in Swedish. Not until 1924 was an English translation available, published in a journal founded and edited by Frederick Allen. Petrén's approach had attracted so "much attention in Europe," a footnote explained, presumably written by Allen, and the high-fat nature of the diet was so extreme, that he deemed it important to have an English translation published, along with a preface written by Naunyn himself to put it in context.

Petrén evoked the same philosophy for treating diabetes that Newburgh and Marsh had arrived at independently: limit the protein and carbohydrate content to minimize the amount of glucose to which the pancreas has to respond, while doing everything possible to avoid underfeeding the patient. Petrén believed in the "strictest limitation of protein" when necessary, which meant he fed his patients fat almost exclusively—"butter and pure fat of swine," allowing not even bacon because its muscle fibers contained too much protein—and green vegetables. Petrén believed that the vegetables and the fat between them would provide enough protein to keep his patients healthy.

In extreme cases, in "grave diabetes," Petrén would occasionally feed his patients only butter for two or three days (900 to 1,800 calories each day). He argued that it was eminently reasonable to feed a patient only butter, if the alternative, Allen's, was to fast them and so feed them little if anything at all. As his patients' conditions improved, Petrén might allow them to eat meat and eggs, but only after the danger of ketoacidosis had passed. Once Petrén began using insulin therapy in his Lund clinic, he reported that his patients with severe diabetes—type 1—did almost as well on the fat-rich diet alone as they did with insulin therapy. The implication was that if the insulin wasn't needed to maintain stable blood sugar, which might be the case without carbohydrates in the diet and only a bare minimum of protein, the patients could still be effectively healthy. It was a radical idea.

What diabetologists thought of these competing dietary therapies, and underlying biochemical rationales, seemed to depend as much on geography as the clinical evidence. Medicine was that balkanized, and it would remain so for another half century. In Germany, for instance, Erich Grafe, director of the Clinic of Medicine and Neurology at the University of Würzburg, published the seminal textbook on metabolic diseases and their treatment in the early 1930s. (The book was translated into English in 1933, at the request and under the supervision of Eugene DuBois, who was then considered the leading American authority on metabolism.) In a lengthy chapter on diabetes, Grafe wrote that he had no doubt that Allen's fasting/undernutrition approach kept patients alive. But it also required "martyrdom" on the part of the patient "who was reduced to a mere skeleton." Grafe couldn't imagine European patients having sufficient faith in their physicians to be persuaded to undergo this level of deprivation. "It required all the typical American patience and trust in medicine to proceed with these cures," he wrote.

As for the fat-rich approach of Newburgh, Marsh, and Petrén, clinicians throughout Europe had demonstrated how well it worked, Grafe wrote, and his own experience and that of other European diabetologists "proved that the principle is right. . . . I know of no method by which the blood sugar may be reduced clearly to normal with as great a degree of certainty." Even with insulin therapy now available, Grafe added, he praised Petrén's nearly all-fat diet, saying he "should not care to live without" it, particularly for those patients whose blood sugar remained stubbornly high.

In the United States, though, Allen and Joslin's opinions dominated, and the fact that Allen had beaten Newburgh and Marsh into print by half a dozen years may have been the deciding factor. By 1920, when Newburgh and Marsh published their first paper on the high-fat diet for diabetes, Allen and Joslin had already published and spoken repeatedly about the dangers of fat in the diabetic diet. Having publicly committed themselves to the belief, they continued to defend it.

In 1921, when Woodyatt presented his thinking at the annual meeting of the Association of American Physicians and discussed the "striking success" of Newburgh and Marsh's high-fat approach, Allen was in the audience and said he found the science interesting and thought far more experimentation would be necessary to settle these questions. He also added, though, that "the only two juvenile patients who are alive today" from his years treating patients at the Rockefeller Institute for Medical Research were the two who escaped being fed high-fat, high-calorie diets. "The sole known difference between them and the others [who died] is that their diets were kept relatively low in fat and calories . . . ," he said. "Other clinical experience convinces me firmly that the attempt to give a high caloric diet or to build up the body weight too high with fat or any other food is injurious and leads to a fatal result in every genuinely severe case of diabetes." One of Allen's colleagues at his New Jersey institute, Frederic Leclercq, later reported that the clinic staff would try Newburgh and Marsh's high-fat diet on three patients and it would fail all three. "The statement that severe cases of diabetes which are resistant to undernutrition or fasting can be cleared up by high fat diets is contrary to all experience in this Institute and in our opinion also contrary to fact," Leclercq wrote in 1922.

Joslin was also in the audience for Woodyatt's talk and said he had experimented with the Newburgh-Marsh approach in the four most severe cases who were not doing well under his clinic's care—four young boys—and it had failed in all four. "In one acidosis became so severe that coma threatened." He noted that the two Michigan physicians had lost track of many of their patients, once they discharged them from the hospital supposedly restored to good health. The four patients in his clinic, he said, "felt that they deserved to know" what happened to those Michigan patients as well. "What shall I do with two boys with diabetes beginning at the age of twelve, but now at the end of three or four years of treatment by the old method? . . . Shall I put these boys on lower protein and push up their fat? Or shall I continue by those methods which have helped them for these three years? With these cases I shall not adopt new methods. We must be positive not to lose the advantages gained."

As Joslin's influence among diabetologists continued to grow through the 1920s, with each subsequent edition of his textbook and his numerous articles, he clearly wrestled with the cognitive dissonance presented by Newburgh and Marsh's fat-rich diet. Here, reading the various editions of Joslin's textbook is again an exercise in dealing with contradiction. As Joslin added new material to each subsequent edition to update physicians on his latest thinking—and as the number of cases on which he came to these conclusions inevitably grew, from one thousand to six thousand and ever upward—his textbooks retained older material unchanged, even if it seemed diametrically opposed to his newer thinking.

After spending his professional career watching children and adolescents die of diabetic ketoacidosis and coma under his care, Joslin could not shake his belief that high-fat diets may have precipitated these "unnecessary" deaths. This remained his primary concern, even after insulin therapy provided a reliable method to prevent them. "Diabetic patients need fat," he explained, "it forms the chief constituent of their diet; but they must not be poisoned with it, they must be gradually accustomed to it," which is precisely what Newburgh and Marsh claimed to have done.

But Joslin had also decided as early as 1916 that what he called the "fundamental principle upon which all treatment has been and is rightly based" is how well that treatment could increase the patient's ability to metabolize carbohydrates. That meant feeding them as much carbohydrate as they could safely consume. He believed he could do that, as Allen had preached, so long as he kept his patients underfed. By including carbohydrates in the diet and restricting fat to keep the patient undernourished, he believed he could always avoid ketoacidosis.

"The whole aim of diabetic treatment . . . is to protect the patient by promoting by every means in our power the combustion of carbohydrate," Joslin wrote in the 1923 edition of his textbook. "Since in diabetes there is little carbohydrate in the body [stored in the liver as glycogen], and that which is eaten is often lost to the metabolism, one should never cease to strive to furnish it and to arrange treatment so that some of it will be burned."

That same year, Joslin praised Newburgh, Marsh, and Petrén for having "courageously demonstrated that patients did not die *in the hospital* [my italics] while living on a high-fat, but low carbohydrate and very low protein diet." Their "diabetic creed," he wrote, "is sound." As a result of their work, Joslin said he had even restored to the diets his clinic was using "a portion of the fat taken away." Although he was also certain that the diet worked because the patients remained underfed and underweight, which was his way of saying that Allen had also been right.

Joslin would continue to cite examples of patients, like his mother, Case No. 8, who had thrived on fat-rich diets, but then he would enumerate the many reasons why he believed patients should be fed carbohydrates, as much as they could tolerate, and why he remained "unalterably opposed" to the high-fat diet as proposed by Newburgh, Marsh, and Petrén. He objected because of the "inflexibility" of their approach. He called it a "life-sentence." Why should diabetics with mild cases be sentenced to a lifetime of abstinence from sweets, grains, and starchy vegetables if it wasn't necessary? "A life sentence takes away all hope, ignores the possibility of the patient's having a tolerance for a much higher quantity of carbohydrate as well as a restoration of tolerance for carbohydrate."

He also thought it was too risky. Yes, it might work in a hospital setting, but what happened when patients were sent home to the care of their local physicians, who might have very limited clinical experience? And Joslin objected because diabetologists, he believed, had to consider what happens not just when the patient adhered to the diet, but when the patient strayed. Patients who were following his guidance, undernourished yet eating to their tolerance of carbohydrates, might err by eating too many carbohydrates, but it wouldn't kill them. At least, not immediately. Eating too much fat or protein on the Newburgh, Marsh, Petrén plans might precipitate a coma and death before either the patient or the physician could prevent it. "The patient who breaks over only in carbohydrate pays an immediate penalty and is warned by increased urination," he wrote; "the patient who breaks over only in fat and protein is not warned and dies."

Now that insulin therapy was preventing diabetic coma and virtually all of Joslin's patients were living longer with the disease, when they did die, more often than not, it was from arteriosclerosis, "disease of the arteries of the lower extremities, heart, brain or kidneys." These were the chronic complications of diabetes now beginning to manifest themselves. In autopsies of diabetic patients who had died, Joslin reported, "whether old or young," the great majority showed significant plaques in their arteries. Joslin had young patients, he wrote, for whom insulin had saved their lives, only for heart attacks to then kill them: Elizabeth J., for instance, who was "in her early twenties while driving with her fiancé had a pain in her heart and died in 5 minutes." This was the new reality. "Arteriosclerosis today in diabetes," he told an audience of physicians in January 1928, "stares us in the face and we must attack it from all angles."

The reason for the arteriosclerosis, Joslin believed, was also the high-fat diets physicians had fed those patients in the pre-insulin era. To demonstrate his point, he invoked a case reported by physicians from Montreal General Hospital in *The American Journal of the Medical Sciences*. The Montreal patient had been diagnosed with diabetes in 1913, had survived almost fourteen years, and now the Montreal physicians were reporting on his death in 1927. It was a bizarre case for Joslin to invoke as a reason to avoid the fat-rich diet, but he may have just read the case report as he wrote up his talk—it had appeared in the most recent edition of the journal—and it had resonated with his many anxieties about high-fat diets.

For most of his first decade with the disease, the Montreal physicians had reported, their patient's therapy was "based upon the principles of undernutrition." In November 1922, with his health failing and his weight having dropped from 160 pounds to under 100, they had sent him to Toronto for insulin therapy. That's when the University of Toronto physicians also advised him to eat a higher-fat diet similar to what Newburgh and Marsh had been proposing. The insulin restored him to health. But five years later, he was dead with "large golden yellow nodules" of fat in his blood vessels and fat-laden plaques throughout his arteries.

The Montreal physicians had not connected their patient's arteriosclerosis to the relatively fat-rich diet he had been prescribed along with insulin therapy. Joslin did. Diabetologists had been observing high levels of fat in the blood of their diabetic patients with justifiable anxiety for years. The early editions of Joslin's textbook had opened with a color plate of two test tubes: one, immediately after being taken from a patient with diabetes, and the other, the same blood after having sat for a half hour, showing the milky white fat that rose to the top.* Joslin was not alone in speculating

---

* It looked "very much like thickened cream," as the New York City diabetologist Rawle Geyelin described it in 1917, after listening to a presentation by Joslin.

that this phenomenon might be a direct result of the fat-rich diets these diabetic patients were eating. Now Joslin convinced himself that was the case.

The Montreal patient "did not live in vain," Joslin told his audience of physicians:

> With this portrayal before us I doubt if ever again anyone will expose a diabetic to a low-carbohydrate high-fat diet for so long a period. Without such treatment as this patient received, in the decade before the discovery of insulin, undoubtedly he would have died, but who of us believe with the evidence now available that if in the earlier years when the diet was restricted in carbohydrate, moderate undernutrition had been continued and obesity avoided, such excessive arteriosclerosis would have developed? The body of this patient seems literally to have been steeped in fat. It is from an excess of fat in the tissues, I believe diabetes most commonly begins and from an excess of fat in the tissues or diet, diabetics, whether human or canine, died formerly of coma, but today, at least so far as humans are concerned, of premature arteriosclerosis.

By 1928, Joslin was defining an "ideal subject for treatment" with insulin therapy as a patient whose diabetes had been diagnosed early "and not damaged with a low-carbohydrate high-fat diet." To the question of why a young patient who had survived ten years with diabetes had arteriosclerosis, Joslin's answer was not "on account of his diabetes" but "because we doctors give it to him. . . . I believe the reason why we see premature arteriosclerosis in young diabetics is, because formerly we were only able to keep them alive with a low-carbohydrate and high-fat diet." Now he said his first rule to avoid arteriosclerosis in diabetic patients was to never overfeed them, "and least of all with fat." He assumed that as they fed their patients ever more carbohydrates—starting with 25 grams a day in the pre-insulin era and up to 125 grams in 1930—they were cutting down the fat they were consuming, preventing arteriosclerosis, and extending their patients' lives by doing so. "If our diets approach the normal," he wrote, "while diabetic symptoms are held in abeyance by insulin and exercise, our diabetics should live about as long as the average individual."

Joslin's admonitions against dietary fat had their intended effect. Physicians shied away from using the high-fat diet, while the debate about dietary therapy for diabetes shifted to accommodate the seemingly remarkable efficacy of insulin therapy: not how much fat could be eaten by diabetic patients without succumbing to coma—insulin therapy had mostly resolved that, in any case—but how much carbohydrate they could now eat, considering the extraordinary job that insulin was doing.

5

Insulin

Insulin does not cure diabetes. Insulin does not allow a diabetic to eat anything he desires. It is a potent preparation alike for evil and for good.

—ELLIOTT JOSLIN,
The Treatment of Diabetes Mellitus, third edition, 1923

The diet in health is made up chiefly of carbohydrate; the diet in diabetes before the discovery of insulin was made up chiefly of fat. Insulin has changed all this. The task of the modern diabetic is not so much to learn how to live comfortably upon less carbohydrate and more fat, but rather to balance the carbohydrate in his diet with insulin so that he can utilize it and thus keep his urine sugar-free.

—ELLIOTT JOSLIN,
Diabetic Manual for the Doctor and Patient, eighth edition, 1948

Leonard Thompson was thirteen in the autumn of 1919 when he first showed the symptoms—swelling ankles, wetting the bed at night—that led to a diagnosis of diabetes. Thompson was put on a diet and then fasted, but his condition deteriorated. On December 2, 1921, he was admitted to Dr. Walter Campbell's diabetes ward at Toronto General Hospital so weak and listless that his father had to carry him to his bed. The resident interns and medical students, as one recalled many years later, "all knew that [Thompson] was doomed." He had the symptoms of both ketoacidosis and a starvation diet that was failing: his hair falling out, his stomach distended, his breath smelling of acetone. He weighed sixty-five pounds. Campbell's secretary later said of Thompson that she'd "never seen a living creature as thin as he was."

Thompson, though, would be the first human ever to receive the pancreatic extract that would come to be known as insulin.* The Toronto physicians had no clinical experience with the drug, and so no reliable idea of the appropriate dose or what might go wrong, but with a patient like Leonard Thompson they had little to lose. If the insulin brought Thompson and patients like him back from the abyss, that would be compelling evidence that this was a significant breakthrough. Thompson's parents allowed the decision to be made by their son. The situation was carefully explained to him—a new experimental drug that might hasten his death or might save him. He readily acquiesced.

On January 11, 1922, Thompson was given an injection of the extract purified by Frederick Banting and Charles Best, the two young researchers who were driving the effort to isolate the hormone at the University of Toronto. The injection had little effect. Over the next twelve days, Banting and Best's biochemist colleague, James Collip, worked to produce a more concentrated extract. By January 23, Collip had made sufficient progress that the Toronto physicians were willing to try again. Thompson was then given two injections daily through February 4. The sugar in his urine nearly vanished; the ketone bodies did. "The boy became brighter, more active, looked better and said he felt stronger," Banting and his collaborators announced a month later in the *Canadian Medical Association Journal.* Their pancreatic extract—not yet officially known as insulin—had a seemingly remarkable ability to reverse the clini-

---

* Thompson was not, apparently, the first patient ever to receive a pancreatic extract that could lower blood sugar levels, just as Banting and Best and their Toronto colleagues may not have been the first to discover what they called insulin. In 1908, a Berlin physician named Georg Zülzer had infused pancreatic extracts into several patients and claimed that it decreased glycosuria "for some time." Zülzer went on to work with the Swiss company Hoffmann–La Roche to develop the treatment further. As the story was recently told by Viktor Jörgens of the European Association for the Study of Diabetes, Roche declined to pursue it in 1914, considering it "unlikely that patients would ever consider injecting themselves multiple times per day."

cal symptoms of diabetes. (Thompson lived another thirteen years before dying in 1935 of pneumonia at the age of twenty-six.)

Banting and the Toronto team of researchers and physicians treated six more patients before publishing their first report. With the new pancreatic extract, they wrote, blood sugar could be "markedly reduced even to normal values." Glycosuria, sugar in the urine, could be "abolished." Ketone bodies "can be made to disappear from the urine." Using a measurement known as the respiratory quotient, they demonstrated that these patients with severe diabetes once again began metabolizing carbohydrates for fuel. And, perhaps most important, the patients experienced "a subjective sense of well being and increased vigor" that was undeniable.

That was the beginning of the insulin era and the remarkable revolution in diabetes therapy that ensued. Purifying the insulin and then mass producing it for physicians worldwide was an extremely difficult process, but by August 1922 the University of Toronto had enough insulin, courtesy of Eli Lilly and Company in Indianapolis, to ship it to other physicians to use for their patients. An informal committee had been formed of the leading diabetologists in the United States—Woodyatt in Chicago; Williams in Rochester, New York; Wilder at the Mayo Clinic; Joslin in Boston; Rawle Geyelin in Manhattan, and Allen at his institute in New Jersey—and all of them would receive shipments of insulin to test on their patients closest to death.

Joslin's first patient was Elizabeth Mudge, a nurse who had been diagnosed five years earlier. Mudge weighed less than seventy pounds on August 7—"just about the weight of her bones and a human soul," as Joslin poetically described it—the day that Joslin's young colleague Howard Root injected the first dose. These early doses were purposely small, a precautionary measure as Joslin and his colleagues, like everyone else at the time outside Toronto, were unsure of the potency and had received very little insulin. Over the course of a week they slowly ramped up Mudge's dosage until she was finally getting enough to clear her urine of all traces of sugar. By then Joslin and his colleagues could see the improvement in Mudge herself. Within six weeks, she was transformed: previously an invalid who had left her apartment only once in nine months, she could now "walk with ease 4 miles daily." By the end of the year she was acting as caretaker to her own invalid mother. By then she had regained twenty of the pounds that she had lost with the disease. Joslin reported in 1928 that she had regained sixty pounds and was "active in her occupation as head nurse."

Allen gave his first insulin injections three days after Joslin, treating six critically ill patients, also starting with very small doses. When Allen wrote to Banting shortly afterward, he called the results "marvelously good" and said they'd already witnessed their patients regaining their strength and vigor.

Allen's most famous patient, Elizabeth Hughes, traveled to Toronto to be treated by Banting himself. She arrived on August 15, three days shy of her fifteenth birthday, "extremely emaciated," wrote Banting of her condition in his notes: "skin dry & scaly, hair brittle & thin, abdomen prommt [prominent], shoulders dropped, muscles extremely wasted, subcutaneous tissues almost completely absorbed. She was scarcely able to walk on account of weakness." She weighed forty-five pounds.

While the other physicians were trying to keep the diet low in calories and carbohydrates, calibrated carefully to the insulin usage, Banting had more insulin available. Hughes gained seven pounds in the first ten days, eating 1,100 to 1,200 calories daily, eating carbohydrate-rich foods like peaches, tomatoes, and toast for the first time in years. Banting let Hughes eat to her heart's content, and he would give her enough insulin to "cover" it, as would become the common terminology. That he had some doubts about his approach is evidenced by the fact that he asked Hughes to keep it to herself, and specifically not to tell Allen.

On August 25, after ten days on insulin, Hughes began eating a variation of the kind of high-fat, calorie-rich diet that had been the standard of care before Allen and his undernutrition took precedence, but now she was consuming it with the benefit of the insulin. She began eating over 2,000 calories a day and, as she wrote her mother, "drinking a pint of heavy cream a day if I can find room for it." After several days, Banting pressed her to eat still more. She then had a breakfast of "two peaches, almost a whole shredded wheat with 4 ounces of heavy cream, an egg, bacon, cream cheese, lots of butter and best of all a whole slice of bread toasted, on which I spread my butter and cheese." After four weeks, Hughes had regained twenty pounds. After six weeks she had even grown a half inch. "I simply don't recognize myself as the same person when I look in the mirror," she wrote to her mother. "Its [*sic*] simply killing too when I go out anywhere and see people I haven't seen for a week . . . they simply stare at me in perfect wonder."

These resurrections would happen throughout the world, wherever and whenever the Eli Lilly insulin became available or the local physicians could make insulin themselves using the techniques that Banting, Best, and their collaborators described in their publications. At William Sansum's clinic in Santa Barbara, California, Sansum and his colleagues used insulin they had extracted and purified themselves. Their first patient was fifty-one years old and had been living for six months in Sansum's clinic on a starvation diet, his weight dropping to under 100 pounds. After three days on insulin, his urine was free of sugar. After a year he weighed 125 pounds, after two years, 153 pounds, and he continued taking the same dose of insulin for the next thirty-eight years. He died at ninety in 1958. At Highland Hospital in Rochester, New York, Williams wrote of one of his patients—a twenty-two-year-old who had been confined to his bed "rapidly approaching death"—that "the restoration of this patient to his present state of health is an achievement difficult to record in temperate language. Certainly few recoveries from impending death more dramatic than this have ever been witnessed by a physician." It was a common sentiment.

The most remarkable recoveries were in those patients already in a coma when they first received insulin. Until 1922, physicians treating patients with diabetes considered coma to be essentially a death sentence. The two conditions insulin definitively, unambiguously cured were diabetic ketoacidosis and coma. The New York diabetologist Nellis Foster reported in 1923 that he had never before seen a patient with diabetes recover from coma, but he had tried insulin therapy on fifteen coma patients, successfully reviving eight of them. Joslin called the treatment of coma with insulin "spectacular." By 1925, he believed that if physicians acted quickly and appropriately, if they had insulin available and knew how to administer it, deaths from diabetic coma should be rare to nonexistent. By then, Joslin's clinic had used insulin on thirty-three patients in diabetic coma, he reported, and all but two had been revived.

Until 1921, the only drugs widely available to physicians that had any true therapeutic qualities were morphine and aspirin (for pain), thyroid extract (for treating myxedema), phenobarbital (a sedative and hypnotic synthesized by German chemists, most notably Minkowski's former colleague von Mering, in 1911), digitalis (for regulating the heartbeat), laxatives, and sleeping pills. "Insulin was totally different from existing medicines," as Robert Tattersall wrote in *Diabetes: The Biography*. Not only did it have to be injected—daily, and almost assuredly for the life of the patient—it was extremely potent and came with exceedingly dangerous side effects. "Its use raised as many questions as it answered, and every part of every answer raised more questions."

It was those dangerous side effects that had Joslin comparing insulin to morphine in the early years. He wasn't the only physician to do so. Both drugs taken at too high a dose can kill. With insulin, though, the progression can be seen coming, at least if the patient isn't already sleeping when it happens; as Joslin wrote, "a warning train of symptoms beginning with nervousness and extreme hunger as the blood sugar drops . . . progressing to sweating and tremor, subconscious or evident, and ending with unconsciousness. . . . Death is possible. [And] unlike morphine poisoning, recovery follows the simplest of measures, the juice of an orange or 1 to 3 teaspoonfuls of sugar, and takes place promptly within five or ten minutes."

Joslin was describing the condition of dangerously low blood sugar known as hypoglycemia. It was effectively a new medical condition in 1922, a new way to die, and it was a side effect of insulin therapy. Physicians of the era called it "insulin shock" or an "insulin reaction." Until the autumn of 1922, the only times such dangerously low blood sugar had been reported were in experiments with animals in which the liver had been removed, and perhaps half a dozen times in case studies of patients with Addison's disease, a disorder of the adrenal glands. In the 1923 edition of his textbook, Joslin said that he had seen it in four patients even before "the introduction of insulin," after "prolonged undernutrition"; four other cases had "come to [his] attention in the clinics of [his] friends."

With insulin therapy, hypoglycemic episodes became unavoidable, all too common. The Toronto researchers had seen them in their lab animals as soon as they had injected the first semi-purified pancreatic extracts. It was hypoglycemia in these animals that convinced them they were injecting the active hormone from the pancreas. Because the Toronto researchers had seen it in the lab, their physician colleagues in the early days knew from the outset what to watch for in their patients. Walter Campbell described it vividly, with his colleague Almon Fletcher, in one of the first articles they published on their clinical experience with insulin therapy at Toronto General Hospital:

> The initial symptom may be a feeling of nervousness or tremulousness, sometimes a feeling of excessive hunger, at other times a feeling of weakness or a sense of goneness. The level at which a patient becomes aware of the fall in blood sugar is fairly constant for that individual, although this is not always the case . . . it is rapidly followed by objective signs—most frequently a sweat which may be very profuse; pallor and flushing is common, sometimes a change in pulse rate. . . . At the same time, the subjective feelings become more severe; the feeling of nervousness may become definite anxiety, excitement, or even emotional upset. . . . Patients have shown a loss of power to perform fine movements with their fingers. . . . Much more severe manifestations are observed with further lowering of the blood sugar. Marked excitement, emotional instability, sensory and motor aphasia, dysarthria [a motor speech disorder that causes slurring of words], delirium, disorientation, confusion, have all been seen.

The simplest way to think of hypoglycemia is that it's the result of an overdose of insulin. Everything after that can be mind-numbingly complicated because what constitutes an overdose of insulin will always depend in turn, first and foremost, on how much food is consumed and, of course, what kind. This means that the appropriate dosing of insulin—both the amount of insulin in each injection and the timing of the injections—is dependent not just on how much carbohydrate and protein and fat is eaten at the next meal (and maybe for the remainder of the day), but when the meals are consumed, how quickly and thoroughly the food is digested, or any other gastrointestinal issues that may be occurring—"In the presence of diarrhea," Joslin wrote, "a very little insulin, even a single unit, may lead to unconsciousness"—any physical activity the patient might do before or after eating, and a host of other unpredictable factors.*

What's more, every patient might respond differently to the same dose of insulin, even patients who appeared to be identical in the severity of their disease. The same patient might even respond differently to insulin at different times of day or at different stages of their disease. In a talk to physicians in late 1925, the Presbyterian Hospital and Columbia University diabetologist Rawle Geyelin explained that this was part of his very challenging learning experience with insulin: severe cases of diabetes could appear mild at times, and so appear deceptively to require very little insulin; mild cases could appear severe—if an infection was present, for instance, like the flu or even a bad case of dental caries (cavities), or an abscess caused by the insulin injections themselves—which meant they could appear to require far more insulin than would otherwise be necessary. "Tolerance for food varies considerably from year to year, from month to month, and in some instances from day to day," Geyelin said.

Even such a seemingly simple issue as the duration of the hormone's action—how long a single dose of insulin would work after injection—could be maddeningly capricious. In most cases, a dose of the Eli Lilly insulin would start to work in the blood half an hour after injection and continue to control blood sugar for the next four to eight hours. Then another injection would be necessary to cover the next meal. So a dose given before breakfast could be expected to last through lunch and perhaps even into the afternoon hours.

---

* When two diabetologists from the Vienna Hospital for Children reported on the fiftieth anniversary of the insulin era on those young patients who had been treated with insulin in the first decade, they noted that two of them later died of hypoglycemia after extreme physical exertion: in both cases, mountain climbing.

But this, too, depended on the patient. Geyelin told the story of two patients to whom he gave insulin at six in the morning only to see them eat their three regular meals without issue and then suffer "severe shocks with convulsions" after nine o'clock at night, more than fifteen hours later. If the insulin reaction occurred later at night, when the patients and their family were asleep, the results could be fatal. Several of the prominent diabetologists of the era lost at least one patient to insulin shock before they established a protocol that prevented it.

Geyelin, whose experience was primarily with childhood diabetes, said that one saving grace was that once the patient experiences "a few severe hypoglycemic shocks," ideally while still in the hospital and so under his care, "the parents become more careful in the correct administration of food and insulin." Walter Campbell and Almon Fletcher in Toronto had recommended in their very first papers on insulin therapy that patients should be made to experience an insulin overdose in the hospital or clinic while first working out the correct dosing. That way the patients would more likely recognize the characteristic symptoms when it later happened by accident. The remedy was "oranges, glucose candies, corn syrup, or cane sugar."

A further complication was the simple fact that no technology existed at the time for patients to test their blood sugar at home. Such tests could be done in hospitals but they were prohibitively expensive: "The cost of one such test," Joslin wrote in 1923, "would probably supply [my patients] with insulin for a week and they prefer the insulin." Urine tests, which were available, only reflected the state of blood sugar roughly six hours earlier. If physicians wanted to get their patient's urine sugar-free, they had little idea whether they were, by necessity, overdosing the patient to do it and so making hypoglycemic episodes that much more likely.

All of this justifiably worried physicians when they contemplated the use of insulin on their diabetic patients. They were thrilled by the promise, but anxious nonetheless, particularly once their patients left the clinic or hospital and would have to administer the insulin injections themselves or, for children with diabetes, have their parents do the proper dosing and injections for them.

It was one thing for experienced diabetologists in major medical centers to learn from their copious clinical experience how best to use a drug this powerful and quixotic and how best to train their patients and the patients' families, but what about physicians with only one or a few diabetic patients in their practices and little (or no) training themselves? While diabetologists almost immediately began experimenting with formulations and methods that would allow their patients to take insulin without injections—all of which would fail—Joslin saw it as a blessing that "its use is restricted to a syringe." In the 1928 edition of *The Treatment of Diabetes Mellitus*, after having used insulin therapy on more than 1,500 patients, Joslin said that insulin had not simplified the treatment of diabetes, but had made it more complex. "The diabetic taking insulin is like a rapidly moving machine which a slight swerve of the wheel will bring to disaster."

Joslin, nonetheless, had also come to believe that reasonably competent patients could easily learn to inject insulin and do so safely. He also believed and made the case that virtually every diabetic patient should be on insulin or at least trained and prepared to use it when necessary. "If a diabetic is not happy, energetic and a joy to himself and his family, and his urine sugar-free," he wrote in his *Diabetic Manual*, "he had better take insulin. The chances are overwhelming that it will do him good." Even those who could control the disease by diet, Joslin believed, might find that diet fails to work with time, or fails during the presence of a flu or other infection. So even these patients should at least know how to use insulin and have syringes and insulin available at home.

Other physicians in the early days of insulin therapy were less sanguine, debating publicly whether it was too dangerous, too potent to be used by any but the most experienced physicians and the most intelligent of patients. "There are certain patients other than those with very mild diabetes for whom insulin is unsuitable," one physician wrote in the *British Medical Journal* in 1924. "Chief of these are the fools."

Insulin therapy is so challenging because of a simple but unavoidable fact: the appropriate dosing depends on diet, and the appropriate diet depends on the insulin dose. This is what mathematicians call a multivariable equation. Such equations do not have a single correct solution, but a range of solutions that can all be considered equally right (or equally wrong). Unlike other pharmaceuticals in which dosages can be standardized, based, if nothing else, on the age and maybe weight of the patient and the severity of the disorder, that is not the case with insulin. There is no obvious or ideal dose for any single patient, only a range of options depending on what the patients eat, when and how often they eat, their level of physical activity (and the energy they expend in that activity), and that host of other mostly unpredictable variables.

While insulin therapy made it possible for those with diabetes to eat carbohydrate-rich foods, the critical question that has yet to be answered is whether eating carbohydrate-rich foods in the long run diminishes their health or shortens their lives. Diabetologists who had accepted without question until 1922 that their patients should be kept undernourished or abstain religiously from carbohydrate-rich foods now had to counsel their patients to eat some carbohydrates throughout the day, not just at every meal but as snacks between meals and even a snack before bed—crackers and a glass of milk, for instance—to cover any insulin that might still be in their system overnight and so prevent what these physicians called nocturnal hypoglycemia.

When Geyelin lectured to an audience of Pennsylvania physicians in October 1925, he described some of the challenges of using insulin therapy to control a patient's diabetes while avoiding hypoglycemic episodes:

In some cases, the proper adjustment between hypoglycemic reaction on the one hand and glycosuria [sugar in the urine] on the other hand has not been obtained until the food has been divided into six meals per day. In other instances, the ideal adjustment has been brought about by giving as many as four doses of insulin per day, one before each meal and one at midnight. In other cases, the relief of glycosuria has been brought about by the simple expedient of transferring the night dose of insulin from one half-hour to two hours after supper. Where a patient has been on morning and evening doses of insulin only, and even when he has been on doses morning, noon, and evening, we have experienced the greatest difficulty in keeping the urine sugar free in the over-night urine or in the first specimen voided after breakfast. In the former instance, moving the night dose to a later time in the evening, or inserting a midnight dose has sometimes proved effective. In the latter type of case, we have usually been able to eliminate the morning glycosuria by moving the breakfast dose to a time from one to three hours preceding the morning meal.

After establishing in the hospital or clinic what appeared to be the appropriate dose of insulin and the appropriate schedule of injections, the expectation was that the patients would then eat the necessary set amounts of carbohydrates at every meal—weighing the foods, ideally, to assure the correct dose of carbohydrate would be consumed for the given dose of insulin—and do so every day for the rest of their lives. The physicians would also have to counsel their patients that when hypoglycemia did occur, as it inevitably would, they had to have the remedy on hand to prevent unconsciousness.

For patients with type 2 diabetes, for whom insulin injections might not be necessary and might even, ultimately, do more harm than good (as we'll discuss), they and their physicians would also have to establish some compromise. The New York City cardiologist Blake Donaldson wrote in 1962 that he said to his diabetic patients, "You are out of your mind when you take insulin in order to eat Danish pastry." But Donaldson's would be very much a minority opinion by then.

6

Rise of the Carbohydrate-Rich Diet

Insulin cannot act with greatest efficiency unless given an opportunity to display its power. An orator needs an audience to show what he can do, and insulin needs carbohydrate. Thus it has been claimed repeatedly by those advocating the higher carbohydrate diet that the carbohydrate can be doubled with little or no additional requirement for insulin provided total calories are controlled. But we are still somewhat skeptical.

—ELLIOTT JOSLIN,
The Treatment of Diabetes Mellitus, eighth edition, 1946

The most important advantage claimed for this form of treatment is the psychological benefit obtained through freedom from irksome regulations. Diabetes is a chronic disease, at present incurable, and to minimize its hardships the fewer the restrictions imposed upon the patient's daily life the better. The advocates of "free" diets reason that if insulin is required at all it does not matter very much to the patient whether the dose is 30 or 50 units a day, whereas it does matter to him a great deal that he should be able to take a normal diet—"that" (as a diabetic doctor has said) "each meal should be an elegant satisfaction of appetite rather than a problem in arithmetic and a trial of self-abnegation."

—C. C. FORSYTH, T. W. G. KINNEAR, AND D. M. DUNLOP,
"Diet in Diabetes," 1951

From the early days of insulin therapy, diabetologists fell into two general schools of thought on how to handle its very considerable challenges. There were those who believed the best approach would be to establish the minimal dose of insulin on which their patients could be healthy, implying that the patients would be eating a diet that minimized the need for insulin. This was the approach that Banting and his physician colleagues in Toronto took (although, as noted, not with Elizabeth Hughes), and that Joslin and his colleagues embraced as well. These physicians believed that the healthier their patients, the happier they would be as a consequence, and so maximizing health and minimizing risk should be the first and dominant priority.

On the other side of the divide were those who believed that their obligation was not just to keep their diabetic patients as healthy as possible, but to minimize the psychological and social burden of living with the disease, and anything that could do that, within reason, was worth trying. These diabetologists, of whom Rawle Geyelin in New York was among the first, argued that the approach taken in Boston and Toronto was unnecessarily puritanical (a word often used to describe the thinking of Joslin with his Massachusetts roots) and would fail in practice anyway because patients would not adhere to it. They came to believe that they could let the insulin therapy do the hard work necessary to control blood sugar, and let the patients reap the benefits by eating much as they wanted. When Adolf Lichtenstein, a professor of pediatrics at Stockholm's famous Karolinska Institute, advocated in 1938 for using this "free diet" for children with type 1 diabetes, he described it as "the first time full advantage has been taken of the introduction of insulin in the therapeutics of diabetes."

The remarkable aspect of the decisions made in this early era about this complex equation of dosing versus diet is that they were made with a complete absence of information about the long-term risks and benefits of the drug itself. Today, if a drug's effects were as immediately remarkable as that of insulin in acute cases—say a miracle anticancer drug that made tumors vanish—it might be approved quickly, but only with the establishment of monitoring protocols to determine whether there are complications that might be worse than the disease itself. Before such a drug could be approved for patients with a chronic disease—one that takes years or decades to do its damage—the researchers would have to demonstrate in exorbitantly expensive randomized-controlled trials that it unambiguously did these patients more good than harm. And once (or if) it was approved, use of that drug, too, would be monitored closely to look for uncommon side effects that these initial trials might not have been large enough or sustained long enough to detect.

When insulin therapy began in 1922, with all insulin's power "for evil and for good," as Joslin had described it, neither the tests, the ability to conduct the tests, nor even the thinking that such tests were necessary existed. The physicians knew only that insulin worked to reverse the symptoms of diabetes. For the more severe cases, any long-term risks of insulin therapy were worth the potential benefits. But the milder the diabetes, the further from death the patient, the more chronic the disorder, the better it responded to dietary approaches, the more important these long-term trade-offs would be. With the advent of insulin therapy, both patients and physicians had to make a risk-benefit analysis with hardly any information on the risks. And as insulin transformed acute cases of the disease into chronic cases, as was its great triumph, these unknowns would become the critical questions, as they still are.

Even the chronic, long-term effects of elevated blood sugar in patients with type 1 diabetes—of having sugar in the urine or having a little sugar as opposed to a lot, whether the blood sugar, the glycemia, was well controlled or not—were mostly unknown, because those patients in the pre-insulin era who had the highest blood sugar or the most sugar in their urine had tended to be those who died quickly. Joslin assumed that high blood sugar caused long-term, chronic damage in and of itself, but he couldn't know for sure. ("The strongest argument against hyperglycaemia [high blood sugar] is physiological," wrote the British diabetologist R. D. Lawrence, who had type 1 diabetes himself, "in that nature has perfected in the normal such a wonderful mechanism to prevent it.")

He also couldn't know if any damage was being done by the insulin that was used to keep diabetic blood sugar under control and/or the *hypo*glycemia that came with that.

As for the insulin therapy itself, nothing was known about long-term effects. Diabetologists seemed to assume it was benign because they thought of insulin therapy as merely replacing a hormone that was missing to begin with, but that was, indeed, just an assumption. For those patients who would clearly die without insulin, these long-term effects, again, were justifiably considered irrelevant. The long-term effects of insulin therapy, if any, would be the price that had to be paid to keep the patients alive. But, even then, would the patient do better, live longer, if the minimal therapeutic dose was used? Large doses would do a better job of controlling glycemia but they would also, most likely, prompt more severe episodes of hypoglycemia. Did frequent or even occasional episodes of hypoglycemia create health issues of their own? Did they injure the brain, for instance, or cognitive abilities? No one knew. No one would know until these patients lived long enough with the disease and, even then, without randomized controlled trials to compare the efficacy and safety of different dosing levels, diabetologists would only be guessing. Not until the 1940s did this possibility even become a topic of discussion in the medical journals. (Such trials would begin in the 1990s, and the results were not encouraging about the benefits of using what researchers took to calling "intensive insulin therapy" for glycemic control.)

When Lawrence discussed these issues in the *British Medical Journal* in 1933, he said with characteristic frankness that some conclusions about the insulin-diet-dosing-health equation seemed "clear and I believe true and ultimate, at any rate until further progress shows up our fallacies." That was the best that could be said.

With all these questions unanswered, clinics would develop their own general protocols for solving the diet-dose equation, all relying ultimately on trial and error: finding first a level of calorie and carbohydrate consumption on which the patients' urine was sugar-free, and then increasing either the carbohydrate content or the calorie content, along with varying doses of insulin, to find a balance in which the patients were getting sufficient food to live an active life with sufficient insulin to make it possible and, ideally, the bare minimum of hypoglycemic episodes. Once the patients left the hospital or the diabetes clinic, their local physicians would be expected to guide them in adjusting the doses as necessary to accommodate the added unpredictability of real life. Textbooks and journal articles would rely heavily on case studies, examples of how different patients handled a particular challenge, with the awareness that these anecdotal experiences were the best tool they had for communicating to patients and other physicians how to approach the diet-dose dilemma. Exercise, for instance, as Joslin noted in the 1928 edition of *The Treatment of Diabetes Mellitus*, exerts a marked effect on insulin needs:* "Case No. 632 says a game of golf is worth 5 units. Days without golf he takes 20 units in the morning and 5 units at night, but on days with golf he omits the night dose."

Three fundamental principles ultimately guided the approach these diabetologists took. Unfortunately, satisfying any one of these principles often required nullifying or transgressing on another. Diabetologists would have to make decisions on a patient-to-patient basis, and these decisions would be based largely on the physicians' preconceptions and beliefs from before the insulin era.

The first principle followed directly from Brown-Séquard's thinking on internal secretions half a century earlier. "Diabetes mellitus is due to a deficiency of the internal secretion of the pancreas," as Banting described it in 1924. "The main principle of treatment is, therefore, to correct this deficiency. If it is found that the patient is unable to keep sugar-free on a diet that is compatible with an active, useful life, sufficient insulin is administered to meet this requirement."

Even this simple idea would be fiercely debated: specifically, was sugar-free urine really necessary, considering how difficult it would be to achieve in many patients without triggering frequent insulin overdoses? Physicians would quickly learn that their patients

---

* Today we would say exercise temporarily increases *insulin sensitivity*, and so a dose of insulin will be more effective in controlling blood sugar after exercise than the same dose given before.

could feel healthy and regain weight while still excreting sugar in their urine. That's true today. But sugar-free urine seemed like the obvious goal, evidence that the disease process had been (mostly) brought under control. The distinct possibility also existed that keeping the urine sugar-free would result in a partial or even complete cure of the disorder. Banting had said so explicitly after he won the Nobel Prize in 1923 (sharing it with John Macleod, in whose University of Toronto laboratory Banting and Best did their research).* At his Nobel lecture, which he gave two years later, Banting had asserted that insulin injections could not only render urine sugar-free in every case of the disease, no matter how severe, but that this should *always* be the goal of insulin therapy.

They had "abundant evidence," Banting said, that patients improved their carbohydrate tolerance while on insulin, meaning they could eat greater amounts of carbohydrates and use proportionally less insulin without sugar appearing in their urine. Of the first fifteen patients they treated with insulin in Toronto—"all extremely severe cases for whom diet had done its best"—Banting said, all fifteen had seemingly gained in carbohydrate tolerance with insulin therapy. As a result, they'd been able to reduce their insulin doses by half to a third of what they had originally required. Among the handful of cases that Banting discussed in his Nobel lecture, his most dramatic example was of a patient whose diabetes after six years was so severe that she had required surgery to amputate a gangrenous leg. She had been started on insulin therapy in preparation for the surgery and it had cleared her urine of the "large amounts" of both sugar and acetone (ketone bodies), Banting said. She then had the scheduled operation, "the stump was entirely healed in three weeks," and after six weeks her "insulin was discontinued and her diet was increased without the return of diabetic symptoms." Nearly three years later, according to Banting, she was still sugar-free and "on a liberal diet without insulin."

There are many possible explanations for what the Toronto

---

* The award was famously controversial. Banting was furious when he learned that he had shared the prize not with Best but Macleod. He split his share of the award money with Best. Macleod split his with the biochemist James Collip. Neither Banting nor Best appeared at the Nobel ceremony.

researchers were witnessing. The most exciting was that insulin therapy was actually restoring their patients to health in more ways than one. Perhaps, Banting suggested, if the pancreatic beta cells that normally secrete insulin are given a chance to rest they can recover or regenerate, just as a strained muscle or a broken limb will heal if allowed to rest. This was based on the belief that the high blood sugar of diabetes never gives these weakened or injured pancreatic cells the respite they need. Diabetes, by this logic, is a vicious cycle: the defect in the pancreas results in high blood sugar, which in turn keeps the pancreas overworked and unable to heal itself. Interrupt the cycle for long enough with insulin therapy and the diabetes itself might resolve. "The condition of the pancreas then corresponds to that of a heart with broken compensation," as Rollin Woodyatt had written in 1909, "and as the treatment for such a cardiac condition is rest, so in diabetes rest is needed for the pancreas." Allen had advocated for undernutrition in part with the hope that by keeping both food consumption and body weight to a bare minimum, the pancreas will have to do less work and so can rest and recover its insulin-secreting ability. With insulin injections doing the work that the pancreas should have been doing naturally, Banting and others now proposed, maybe this cycle could indeed be broken. According to Joslin, the prominent German diabetologists Naunyn and von Noorden had both suspected that diabetes might be curable. Maybe it was.

While Joslin himself was skeptical—he remained skeptical that *any* patients had been cured—he nonetheless argued for sugar-free urine on the same basis that Banting did. "The tendency of the diabetic patient to gain in tolerance for carbohydrates when the urine becomes sugar-free," he asserted in numerous editions of his textbook, beginning even before insulin therapy with the first edition in 1916, "is the fundamental principle upon which all treatment has been and is rightly based, and that by which the value of all therapeutic measures is determined." If that was the case, then sugar-free urine had to be the therapeutic goal, and enough insulin would have to be used, in harmony with the diet, to make that happen. When diabetologists took to arguing for higher-carbohydrate diets and so more insulin rather than less, they believed that their patients thrived in part because the added carbohydrates could seemingly "whip up intrinsic insulin production," evidence that the pancreas might be recovering.

The second principle was that insulin therapy should be layered on top of the best thinking on dietary therapy, not the other way around. In the 1923 edition of his textbook, written when Joslin had accumulated over a year's worth of clinical experience with insulin therapy—more than 350 patients—he described the thinking this way: "With insulin the principles of the diabetic diet are utilized, not replaced" and "the treatment of diabetes with insulin should follow the best lines of treatment of diabetes without insulin."

This logic is difficult to challenge. It has the expectation of doing the least harm, as the Hippocratic oath dictates. These physicians would not be reinventing dietary therapy to work with insulin, at least not at first (as we'll see); they would be using insulin cautiously while maximizing the potential of what they had learned before insulin had been available. This had an unintended consequence, though: it guaranteed, as Russell Wilder observed in 1940, that the thinking of the pre-insulin era would be woven deeply into the thinking that followed, whether that thinking was correct or not. In the face of enormous ignorance about the long-term effects of both high blood sugar and insulin therapy, these diabetologists would continue to believe what they had always believed about the nature of a healthy diabetic diet. And they would do so, regardless of evidence to the contrary.

The third principle is similar in its intent and related in its outcome: the less insulin used the better. It was best for physicians to establish the proper dose by working up from very small doses rather than down from larger ones. The reasons were many— Joslin enumerated half a dozen in his textbooks—including the unknowns about long-term complications of insulin therapy, the ever-present danger of hypoglycemia from insulin overdoses, and the fear that ketoacidosis could still kill quickly should insulin therapy be suddenly discontinued. Hence, the ideal approach to insulin therapy, the one with the least risk, used the smallest dose possible to render the patient's urine sugar-free.

Joslin was particularly sensitive to the problems engendered by relatively high doses because of the second patient he had treated with the drug. In the 1923 edition of his textbook, he referred to the case briefly: "Thomas D., Case No. 1305, omitted insulin for five days, continued his relatively high[-calorie] diet, developed a mild infection, and entered the hospital to die seven and a half hours later of coma." By the 1928 edition, Joslin was taking responsibility for allowing Thomas to eat a diet that required relatively large doses of insulin to cover it. Had Thomas been on lower doses and suitably undernourished, Joslin implied, he might have survived five days without insulin (during which his family doctor was out of town and his family preoccupied with another illness).

When Joslin and his colleagues first began using insulin therapy in the Boston clinic, they would start the patients off on a single unit of insulin, either once, twice, or three times a day before meals, and then they would increase those doses by a single unit at a time until they hit upon the right balance between carbohydrates consumed and insulin injected. A unit of insulin itself had been defined by the Toronto group. It was a third of the amount of insulin necessary to drop the blood sugar of a 2 kilogram (4.4 pound) rabbit (which had not eaten for twenty-four hours) to the level at which it experienced hypoglycemic convulsions. Rabbits were apparently very consistent in that sense. Humans were not. And so, as we discussed, what a single unit of insulin would do in any one patient might have little relationship to what it would do in any other. As Joslin said in the 1923 edition of his textbook, in humans the value is "quite theoretical." By 1928, Joslin and his colleagues were starting patients with 5 units of insulin before each meal.

These small doses also had the advantage of demanding diets that were restricted as well, further strengthening the second of the guiding principles of insulin therapy. From Joslin's perspective and Frederick Allen's, small doses of insulin meant patients with diabetes would *have* to eat thoughtfully still, restricting all calories and particularly fat calories. Ideally, they would still remain at least slightly undernourished, and so under their ideal weight, reducing further the burden on the pancreas. For those who found Newburgh and Marsh's papers compelling, as Russell Wilder at the Mayo Clinic did, small doses of insulin meant diets that were fat rich and still restricted almost entirely in carbohydrates. Either way, insulin would take care of the symptoms of the diabetes, but patients would still have to eat either less than they preferred or very differently from their healthy friends and family. It was reasonable to assume that they would remain healthier, because of both the dietary restriction and the minimal insulin doses.

These three principles could be thought of as an ideal of insulin therapy, but they would not hold up to the realities of treating patients, particularly children. Physicians began chipping away at them, if not discarding them entirely, as they realized how much of the job of controlling blood sugar with insulin was still beyond their control.

Among the more revealing articles in the literature from this era is one published in a Scandinavian pediatrics journal in 1932, in which local physicians reported that over a third of their young diabetic patients had died within a few years of diagnosis even with insulin therapy available. This was a death rate more than three times higher than the Mayo Clinic was reporting at the time and many times higher than Joslin's clinic in Boston. The reason, according to the Norwegian pediatrician Kirsten Utheim Toverud, was that some parents either refused insulin therapy for their children outright or didn't give their children injections on the strict schedule necessary. The farther the families lived from a major city and/or a hospital itself, the worse the children did, suggesting that their parents were not prepared or sufficiently well-educated to accept a reality of what had now become modern medicine. The director of the Gothenburg Children's Hospital in Sweden, Arvid Wallgren, told the story of one child who had been *living* in his hospital for eight years because the parents "positively refused" to give the necessary insulin injections themselves.

Pediatricians, dealing with the most lethal form of the disease, type 1 diabetes, in their young patients, became the greatest proponents of making life as easy as possible for them, liberalizing the carbohydrate content of the diet and letting high insulin doses cover what would have been considered, pre-insulin, a litany of unacceptable dietary transgressions. Once this free or liberal diet philosophy was justified for children, it became much easier for physicians treating their adult patients to embrace it as well. Physicians assumed their patients were not always following their dietary advice anyway, so why not let them eat whatever they desired and use however much insulin was necessary to assure that they were free from coma and the conspicuous symptoms of the disease?

Some diabetologists even argued that the previous focus on desugaring the patients' urine was rendered obsolete by insulin. Even R. D. Lawrence, speaking from his personal experience, felt sugar-free urine was too much trouble to accomplish without obvious benefit. Sugar-free urine, after all, was simply the means by which diabetes was diagnosed, not the reason why diabetic patients could not lead a normal life. It was the thirst, the hunger, the urination, emaciation, and the infections, these physicians argued, that discomforted and disabled their patients and that the insulin would treat. If diabetic patients could be protected from ketoacidosis and feel healthy with more insulin and eating a relatively normal diet, wasn't that sufficient? "The common denominator of all these reports is the increased 'feeling of fitness,' mental alertness and physical vigor," wrote the Santa Barbara diabetologists William Sansum and Percival Gray in *The Journal of the American Medical Association* in 1933. "That such desirable subjective values can be purchased at a moderate cost in insulin, our own observations and those of other clinicians demonstrate."

As early as the spring of 1923, after only eight months of experience with nine children, all with severe cases of diabetes, Rawle Geyelin and his colleagues at Presbyterian Hospital in New York City were allowing these children to have sugar in their urine, so long as they otherwise felt healthy and were regaining weight. Keeping the urine sugar-free without causing the "disagreeable and sometimes alarming symptoms" of hypoglycemia was proving too difficult, so they settled for "moderate glucosuria in all cases," and said this approach allowed them to use smaller insulin doses and reduce the likelihood of insulin overdoses.

By the autumn of 1925, though, Geyelin and his colleagues were feeding their young patients more carbohydrates and using however much insulin was necessary to keep the urine sugar-free. They had patients who had been failing to gain weight and strength and complaining of hunger, even with insulin, Geyelin reported, and increasing the insulin dose without allowing these kids to eat more carbohydrates would mean more insulin overdoses. The argument would later be made that for children with diabetes to grow and develop normally, just as with healthy children, "they need nutritionally adequate diets." That meant any approach that restricted either calories (undernutrition) or protein (the Newburgh-Marsh program) could result in children whose growth would be stunted despite insulin therapy.

With Joslin and Allen arguing the dangers of high-fat diets and the potential toxicity of the Newburgh-Marsh approach, Geyelin chose what seemed like the least harmful option: add carbohydrates to the diets, cut fat to balance the calories, and use however much insulin was necessary to keep the patient's urine sugar-free. Geyelin said that he assumed insulin doses would have to be increased significantly to cover the added carbohydrates, but found that often wasn't necessary. So he allowed his young patients to eat "striking" amounts of carbohydrates—"quite shocking [to] several veteran diabetic observers." Some of his patients quintupled the amount of carbohydrates they had been eating, with no immediate increase in the insulin doses necessary for sugar-free urine. For these patients, the insulin appeared to work more efficiently as they consumed more carbohydrates, the opposite of what Joslin had been reporting.

The advantages seemed obvious, Geyelin wrote: for starters, the children certainly appreciated the opportunity to eat foods they liked, and they seemed less likely to cheat if they did. Moreover, some of Geyelin's younger patients had failed to gain any weight or even grow in height for two or three years on the conventional program that still relied on some undernutrition. Once Geyelin allowed them to eat more carbohydrates and increased the insulin dose, he later reported, they "gained as much as 15 pounds in weight and 3 inches in height in one year." When Geyelin visited Montreal, he convinced Israel Rabinowitch of the Montreal General Hospital to start using high-carbohydrate diets, albeit still relatively low in calories, after telling him of a fourteen-year-old patient who had weighed fifty-one pounds when Geyelin first saw her in January 1923, and was only four feet two inches tall. He tripled her carbohydrate allowance to 1,200 calories a day, needing only a little more insulin to cover it, and over the next half-dozen years she gained almost sixty pounds and grew fifteen inches.

When Geyelin described his approach and its results in 1935, he reported that his patients felt better eating this way, they possessed more strength and physical endurance, and their dietary habits were like those of "normal human beings." They weren't as hungry as they had been when they had been eating as Joslin and others recommended. They stuck to their diets, feeling little urge to cheat or binge on carbohydrate-rich foods. They seemed well protected from both insulin reactions and diabetic ketoacidosis.

Several European pediatricians had come to the same conclusions about how best to treat their young patients. What Geyelin had observed and these pediatricians confirmed was that the children could eat considerably more carbohydrates with only small, if any, increases in the doses of insulin necessary to keep urine sugar-free. This clearly seemed to be a good thing: maybe the more carbohydrates they ate, the greater their carbohydrate tolerance. A reasonable assumption, albeit a wrong one, as it would turn out, was that these children were getting healthier.

If nothing else, the psychological benefits seemed undeniable. In 1931, the German pediatrician Karl Stolte, who had treated his first cases of diabetes in the pre-insulin era at Naunyn's institute, outraged the older generation of German diabetologists by describing his young diabetic patients eating "with delight" a birthday cake from the local pastry shop without any significant increase in the sugar they excreted. "The euphoria," Stolte said, "helped as it did so often in life to overcome difficulties." In a 1933 book on diet, the Danish pediatrician Carl Friderichsen put it bluntly: "Psychologically it is better to accommodate the doses of insulin to the food than vice versa." Five years later, when Adolf Lichtenstein at Stockholm's Karolinska Institute described his use of an "absolutely free" diet, he reported that he had used it with fifty children by then, from toddlers to sixteen-year-olds, and the children always embraced it. "It is necessary to witness the joy of the diabetic children at the release [of dietary restrictions]," Lichtenstein wrote, "in order to understand fully the psychologic importance of the free diet."

There were (and still are) practical reasons as well for letting children with diabetes eat *almost* whatever they want (even Lichtenstein cautioned against dietary "luxury, which also ought to be avoided by healthy children," and Stolte against bingeing on "whipped cream and sweets," despite any euphoria they might induce) and then matching the dose of insulin to whatever carbohydrate-rich foods they were eating: they could eat what their family and friends were eating. Parents were not required to cook separate meals for the child with diabetes; children were not required to go hungry while they watched their healthy siblings eat whatever and however much they desired. Those kinds of practical issues, the Norwegian pediatrician Utheim Toverud argued, "can hardly be overestimated in connection with a disease lasting a lifetime."

Karl Stolte could see no reason for physicians to treat diabetes, a disease in which an "internal secretion" was deficient, any differently from their approach to other such endocrine deficiency diseases: "Who thinks of changing the diet with other deficiency diseases?" he wrote. "Adding back what is missing [i.e., insulin] brings healing!" He acknowledged that he lacked the clinical evidence necessary to support the safety of his protocol—ideally, decades of follow-up with his patients—but he also insisted that the benefits were indisputable: children with diabetes experiencing "full health, feelings of safety, security and happiness. . . . The times of unilateral carbohydrate restriction are over," Stolte asserted, "otherwise insulin would have been discovered in vain!"

The same trend toward higher-carbohydrate diets had also started for adults shortly after the availability of insulin. William Sansum, head of what was then the Potter Metabolic Clinic in Santa Barbara and later the Sansum Clinic, said that as soon as he read about Banting and Best's discovery he had hoped his patients would be able to eat something akin to normal diets. A single "discontented patient" convinced him that this might be possible. A successful Colorado businessman, fifty-one years old, had a mild case of diabetes that was being treated by a high-fat diet without insulin.

After six months, Sansum explained, the patient stated "frankly that he had gained nothing by the six months of treatment, and that there must be something wrong either with him or with our diet." Sansum and his colleagues suggested he eat more carbohydrates and less fat, and they now used small doses of insulin to keep his urine sugar-free. Over the course of two months in the fall of 1925, they repeatedly upped his carbohydrate consumption until opting for what Sansum called "a radical experiment": they more than doubled the amount of carbohydrates the patient ate each day and cut his fat by almost half, now allowing him white bread, potatoes, milk, and large servings of fruit. Within twenty-four hours the patient said he felt better. Eight days later he returned to work, now taking large daily doses of insulin—more than 100 units—but feeling "fully restored . . . to his former prediabetic state of mental and physical activity." Eight months later, as Sansum told it, he was still eating the same way, but needing only a third as much insulin.

Sansum reported in January 1926 that he had used the high-carbohydrate diet on more than 150 of his patients and he suggested that all were thriving. Some of his patients needed as much as 200 units of insulin a day to keep their blood sugar controlled—a dosage that Joslin and others could not imagine giving to their patients—but Sansum thought it was worth the trade-off. His patients showed no signs of diabetic ketoacidosis and had a "margin of safety" with all the carbohydrates assuring the status quo. The patients got to eat the foods they craved and felt better doing it. In 1933, Sansum reported that he had used the diet on more than a thousand patients with outcomes that could compete with that of any other clinic, Joslin's included. (Sansum clarified that his high-carbohydrate diet was not a "free diet" like Stolte and others were using in Europe, but rather a doubling of the allowed carbohydrate content and a halving of the fat.)

Two years later, the diabetes world would change once again. This was the beginning of what Joslin called the "Hagedorn era." Joslin announced its arrival in January 1935 in a single paragraph added to the first chapter of the fifth edition of *The Treatment of Diabetes Mellitus*. Hans Christian Hagedorn, founder of the Nordisk Insulinlaboratorium (now the pharmaceutical giant Novo Nordisk) in Copenhagen, had added a compound called protamine to insulin, causing it to dissolve only gradually in the bloodstream and therefore remain active for twice as long after injection than regular insulin. Tests at Hagedorn's institute suggested that protamine insulin could control glycemia for eight hours, making it appear ideal for overnight use. The following year, Toronto researchers added zinc to the compound, now called protamine zinc insulin (PZI), which could remain active in the circulation for twenty-four hours. Within a year, major insulin manufacturers worldwide were producing protamine zinc insulin and shipping it to diabetologists for trials.

Patients could now ideally control their diabetes with a single daily injection of insulin, even those with relatively severe disease. Joslin described the diabetic control brought about by PZI as "extraordinary . . . less coma, less tuberculosis, fewer insulin reactions . . . fewer injections and there is a feeling of confidence on the part of the patient which allows him to live with comfort." Patients with mild diabetes who had "previously [been] appalled at the thought of multiple injections" might now be enticed to embrace this new once-a-day insulin therapy.

Like all advances in diabetes, though, PZI came with consequences and complications that could not be foreseen. While it could control blood sugar for twenty-four hours, having some insulin working to lower blood sugar all day and all night is a very different physiological condition from having a pancreas that secretes insulin as needed and only, ideally, as much as needed. Patients might have had fewer insulin reactions in total, but when they did have them the episodes tended to come on slowly, often unrecognized. Russell Wilder described going to a conference with a physician colleague and a dietitian, both of whom had diabetes and had been among the first at the Mayo Clinic to use the new insulin. The physician was later found wandering the streets as though drunk, and the dietitian, "too long without food" on the drive home, was unable to write her name on a hotel register during check-in. "I don't have diabetes anymore," one of Wilder's patients on PZI supposedly said to him; "I have insulin reactions." Now physicians worried about possible long-term effects of these prolonged hypoglycemic episodes as researchers studying them in animals suggested that they could lead to permanent brain damage if they weren't treated quickly.

One unintended consequence of PZI was the further liberalization of the diabetic diet. Avoiding hypoglycemia on the new insulin could be even trickier than it had been, as was avoiding high blood sugar or sugar in the urine immediately after meals. The result was another trial-and-error scramble. Here's Russell Wilder in 1938 reviewing the new state of affairs:

> It is advantageous under such circumstances, and at times desirable even when less carbohydrate is given, to spread the meals, as [Garfield] Duncan [of Pennsylvania Hospital in Philadelphia] has proposed, by giving the breakfast early, saving a portion of it to be taken in the forenoon and taking a lunch at midday, an afternoon snack, a late supper and food at bedtime. Joslin also has recommended more frequent supplying of food in meals and lunches. [Barnett] Greenhouse [New Haven, Connecticut] has stated that he subtracts from the day's dietary prescription the value of three glasses of milk, one of which is given at 10 a. m., one at 3 p. m. and one at bedtime. [Henry] Ricketts [of the University of Chicago Pritzker School of Medicine] has subtracted a small amount of carbohydrate from the breakfast for a midmorning feeding, and in two thirds of his cases finds a bedtime meal to be a necessity. [Herbert] Pollack [of Mount Sinai Hospital in New York] has said that he gives no fruit other than banana at breakfast and some other fruit later in the morning. He has found the carbohydrate of banana to be absorbed more slowly than that of other fruits. Also, he has said that he gives two thirds of the protein of the dietary prescription at the evening meal and extra protein-containing food, such as cheese or meat, at bedtime.*

---

* Some patients just gave up and went back to the short-acting regular insulin. Charles Fletcher, for instance, renowned for supervising the first clinical trial of penicillin, diagnosed himself with diabetes in 1940. He considered protamine zinc insulin to be "socially intolerable" because it required eating dinner at the same time every night, which was incompatible with his lifestyle.

One obvious solution was to let patients live with sugar in the urine and however much insulin was necessary for them to feel healthy. Sansum had admitted that his patients rarely satisfied the criteria for sugar-free urine, but he was unconcerned because patients felt so good. "Physical vigor, mental alertness and social usefulness do not always register on metabolic progress records," Sansum wrote.

Edward Tolstoi, of New York Hospital and the Cornell University Medical College, became the leading proponent of the idea that patients with diabetes should get to eat what they wanted and let insulin cover it. If they still had sugar in their urine, it was unimportant so long as the patient felt healthy. Tolstoi's articles describing his logic are both a vivid description of the realities of dealing with diabetes at the time and a case study in physicians acting on what they were witnessing in their own clinics, regardless of the conventional thinking about standard of care. Tattersall, in his history of diabetes, describes Tolstoi as the most notorious of the physicians advocating for free diets, but the logic that powered Tolstoi's notoriety is nonetheless compelling.

Tolstoi and his colleagues were excited by the availability of PZI, he wrote, but found it difficult to live up to the "established and conventional criteria" of sugar-free urine and normal blood sugar. They tried to supplement long-acting insulin given in the morning with the original four-to-eight-hour insulin, now called regular insulin, before meals. They tried "juggling the diets . . . as recommended by some workers in the field," delaying most of the carbohydrates until later in the day, withholding fruit juices at breakfast. When that failed, they tried increasing the insulin doses and "our patients developed most alarming and prolonged hypoglycemic reactions which were extremely subtle in onset." At this point, Tolstoi wrote, their patients were still burdened with multiple daily injections of insulin and ever more complicated dietary advice.

But all wasn't lost. "Some of the patients treated with one daily dose of protamine insulin failed to report at weekly intervals as was their routine," Tolstoi writes:

While away they kept a record of their urine analyses. . . .
When they visited the clinic after a two, three or even four

week absence, the reports revealed [considerable sugar in the urine] at all times. The patients stated that they "never felt better and stronger." They had to "force themselves to drink water," they enjoyed their food but the hunger was not extreme, and if it were not for the "terrible" urine tests they would not have come for a checkup. They certainly enjoyed the new freedom of one injection a day. It was most impressive to have a record of a continuous heavy glycosuria and a singular freedom from any and all the recognized clinical symptoms of diabetes.

Tolstoi had no explanation for why his patients were maintaining a healthy weight, "amazingly free from any and all symptoms of diabetes mellitus," despite secreting as much as 600 calories a day of sugar in their urine. "Such facts were startling and certainly unorthodox," he wrote. "They contradicted all established concepts," but they also suggested that the PZI on its own was enough to keep patients healthy.

Tolstoi and his New York Hospital colleagues took to counseling their patients to take a moderate dose of PZI every morning and then "put the equipment away until the following morning." They could eat whatever their friends and family were eating. Tolstoi and his colleagues considered treatment to be "satisfactory," even if the patient had sugar in the urine. More important was that he be otherwise "symptom free, maintaining his weight or is gaining, and has no acetone in the urine on the prescribed diet and insulin, and furthermore, is socially and economically useful."

Tolstoi's approach seemingly solved the problem of poor adherence to diabetic diets. In his experience, Tolstoi wrote, the patient inevitably became "impatient with all the fuss and lapses into diabetic sins," sometimes immediately. "We learned quite frequently that a patient would leave the clinic," he wrote in 1950, "after having been complimented on his excellent co-operation, and would go at once to the hospital cafeteria for coffee and doughnuts or chocolate cake and sometimes we found him enjoying an ice cream soda. We then looked the other way while saying to ourselves, 'Oh well, he was sugar-free.'" With the free diet and the relative lack of concern about sugar in the urine such indulgences would now be allowed. Only "overindulgence" and "concentrated sweets" would be frowned upon. Tolstoi's concept of "modern diabetic care," he wrote in 1950, came down to this: "Why not let them eat what they want as long as they gain or maintain their weight? Why be concerned about their urinary sugar?"

As Tolstoi and others argued for ever more carbohydrate-rich, fat-restricted diets, even Joslin and his colleagues in Boston had been relaxing their dietary restrictions. In the eighth edition of Joslin's textbook, published in 1946, he pointed out that what Gray and Sansum had considered a radically high-carbohydrate diet and argued for over a decade earlier—180 grams of carbohydrates per day, or 720 calories' worth—was now perfectly appropriate. Nonetheless, he continued to worry about his patients eating too few carbohydrates and too much fat, which he still assumed was the likely cause of atherosclerosis.

By the late 1940s, though, the epidemic of diabetic complications was becoming inescapable, and a likely answer to Tolstoi's rhetorical questions—Why not let patients eat what they want? Why be concerned about urinary sugar?—was that the cause of these complications was indeed what they were eating (and perhaps the excessive insulin needed to cover it, and the resulting poorly controlled blood sugar).

Tolstoi believed such speculation had been refuted. He cited a report by Joslin and his colleagues in which they had surveyed two hundred patients who had been diagnosed under the age of fifteen and had lived more than twenty years with the disease. "These patients were treated by calculated diets," Tolstoi wrote, "and every effort was made to keep their urine sugar free and their blood sugar at near normal levels. Yet 184 patients revealed disease of the blood vessels in one form or another." Tolstoi's own patients, he said, were doing no worse.

One obvious possibility, as Tolstoi (and many other diabetologists) argued, is that the diabetic complications were the result of the duration of the diabetes. In other words, the complications were inevitable, unavoidable. If that was so, then those with diabetes might as well eat whatever they liked and enjoy their lives as long as they could. In 1954, when the University of Edinburgh diabetologist and physiologist Derrick Dunlop asked in the *British Medical Journal* whether these degenerative diabetic complications could be prevented, he quoted Arthur Mirsky, a renowned physiologist, physician, and psychoanalyst, saying eight years earlier that "it makes little difference how the diabetic is treated; if he lives long enough he will develop one or another form of vascular disease." Dunlop described the comment as "a profoundly pessimistic profession of therapeutic nihilism." But that didn't mean it wasn't true.

Dunlop described his clinical experience with diabetic patients as similar to Tolstoi's and the other proponents of free diets. In the 1920s, when he was a young physician, he had had "no doubt" that "if you were a good diabetic and kept the commandments you not only remained well so far as good nutrition and the avoidance of ketosis were concerned but you avoided the later complications of the disorder." He came to question this thinking over the years and "flirted" with the free diet and the possibility that it didn't matter "how great the quantity of sugar lost in the urine so long as enough insulin was administered to ensure an efficient carbohydrate utilization from the abundant intake."

Beginning in 1946, Dunlop and his Edinburgh colleagues conducted what may have been the first clinical trial ever to test the notion that free diets were as good as any other for keeping diabetics healthy. They selected fifty newly diagnosed patients from the Royal Infirmary in Edinburgh and counseled them to eat "as their appetites dictated" (with the exception of "table sugar, jam, chocolates and sweets"). They were given one or two shots of insulin a day and once a month the dose might be adjusted to help them maintain an adequate weight and keep them free from the symptoms of diabetes and hypoglycemic reactions. They measured sugar in the urine during these checkups, but didn't use it to regulate treatment. These patients were then compared with forty patients from the clinic who had shown that they were faithfully following their calculated, restricted diabetic diet and keeping their blood sugar under relatively good control.

The Edinburgh physicians followed their subjects for five years and reported in 1951, in the *British Medical Journal*, what they had learned. With the possible exception that the patients who had been obese prior to developing diabetes tended to go back to being obese on the free diet, these patients, eating as they desired, seemed to be staying healthy. "Results were uniformly good" for patients over thirty-five. The growth of the children eating the free diet was rapid but not unusually so. Seven young patients showed a progressive worsening of the diabetes and had to be switched to restricted diets, but otherwise the free diet seemed perfectly adequate. Dunlop and his colleagues did stress, however, that five years was not long enough to establish the effect on the potential long-term complications of the disease.

The caveat was justified. The results in the succeeding years, Dunlop reported just three years later in the *British Medical Journal,* "have been disastrous." Only nine of his original fifty patients on the free diet were in what he described as "good shape" at the end of nine years. When Dunlop now divided the patients into those who had good control of their diabetes and those who didn't, the latter had three times the rate of complications, of what he called "diabetic degenerative lesions—endocrine, infective, or metabolic." He described himself as having "returned to [his] original simple diabetic faith . . . the careful control and aggressive treatment of the disorder over the years is the most important factor in their prevention or postponement."

Dunlop also suggested a likely reason why he and others had failed in their research to confirm the benefits of good blood sugar control. Whether it delayed complications or not, *none* of the patients they were seeing were doing an adequate job. These diabetologists were not studying patients who had good control of their diabetes and comparing them to patients who had poor control, as they thought they were, he wrote; rather they were dealing with "individuals whose diabetes has been under varying degrees of inadequate control and [speaking] of the best of them as having good control."

Diabetic patients would see their doctors and be told they were doing fine because they had no sugar in their urine and their monthly blood test was acceptable. But they felt other than fine; whether it was from the inexorable emergence of the diabetic complications or simply the constant sense of exhaustion, fatigue, and either entering or recovering from a hypoglycemic episode. What their urine tests were showing seemed to have little relationship to how they felt.

"The problem, of course," as the 1985 edition of Joslin's textbook made clear, "is that it is almost impossible with injected insulin to mimic the exquisitely accurate on-off feedback system that so closely regulates glucose concentration in the normal. The analogy can . . . be made to a refrigerator in which, instead of being regulated normally by a thermostat, the cooling unit is timed to run an arbitrarily selected 15 or 20 minutes each hour independent of whether the machine is too hot or too cold, resulting obviously in wild swings in one direction or the other." Patients whose urine was sugar-free might have better control of their diabetes than patients whose urine was not, but whether anyone had truly good control was unknown. In the 1950s, the question of what causes the diabetic complications and whether legitimately good control of the disease—whatever that might mean—could prevent them was still unanswered.

7

Good Science/Bad Science, Part I

The use of insulin to fatten patients unduly or to enable an obese diabetic to remain obese, or for gratifying mere gluttony or craving for excessive sugar and starch . . . must be considered inexcusable and dangerous. These forms of misuse are to be anticipated on a large scale when insulin becomes generally available, and furthermore will probably be hard to combat because the harm may not be immediately evident.

—FREDERICK ALLEN AND JAMES SHERRILL,
"Clinical Observations with Insulin," 1922*

If these favorable reports are true, we must no longer limit the use of insulin to the treatment of diabetes alone. We may look upon it as a biologic product unrestricted in its action to one system or organ in the body. Its influence apparently extends over the entire organism to a degree which we do not sufficiently understand or appreciate.

—ROY METZ,
"Insulin in Malnutrition. Further Observations," 1932

---

* Allen and Sherrill published their analysis in a 1922 edition of Allen's *Journal of Metabolic Research* that was not actually published until the spring of 1923. As Allen explained, he had held out on publishing the volume until he could include reports on the clinical experiences of all the major diabetologists who had been given the opportunity to test the new insulin therapy.

After the discovery of insulin, with its remarkable ability to rescue diabetic patients from the brink of death and return them to health, physicians throughout the medical community asked a seemingly obvious question: Would these same restorative powers work on other diseases? Within a decade, they had experimented with insulin therapy on a wide range of disorders. One 1933 accounting published in *The Journal of the American Medical Association* included "cardiac decompensation, scleroderma, cancer, toxemia of pregnancy, certain types of uterine hemorrhages, morphinism . . . rickets, pellagra, chronic ulcers and circulatory disturbances of the extremities, and even to hasten the healing of experimental fractures," all, so the author claimed, obtaining "favorable results from its use."

The most obvious potential application for insulin was for disorders associated with emaciation, as Frederick Allen prophesized in 1923, with physicians hoping that the hormone would either fatten up their patients or stimulate their appetite. If the patients were malnourished, as was often the case, stimulating their appetite would seemingly solve that problem. But could insulin replicate in these conditions one of its most remarkable effects in diabetes: the striking rapidity with which patients would "fill out" and regain their health.

Emaciated diabetic patients, as we've discussed, upon the very first injections of a therapeutic dose would begin to lose their skeletal appearance, gain weight, and quickly take on the appearance of health. Their tissues would seemingly soak up fat like a sponge, though no one knew why. Researchers working with pancreatectomized animals had witnessed the same thing. "The fact that insulin increases the formation of fat," as Charles Best and his University of Toronto colleague Reginald Haist wrote decades later in *The Physiological Basis of Medical Practice*, "has been obvious ever since the first emaciated dog or diabetic patient demonstrated a fine pad of adipose tissue, made as a result of treatment with the hormone."

In the spring of 1923, when Allen published his analysis of nine months of insulin therapy at his New Jersey institute, having used the drug on more than 160 patients, he described this fattening phenomenon as "one very noticeable early effect of insulin treatment." Often within a day or two of their very first injection, they would see "a filling out of the face [that] is out of proportion to the general gain of weight."* But it had a discomforting aspect. The "patients seem to tend readily to become obese," Allen wrote, with a disproportionate amount of fat accumulating around the waist.

This was happening even on the very small doses of insulin they were using in those early months on both their older patients, who might have been obese prior to their diabetes diagnosis, and their young patients, who had the acute form of the disease and may have always been lean. The danger of diabetic patients becoming overweight or obese with insulin therapy was one primary reason, Allen said, that "physicians should be on their guard against the possible consequences of flooding the body with one of its most powerful hormones."

Joslin occasionally witnessed the same phenomenon in his patients. This was another reason why he argued to keep insulin doses low and patients slightly undernourished, by which he meant slightly beneath their ideal weight. By 1928, Joslin and his colleagues were starting patients off with 5 units of insulin before each meal, up from the 1 unit they started with in 1923. But he was still trying to keep his patients' average daily dose below 15 units a day. The highest dose he ever prescribed to a patient, he wrote, was 63 units daily, and even that decreased with time. He had a few other patients who took as much as 45 units, but at least one of those died in coma. He also admitted to having one patient, No. 2476, who "did not even measure his insulin doses, ate liberally and gained 100 pounds in weight." Patient 2476, Joslin wrote, had driven his family to distraction because of his hypoglycemic episodes. He had become so large, though, that when he had an insulin reaction and convulsions in the hospital under Joslin's care, Joslin and three other doctors couldn't safely get near enough to

* This effect within hours may likely have been from water retention, more so than restoration of fat.

feed him the carbohydrates or inject him with the glucose or adrenaline necessary to return his blood sugar to a normal level. Rather they waited two hours until his convulsions subsided and then gave him oral and intravenous glucose. "I suppose," Joslin wrote, "the advocates of high insulin dosage will retort—see how strong big doses of insulin make a patient and how much weight can be put on!"

As the free diet movement grew in popularity in the post-insulin years, patients were often injecting 40 or more units of insulin a day. Sansum was starting his patients on 40 units and working up from there. It wasn't unusual for his patients to be injecting 100 or 150 units a day, occasionally even 200 units, massive doses by Joslin's standards. In 1926, Sansum reported in *The Journal of the American Medical Association* that many of his patients were putting on fifty, even sixty pounds—either gaining or regaining—on his high-carbohydrate diet, high-insulin approach, much of it within a single year. R. D. Lawrence, who experimented with Sansum's approach himself, later explained why he found it both enticing and worrisome: "Sansum and his colleagues claimed that their patients felt much better and certainly were more satisfied with the high carbohydrate diet. Most, however, became fat—definitely overweight." He was not an advocate. "I personally fail to see the advantage of jam and marmalade in the high carbohydrate diet if a rasher of bacon and an extra pat of butter are anathema," he wrote in the *British Medical Journal* in 1933.

Rawle Geyelin, who had helped launch the higher-carbohydrate diet for diabetes with his pediatric patients, also talked about preventing patients from getting too fat. But he acknowledged that "it has not always been easy to achieve this goal." Even in children, particularly adolescent girls during puberty, he wrote, putting on excess fat with insulin therapy and higher-carbohydrate diets could seem "unavoidable." For some physicians, as with Lawrence, this was reason enough to temper their enthusiasm for high-carbohydrate diets and the insulin they required to keep blood sugar controlled. Byron Bowen, who ran a diabetes clinic at the Buffalo General Hospital in western New York, explained in a 1930 article that while he was giving his patients higher-carbohydrate diets than he had formerly, he lacked "the courage" to apply Sansum's method widely. "The great tendency for the diabetic patient treated with insulin to become fat is common experience," he wrote, "and obesity as a forerunner of diabetes is an outstanding association."

That was the rub, and still is. Obesity is so closely associated with the chronic form of diabetes—type 2—that it seems part and parcel of the disease. Ever since Allen and Joslin and the pre-insulin era, diabetologists have assumed it is obesity that somehow causes the diabetes in most of their older patients. In 1923, Joslin reported that more than 40 percent of his patients had been obese prior to the diabetes diagnosis. He suggested that if they had "the exact data"— measurements of both weight and height, which weren't regularly recorded by physicians of that era—the real proportion would be "fully twice as great." Adults over fifty, reported Joslin, rarely if ever acquired diabetes if they remained lean. Overweight or obesity seemed almost a prerequisite for the condition. Life insurance companies had already published data supporting this. "Diabetes, therefore, is largely a penalty of the obesity," Joslin had concluded, "and the greater the obesity the more likely is Nature to enforce it. The sooner this is realized by physicians and the laity, the sooner will the advancing frequency of diabetes be checked."

Carl von Noorden, the great German diabetologist, had suggested presciently that every obese adult should get their blood tested regularly for high levels of blood sugar, predicting that this would be visible in a kind of prediabetic state before the glucose appeared in the urine and a definitive diabetes diagnosis would be made. Joslin suggested the same for obese children. By Joslin's calculation, individuals with obesity might be as much as forty times more likely to develop diabetes than lean individuals. He called the prevention of obesity "a splendid opportunity" to prevent diabetes. He would also acknowledge that not all people with obesity developed diabetes, which suggested to him that it wasn't the excess fat that was the problem, but "the process of overeating by which the deposit of fat is acquired." As late as the 1980s and the twelfth edition of Joslin's textbook, the authors were still quoting Joslin's aphorism: "With an excess of fat diabetes begins and from an excess of fat diabetics die."

From the discovery of insulin onward, the one consistent principle in the dietary treatment of diabetes has been that the patient should achieve and maintain a healthy weight. Lifestyle management guidelines will discuss, as the American Diabetes Association does, the targets of dietary therapy as both to "achieve and maintain body weight goals" and "attain individualized glycemic . . . goals," but the implicit expectation is that the glycemia can be controlled through pharmaceutical therapy alone, while the latter, weight management, requires some combination of diet and exercise. Those who are overweight or obese should do what's necessary to lose weight and not regain it. Those who are underweight when started on insulin therapy should watch what they eat so as to prevent excessive weight gain.

This has always seemed simple enough, but it has often proved difficult to achieve. For many patients, the manifestation of obesity has seemed unavoidable. "When diet failed to reduce blood-sugar levels enough," as Robert Tattersall wrote in *Diabetes: The Biography*, "the only other treatment until the 1950s was insulin, which was started reluctantly in overweight middle-aged people, because it caused weight gain."

Diabetologists would offer many possible explanations over the post-insulin decades for the fact that so many of their patients on insulin therapy would struggle (and eventually fail) to maintain a healthy weight. The most likely explanation, that insulin itself is fattening, hence the greater the dose of insulin given, the fatter the patient will become, is the one that has presented the most challenges to how diabetes therapy itself evolved and how we think about diabetes therapy today.

In the very earliest years of insulin therapy, once diabetologists had taken to publishing before-and-after photos of their emaciated, near-death patients filling out on insulin, physicians confronted with other chronically emaciated patients naturally wondered if insulin would have the same effect on theirs. These doctors would read a report documenting one or a few cases of successful insulin therapy and would try it on their own patients and publish *their* observations. As we've discussed, this is how medical practice evolved in the years before randomized controlled trials and the evidence-based-medicine era. It still does, but now tempered by the results of clinical trials.

In August 1923, a year into the insulin therapy era, a Pennsylvania physician named Robert Pitfield reported using insulin on two infants, four and five months old, who were failing to thrive in the hospital. Pitfield had read what insulin could do for young diabetics and so he tried it on these two malnourished babies and wrote a one-page report in *The New York Medical Journal*. He injected the babies with 1 unit of insulin each, he wrote, along with "a prophylactic dose of glucose" to prevent insulin shock. Both did well. Of the four-month-old "Baby D," Pitfield wrote that "from the beginning he at once began to gain in weight and seemed happy and contented." In just sixteen days, the child gained a pound. He also had at least one insulin shock—"he fell into a hypoglycemic collapse"—but Pitfield was able to restore him to health quickly with glucose.

In St. Louis, William Marriott, a pediatrician and newly appointed dean of the Washington University School of Medicine, reported the following year that he and his colleagues had tried insulin on malnourished infants as well, with similar success. Marriott had seen Pitfield's report and had also been influenced, he wrote, "by the observations that diabetic patients on insulin treatment often gain weight at a phenomenal rate, even when the food intake is not excessive." Marriott said they achieved the best results when they gave the infants "considerable amounts" of insulin—15 units—along with enough glucose, also by injections, to cover it. In some cases, Marriott was confident that the insulin therapy had saved the child from death by inanition, perhaps "within a few days or hours."

Orville Barbour, a pediatrician from Peoria, Illinois, was visiting Marriott's hospital in St. Louis when he learned what Marriott was doing. When he returned to Peoria, he tried it on the infants in his practice who were failing to thrive. Barbour, too, started with small doses before meals but then realized that the effect was more impressive when "the insulin is pushed to the limit of the individual's capacity," the point at which hypoglycemic reactions begin. In thirty-eight of forty patients, he reported, some as young as a few days and none older than eleven years, the children improved on insulin, their appetite and weight increasing. Barbour reported that he first gave the children diets rich in both carbohydrates and fat, but soon learned that the children added weight more quickly when fat content was kept low and carbohydrates high. "We now give to all of these children," he wrote, "as little fat as possible, as much protein as possible, and as much carbohydrate as necessary to meet the patient's caloric desires."

Some children did much better than others, gaining as much as two pounds a week on insulin; for others, as Louis Fischer and Julian Rogatz, New York City pediatricians, reported in *The American Journal of Diseases of Children* in 1926, the effect of insulin therapy was either "indifferent" or "confusing." They had no inkling why, or what characteristics might predict success or failure. University of Toronto pediatricians reporting on their experience were unenthusiastic: Only half of their malnourished babies gained weight on the insulin therapy and, as with children with diabetes, the effect of the insulin could vary "tremendously" in different infants; "consequently its administration is not without danger."

The use of insulin therapy quickly spread from malnourished children to malnourished, emaciated adults. Wilhelm Falta of the University of Vienna published the first report of that use in 1925 in a German-language journal. Falta had been a student of von Noorden's and a pioneer of the science of endocrinology in Europe. He had suggested in a 1913 monograph well before the discovery of insulin that a "functionally intact pancreas" was a necessity for putting on fat. Once insulin was available, the idea of using it to fatten up emaciated *nondiabetic* patients seemed a natural extension of his thinking. He called insulin therapy the "fattening cure" and described the experience of three patients who gained significant weight on insulin therapy. One was a thirty-year-old woman who had weighed barely ninety pounds when he began his treatment. In less than eight weeks she gained twenty-four pounds and looked "quite plump." After the insulin treatment was discontinued, she added another nine pounds in three weeks. She had been ravenously hungry while on insulin therapy, but once her weight finally plateaued, Falta wrote, "the bouts of ravenous hunger disappeared."

Falta's report then sparked what Sterling Nichol, a prominent Miami cardiologist, called "a wave of enthusiasm for attacking all types of pathological leanness with insulin and a high calorie diet, including a good supply of carbohydrates." Most of the papers were published in the European journals, Nichol recounted in 1932, but American physicians were paying attention and also reporting their observations. By Nichol's count, over one hundred articles had already been published documenting the use of insulin therapy on chronically emaciated patients. "In reading this literature," Nichol said, "one is impressed with the almost universal improvement obtained by this method of therapy."

Once again, the doses of insulin these physicians felt comfortable using on their patients increased with time. In this case, the goal, after all, was to *induce* weight gain; these physicians didn't care how much their patients ate so long as it was enough to prevent insulin shock. And so malnourished, emaciated, and anorexic (technically meaning "lack of appetite") but *nondiabetic* patients were being given 60 units of insulin a day, a few patients even more, along with the carbohydrates necessary to prevent hypoglycemic reactions. In 1929, Kenneth Appel and two colleagues from Philadelphia's Pennsylvania Hospital reported using what they called "insulin hypernutrition," prompting weight gains of nearly two and a half pounds a week in women who had previously seemed incapable of eating enough to gain weight. "Not a few" of these women, the Philadelphia physicians reported, "were brought to a weight never before reached in their lives." They described the results as "quite remarkable" and, just like diabetic patients on insulin therapy, these women took on the aura of health as well.

In Boston, Harry Blotner, a diabetologist at Peter Bent Brigham Hospital and Harvard Medical School, had been teaching "chronically underweight" patients how to inject themselves with insulin, just as if they had been diabetic, and to eat liberally while taking 10 units of insulin, twenty minutes before each meal. "After insulin injections were begun," Blotner reported in 1933, "each patient began to gain weight and some continued to gain at a remarkable rate." They also tended to have "marked improvement in sense of well being. Some of the patients remarked how wonderfully well they felt, while others said insulin made new individuals of them. Their appetites increased and occasionally became voracious."

One of Blotner's patients took insulin shots daily for twelve weeks and gained thirty-one pounds; another gained nine pounds in a week. Some patients continued to put on weight even after they discontinued the insulin therapy. This could be a problem, though, as Blotner suggested five years later when he reassessed his clinical experience in *The New England Journal of Medicine*: one formerly emaciated patient, he wrote, went on to become "definitely obese," gaining fifteen pounds on the insulin therapy and then another fifty-four pounds after discontinuing it.

As with diabetic patients eating carbohydrate-rich diets, what Blotner had observed of his emaciated patients coming to feel "wonderfully well" on insulin therapy would be another common theme in these reports. "Euphoric" was the word used, for instance, by three Brazilian physicians describing the response of their mal-nourished patients to insulin therapy in 1942 (while also observing that maybe the euphoria was caused in part by "the esthetic satis-faction of gaining in weight").

Physicians also reported instances when insulin had beneficial effects on severe cognitive disorders. Appel and his Pennsylvania Hospital colleagues wrote that four of their patients who had been clinically depressed took to joking, knitting, and playing cards in the ward while on insulin therapy. One patient with dementia prae-cox, a cognitive disorder in young adults that would now be diag-nosed as schizophrenia, "became agreeable and cooperative and has so remained."

In 1930, a young Austrian physician, Manfred Sakel, reported that insulin could ease the symptoms of morphine withdrawal, an observation that would lead directly to one of the most controver-sial therapies in the history of psychiatry. Sakel noticed that when he accidentally overdosed his patients with insulin and induced a severe hypoglycemic episode, the patients appeared particularly lucid afterward. As one historical account described it, Sakel "satisfied himself that hypoglycaemia could safely be reversed, thus permitting deeper levels of induced coma, a step that earlier workers had not taken."

The result was *insulin shock therapy*, also known as *insulin coma therapy*, as a treatment for schizophrenia, pioneered by Sakel at Vienna's University Clinic. Between 1933 and 1935, he published over a dozen papers on the technique. Then Sakel and Joseph Wortis, a young American psychiatrist who was in Vienna undertaking a "learning analysis" with Sigmund Freud, introduced the technique in the United States. While occasionally patients died from the insulin shock, Wortis wrote in 1936, their deaths could still be justified because of "the remarkable results" in other patients in what was "an otherwise almost hopeless disease."

Individual responses to insulin, not surprisingly, varied widely and unpredictably in these patients. The psychiatrists, like the diabetologists, were left to work out doses and methods based on intuition and what Wortis described as "rule of thumb." "Twenty units of insulin may produce convulsions in one patient," he wrote, "while I have seen another walk around hours after an injection of 130 units." Some patients received more than 200 units of insulin a day. Because the goal of the insulin was to induce a coma and, at the very least, prolonged hypoglycemia, the injections were given while the patients were fasting and no food was allowed after the injection to balance out the insulin.

The protocol would be repeated in some cases daily, typically for two or three months, until the patient became sustainably lucid or it was clear the therapy was of no benefit. Another judgment call was precisely how long the patient should be allowed to persist with hypoglycemia or remain in a coma before the attending staff opted to "rescue" them with sugar water or sugar milk, if the patients could swallow, or tube feeding of glucose, or injections of glucose or adrenaline, if they could not. Patients would be left to experience the hypoglycemia even if they became extremely agitated. "An exception is allowed when a patient grows literally wild with hunger," wrote Wortis in the paper that introduced the technique to psychiatrists in the United States. "Such patients should be fed at once."

In his 1986 history, *Great and Desperate Cures*, the University of Michigan neuroscientist Elliot Valenstein wrote about visiting a Veterans Administration hospital in Topeka, Kansas, in 1950 and watching insulin shock therapy be administered to thirty patients at a time. "Patients furthest along in the treatment series," he recalled, "with the highest insulin doses, might be having violent convulsions. With all these people—tossing, moaning, twitching, shouting, grasping—I felt as though I were in the midst of Hell as drawn by Gustave Doré for Dante's *Divine Comedy*."

Hell or not, Sakel and Wortis claimed that the use of insulin shock therapy at Vienna's University Clinic had resulted in remission of schizophrenia in over half of the patients on which it had been used. Psychiatrists would report seemingly miraculous cures like those seen with diabetes, anecdotal accounts of patients who had suffered years of dementia waking from an insulin-induced coma suddenly and remarkably lucid. By the late 1930s, in mental hospitals throughout Europe and the United States, insulin shock therapy had become a standard treatment for schizophrenic patients.

In the 1950s, insulin shock therapy began its precipitous fall out of fashion, as the British psychiatrist Harold Bourne and others began arguing that any clinical benefits might merely be placebo effects—a result of the vastly increased attention that patients with schizophrenia received from doctors and caregivers in the insulin ward. The first effective antipsychotic medications replaced insulin shock. Clinical trials suggested that insulin therapy was at best no better than a drug called chlorpromazine, and the drug was "safer, easier to administer, and better suited for long-term management." Insulin therapy, as Morris Markowe and colleagues from London's Springfield Hospital wrote in *The Lancet* in 1956, "is much more dangerous, even in the most expert hands; it is extremely unpleasant for patients and imposes more restrictions on their freedom, and it taxes the nurses more severely both in numbers and skill."

All of this would have little relevance to a book on diabetes except for the issue of weight. Patients receiving insulin shock therapy got fatter, and the fatter they got, the better the therapy seemed to work. A historical review published in 2000 in the *Journal of the Royal Society of Medicine* reported that "most patients emerged from treatment grossly obese," which may have been hyperbole. A 1955 report on fourteen years of insulin shock therapy at Pennsylvania Hospital revealed that one in every seven patients gained over thirty pounds and these were the ones most likely to benefit. One in three gained more than twenty pounds on a therapy that was used, typically, for only two to three months. Among the memorable patients who had been subjected to insulin shock in the 1950s and gained excessive weight was the Princeton mathematician and later Nobel laureate John Nash. The poet Sylvia Plath experienced a "drastic increase in weight" on the treatment.

Over the years, physicians would offer up a multitude of potential explanations for why their patients tended to fatten when injected with insulin. Perhaps it prompted them to eat regularly to avoid hypoglycemia, low blood sugar. These patients not only ate "much more food" when taking regular injections of insulin, as a 1932 editorial in *The Journal of the American Medical Association* described it, but often ate these meals "with relish and gusto."* So maybe insulin is a powerful "stimulant to the appetite," as some physicians assumed. Physicians reporting on insulin therapy expressed little doubt that their patients who had previously had to be coerced to eat now often developed robust appetites if not voracious hunger.

Laboratory rodents, too. In 1940, three researchers at the Scripps Metabolic Clinic in La Jolla, California, reported that they had made rats obese by injecting them with the new protamine zinc insulin. The animals doubled their food consumption and blew up like little balloons. "The rats got so obese," the Associated Press reported, "that they had difficulty rolling over. They got that way by overeating voluntarily, like man." The researchers, led by Eaton MacKay, suggested that the insulin caused hypoglycemia, which

---

* In Sylvia Nasar's 1998 biography of John Nash, she describes the "insulin patients" as getting "richer and more varied food," too. "They got special desserts. They had ice cream every night at bedtime."

stimulated the animals to eat too much. Others assumed that their diabetic patients on insulin therapy got fatter because they were no longer losing calories in their urine, but were eating more than they would otherwise because insulin made it possible, or because they were treated to prevent a hypoglycemic episode.

When psychiatrists offered explanations for the weight gained on insulin shock therapy, they would do so, not surprisingly, from a psychiatric perspective. Why would the patients suddenly begin eating enough to get fat when they had been anorexic and getting ever thinner, often for years? The Pennsylvania Hospital physicians suggested that the insulin "imposes a situation of dependency and helplessness upon the patient, in which the need for feeding and being mothered becomes a matter of life or death." "The repeated gratification of these needs," they added, might somehow lessen "the need for regression to psychosis." When London physicians reported in 1966 on the efficacy of using both insulin therapy and the anti-schizophrenic drug chlorpromazine on emaciated patients diagnosed with anorexia nervosa, they reported "large and rapid gain in weight"—averaging almost five pounds per week— "however resistant initially the patient was to eating." They speculated that the chlorpromazine "lessens the patient's fear of and resistance to eating," and so they ate more and regained weight. (These British physicians also worried that "compulsive overeating, resulting in the patient becoming overweight, may recur during the recovery phase.")

Another possibility existed as well, and it was the simplest possible explanation: perhaps insulin makes patients get fat because it directly promotes fat storage. This is what Wilhelm Falta had suggested a decade before insulin had been discovered. Whatever pancreatic hormone was missing in diabetes, he had suggested in his 1913 monograph, "must play a role in the development of some forms of obesity."

Falta's logic was based on a simple observation: in acute cases of diabetes, type 1, which is correctly understood to be a disease of insulin deficiency, not only did patients see a depletion of their carbohydrate reserves—the glycogen stored in the liver—but "a melting of body proteins and loss of fat." These patients were typically diagnosed after they had experienced a significant weight loss. The longer they went without insulin therapy, the more emaciated and skeletal they became. Pancreatectomized dogs experienced the same. Hence, Falta's conclusion: "For fattening, one needs a functionally intact pancreas." Give insulin to diabetic patients—or, more accurately, give them too much—and the result is what Falta was calling "insular obesity" or "diabetogenous obesity," a term his mentor von Noorden had used a decade earlier. This was the logic that led Falta to experiment with insulin on his emaciated nondiabetic patients to determine if it worked as he expected. He reported that it did.

The possibility that one of insulin's many roles in the body was to directly facilitate the storage of fat in fat cells could also explain two other puzzling observations of insulin therapy. One was the excessive accumulation of fat, a swelling or tumescence, at the sites of repeated insulin injections. One 1930 report described the swellings in a sixteen-year-old diabetic patient as "halfway up the thighs . . . two symmetrical tumours, about the size of fists" with the skin over them showing the "numerous marks left by the insulin injections." The swellings were not painful to the touch. When the insulin injections were stopped or moved to a different site, the swellings would diminish. A likely explanation is that the injected insulin was having a direct effect on the accumulation of fat at the site of the injection.

The second observation was that patients with uncontrolled diabetes very often had high levels of fat in their blood, a condition known as lipemia. Today it's commonly known as hypertriglyceridemia ("triglyceride" being a technical term for a molecule of fat). "Severe diabetes was, and is, the only disease in which lipemia is frequent enough to be of special appearance," Joslin noted in the 1928 edition of his textbook. "The milky appearance of the serum and the 'cream' which rose from it on standing indicated the fat." The German diabetologist Felix Klemperer had reported the fat content in the blood of a diabetic patient twentyfold higher than that typically observed in the blood of healthy individuals.

The more severe the diabetes, the greater the lipemia, and the greater the lipemia, the more dire the prognosis for the patient. The fat in the blood is associated with heart disease risk and atherosclerosis just as blood cholesterol is. Joslin was particularly worried about lipemia, considering it more insidious than high blood sugar because it required more than a urine test to detect it—the blood itself had to be analyzed, which meant a visit to a physician who could take a blood sample and send it off to a laboratory. The association between lipemia, uncontrolled diabetes, and bad outcomes—coma and death in the pre-insulin era—was another reason Joslin had taken to preaching against high-fat diets; he assumed that the excessive fat in the blood derived from excessive fat in the diet. Later, after Newburgh and Marsh reported that their high-fat diet resolved lipemia in their Michigan patients, Joslin acknowledged that he had probably been wrong about the dietary origin of these fats in the blood.

Considering the intimate relationship between lipemia and uncontrolled diabetes, it wasn't surprising that insulin therapy resolved the condition. The fat would clear from the blood quickly once patients started on insulin, often within hours or a few days. But that raised an obvious question: If fat in the blood disappeared with insulin therapy, where did it go? In 1928, Israel Rabinowitch in Montreal described a patient whose terrible lipemia was resolved almost entirely with insulin. Over the course of eighteen hours, enormous amounts of fat had "disappeared from the blood" of the patient, and the same happened the next day in just twelve hours with another injection of insulin.* Rabinowitch thought it hard to imagine that the patient's body had used that fat for energy—oxidized it—as the patient had been bedridden and comatose at the time.

So where did that fat go? One possibility is that it had come from the fat tissue originally and that insulin had somehow induced the fat tissue to take it back up again. By the early 1930s, researchers were reporting that insulin stimulated the storage of carbohydrates as glycogen in liver and muscle cells. Maybe it did the same with

---

* By Rabinowitch's calculation, some 460 grams of fat—over a pound of fat—had vanished on the first day and 375 grams the next. "This accounted for over 3000 calories," he wrote. These amounts, though, seem an order of magnitude—ten times—higher than physiologically possible.

fat in fat tissue. Insulin is an "excellent fattening substance" and a "splendid builder of reserve material," is how the German authority Erich Grafe described this possibility in 1933 in *Metabolic Diseases and Their Treatment*. This fattening effect, he added, was likely "due to improved combustion of carbohydrate and increased synthesis of glycogen and fat."

Neither Grafe nor anyone else had been able to explain this phenomenon previously, partly, as Grafe himself suggested, because of the way insulin had been perceived by the physician researchers who studied it. This was the same problem Denis McGarry would remark on almost sixty years later in his *Science* article, "What If Minkowski Had Been Ageusic?" and appears repeatedly in this history. Diabetologists were focused so completely on the most conspicuous manifestation of diabetes—high blood sugar and sugar in the urine—that they thought of diabetes as a disorder of uncontrolled carbohydrate metabolism, which implied a failure of the muscles to use that blood sugar for energy. For this reason, that's essentially all they studied. They were ignoring other organs and other effects of insulin, specifically in the liver and the fat tissue and even the pancreas itself, because they did not see the immediate relevance and because these phenomena were far more difficult, if not impossible at the time, to measure.

Another factor was at play here, though, and its influence cannot be overstated. Physicians treating diabetes and those few trying to understand its mechanisms were assuming a way of thinking about fat accumulation that seemed intuitively obvious—that people put on excessive fat because they eat too much and stay lean when they don't—and they assumed, as did physicians throughout medicine, that that was all they needed to know. They came to perceive the accumulation of excess fat as the result of little more than bad behavior or a lack of willpower.

This is where Louis Newburgh of the University of Michigan would have his extraordinary, if detrimental, influence on diabetic therapy through his work on obesity. While diabetologists, led by Joslin, were rejecting the therapeutic relevance of Newburgh and Marsh's high-fat diet for diabetes, Newburgh claimed he had established experimentally the fundamental cause of obesity and, thus, the accumulation of excess fat that played such a seemingly critical role in the diabetic condition.

At a time when the scientific discipline of obesity research constituted perhaps a dozen physicians in the United States and Europe trying (and typically failing) to treat patients burdened by obesity and musing on its causality in the medical literature, Newburgh was the first and only researcher to claim in that literature that he had experimentally tested competing hypotheses about the cause of the condition. In doing so, he gave his work the aura of hard experimental science, which can explain in part why his conclusions were so rarely questioned.

Through the 1920s, the general thinking on the cause of obesity had allowed for two equally commonsensical ways to imagine it, defined essentially by how much the individuals burdened with obesity seemed to eat. If they were big eaters, it was easy to assume that their obesity was caused by how much they ate. But "many stout persons are light eaters," as the 1914 edition of William Osler's *The Principles and Practice of Medicine* observed, suggesting, for them, at least, that other factors were involved. "The medical profession in general, believes that there are two kinds of obese persons," is how Newburgh described this schism in 1930 (writing with his lab assistant, Margaret Johnston), in the first of the half dozen papers he would publish on the subject: "those who have become fat because they overeat or under-exercise; and those composing a second group whose adiposity is not closely related to diet, but is caused by an endocrine or constitutional abnormality."

These competing ideas represented a conflict of nurture versus nature or behavior versus physiology. If people get fat because they eat too much or exercise too little, then obesity can be considered the penalty or the direct consequence of their behavior. Joslin wrote about the "individual patient who *allows* [my italics] himself or herself to become fat," as though it were only an issue of willpower. Diabetologists would effectively believe this, assuming that their overweight or obese diabetic patients could achieve and maintain a healthier weight merely by eating less, restricting their calories as the guidelines from the diabetes associations would inevitably suggest and still do.

The alternative, the endocrine or constitutional disorder concept, may have been described best not by a physician or researcher, but by a fictional character, John Tarleton, in George Bernard Shaw's 1910 play *Misalliance*: "It's constitutional," Tarleton says. "No matter how little you eat you put on flesh if you're made that way." In the medical literature itself, this way of thinking about obesity was a German-Austrian conception, promoted most prominently by Gustav von Bergmann, the leading German authority of the era in internal medicine,* and Julius Bauer of the University of Vienna (the "noted Vienna authority on internal diseases," as *The New York Times* referred to him). Bauer's field of expertise was the application of genetics and endocrinology to clinical medicine, a discipline he had pioneered in a 1917 monograph entitled *Constitution and Disease*.

Bauer and von Bergmann argued that obesity, in effectively *all* cases, had to be a physiological disorder with a large endocrine component, for a host of reasons, not just because it often seemed an intractable condition, but because of their observations about fat accumulation itself. Body fat, for instance, is often localized to particular regions of the body—the breasts or the buttocks in woman, the abdominal area in men (the beer belly)—and in fatty tumors called lipomas. This observation alone suggested that whether or not a particular tissue or part of the body will take up fat, and how avidly it will do so—what von Bergmann referred to as "lipomatous tendency" and Bauer as "lipophilia" and described as "the exaggerated tendency of some tissues to store fat"—must be determined by factors localized to those specific areas. It was hard to imagine how the amount of energy consumed or expended could be relevant to this localized accumulation of fat.

The fact that men and women fatten differently, as Bauer observed, is compelling evidence that sex hormones play a primary role in fat accumulation. And the observation that obesity runs in families, which Bauer had reported in almost 90 percent of the more than four hundred cases he had accumulated in his clinic,

---

* Until 2010, the Gustav von Bergmann Medal was the highest honor awarded by the German Society of Internal Medicine.

and von Noorden had reported in 70 percent of his obese patients, is compelling evidence that the disorder has a large genetic component, and that, in turn, implies the existence of genetically determined neural, hormonal, and metabolic factors that could both influence body weight and body fat distribution *and* be passed on by those genes from generation to generation. "The genes responsible for obesity," Bauer wrote, "act upon the local tendency of the adipose tissue to accumulate fat (lipophilia) as well as upon the endocrine glands and those nervous centers which regulate lipophilia and dominate metabolic functions and the general feelings ruling the intake of food and the expenditure of energy. Only a broader conception such as this can satisfactorily explain the facts."

Newburgh disagreed, though, and as a result, so eventually did the American medical establishment. In April 1929, Newburgh gave a presentation in Boston at the annual conference of the American College of Physicians, reporting that he had experimentally tested the two competing hypotheses and ruled out the possibility that obesity might have an endocrine or constitutional cause. Newburgh had recognized that his diabetic patients were often obese and that their diabetes could be controlled with weight loss. This had convinced him that even obese diabetic patients will lose significant weight when subjected to sufficient restriction of how much they eat, when they are "semi-starved," to use a term that has since fallen out of fashion. The evidence Newburgh presented in Boston was of a handful of obese patients and one single lean volunteer, all subjected to the same semi-starvation regimen and all eventually losing body fat at similar rates.

From this observation, Newburgh turned a loosely linked chain of questionable assumptions into a statement of fact. If obese patients could lose weight by semi-starvation, he concluded that obesity itself was clearly caused by eating too much—a "perverted appetite," he called it—and so *not* by a constitutional or endocrine disorder. "There is no specific metabolic abnormality in obesity," Newburgh stated unequivocally in *The Journal of the American Medical Association* in the third of the half dozen papers he would write on obesity, evoking this same evidence, between 1930 and

1944. "All obesity is 'simple obesity.' The increase in weight merely represents an inflow of energy greater than the outflow."

The logic raised questions, though, that Newburgh's conception had to answer. If the cause of obesity is merely an imbalance in the flow of energy into and out of the body, why don't those who are becoming obese do what lean people apparently do effortlessly, which is eat less and exercise more such that their energy intake and outflow balance? In most cases, the adjustment necessary would be a very small one: fat gain of several pounds a year, the rate at which most children and adults gain excess weight, is equivalent to storing only a few tens of calories every day as fat, a bite or two of food. So why would people allow this imbalance to continue if the necessary corrective steps were so easy?

In providing the answers to these questions, Newburgh catalyzed the transformation of the scientific perception of obesity from a chronic, disabling physiological disorder into, at best, an eating disorder or a character or psychological defect. Children get fat, Newburgh explained, not because of any constitutional, genetic, or endocrine predisposition, as Bauer and von Bergmann were arguing, but because they'd "been deliberately trained to over-eat by their parents." Adults got fat because they suffer from "various human weaknesses such as over-indulgence and ignorance." When the *Los Angeles Times* reported on Newburgh's work in 1932, the headline put it in appropriately blunt perspective: "Just Gluttony Makes Obesity. Michigan Professor Strips Defense of Portly."

By the 1940s, the University of Michigan was already crediting Newburgh with having "undermined conclusively the generally held theory that obesity is the result of some fundamental fault." Newburgh was describing obesity as "an entirely preventable disease," and physicians who considered themselves authorities on the subject took to ridiculing the notion that obesity was caused by anything *other* than eating too much. These were "lame excuses," as Mayo Clinic physicians would write, for "avoidance of the necessary corrective measures," which is to say, what the lean apparently did naturally: eat in moderation and maybe exercise. By the late 1950s, the leading authorities in obesity research, now a burgeoning field of study, included psychologists and psychiatrists trying to explain why people would continue to overeat and get fat—"unresolved emotional conflicts" were theorized, or the need to relieve "the nervous tensions of life"*—even when the physical and psychological burdens of the obesity itself were so manifestly obvious. The two dominant hypotheses of obesity and body weight regulation through the 1970s—the glucostatic hypothesis (promoted by Jean Mayer of Harvard) and the lipostatic hypothesis (promoted by Gordon Kennedy of the National Institute for Medical Research in London)—were both attempts, essentially, to explain why people with obesity might not know enough to stop eating when they should.

The implications of this logic and the triumph of Newburgh's thinking were profound and had long-lasting consequences. One immediate and direct result was that it confirmed the preconception of physicians and diabetologists that their patients had become obese by aberrant eating behavior and they could lose weight by some combination of eating less and exercising more. Diabetes textbooks and guidelines on therapy would invariably insist that dietary therapy had to accomplish two primary goals—controlling blood sugar *and* controlling weight—and assumed the patients could accomplish the latter merely by reining in their appetites, by eating less.

That the results of this approach to dietary therapy were "remarkably poor"—quoting the New York Hospital psychiatrist Albert Stunkard and his colleague Mavis McLaren-Hume in 1959, in one of the most famous papers in obesity medicine—was the unfortunate caveat. The published reports in the medical literature left little doubt that this was true, as Stunkard and McLaren-Hume made clear, and this reality would become widely accepted in the field despite the cognitive dissonance it would provoke.

The diet therapy chapter in the first edition of *The Handbook of Obesity*, for instance, published in 1998 and edited by three of the most prominent authorities in the field—George Bray, Claude

---

* It's still common to have obesity blamed today on "comfort eating," as one 2017 textbook describes it, "to reduce negative emotions, such as anger, loneliness, boredom and depression."

Bouchard, and W. P. T. James—describes "the reduction of energy intake" as "the cornerstone of treatment" and "the basis of successful weight reduction programs." It then proceeds to acknowledge that "90–95% of those who lose weight during a dietary program will regain it within several years" and describes that reality as prompting "charges that a traditional treatment for obesity should be abandoned." The fourteenth and latest edition (as of this writing) of *Joslin's Diabetes Mellitus* acknowledges much the same. The chapter on obesity was written by Jeffrey Flier, an obesity researcher and later dean of the Harvard Medical School, and his wife and research colleague Eleftheria Maratos-Flier. The Fliers also describe "reduction of caloric intake" as "the cornerstone of any therapy for obesity." But then they enumerate all the ways that this cornerstone fails. After examining approaches from the most subtle reductions in calories (eating, say, 100 calories less each day with the hope of losing a pound every five weeks), to semi-starvation diets of 800 to 1,000 calories a day, to very-low-calorie diets (200 to 600 calories), and even total starvation, they note that "none of these approaches has any proven merit."

Neither of these textbooks entertains the possibility that the traditional dietary approach fails because the thinking on which it's based is incorrect. That was the implication of the science of fat metabolism and fat storage that also emerged in the mid-twentieth century, but it emerged, weirdly enough, independent of the science of obesity. Once the medical community reached a consensus that Newburgh had been right and that obesity was an eating disorder, obesity researchers paid little attention to this school of research, all of which suggested that obesity, as Bauer and von Bergmann had insisted, was caused by a dysregulation in these physiological phenomena and that insulin played a critical role, if not *the* critical role. To put this simply, after embracing Newburgh's thinking in the post–World War II decades, those researchers and clinicians who thought of themselves as studying obesity, a disorder of excess fat storage, paid little to no attention to the emerging science of fat storage itself, as though the storage of *excess fat* and the storage of *fat* itself had nothing in common.

The physiologists and biochemists studying fat accumulation

in their laboratories tried to rectify this situation, but they failed. In 1965, the American Physiological Society published an eight-hundred-page *Handbook of Physiology*, subtitled *Adipose Tissue*. The purpose of the book, with its sixty-nine chapters and over four thousand references, was to make available to physicians and obesity researchers in one volume as much as possible—"to present indeed the whole existing up-to-date knowledge"—of what had been learned about the physiological regulation of fat metabolism and fat storage over the preceding decades.

The introductory chapter was written by Ernst Wertheimer, a biochemist who had fled Germany after the Nazi purges of Jewish academics in 1933 and settled in Israel, working at the Hebrew University–Hadassah School of Medicine in Jerusalem, where he had done much of the early research. The literature on the science of fat, Wertheimer explained, had been "meager" in the early 1930s, when this scientific revolution began. To the medical community, fat tissue was a kind of amorphous connective tissue "filled, by chance, with droplets of fat." When physiologists thought about fat tissue at all, they assumed its main task was to insulate the body against heat loss and provide mechanical support and maybe cushioning where necessary. "The fat stores were believed to be purely passive in nature," Wertheimer wrote, "and not in any way involved in the general metabolism of the body."

All of this had changed radically by the 1965 publication of the *Handbook of Physiology: Adipose Tissue*. Researchers had come to learn that fat tissue and the fat cells themselves played vital roles in energy metabolism in the body, and that insulin was the dominant hormone orchestrating this role. Moreover, no tissue in the body was as sensitive to insulin as the fat tissue, and the fat tissue was responding to insulin by synthesizing and storing fat. From a physiological perspective, as Wertheimer wrote, the action of insulin on the human body *could not be understood* without understanding its role in fat storage and metabolism. Most diabetologists, though, remained uninterested, even when much of the research had been done with diabetic animals and often in collaboration with diabetologists themselves.

As is often the case in any science, researchers who might have studied the problem initially lacked the tools necessary to do so. To understand what insulin was doing in the human body required understanding its influence on the fats, protein, and carbohydrates consumed in the diet. That, in turn, required a method to track those macronutrients from their consumption through all the various metabolic processes of the body—the *intermediary* or *intermediate metabolism*, as it would come to be called—until they were eventually used to generate energy or excreted. These were the processes to which Rollin Woodyatt had directed attention in 1921 when he reminded diabetologists that the macronutrients being used for energy in cells are not necessarily those that had just been consumed, but rather those nutrients that had been digested and absorbed, after what could be an endless series of biochemical processes, of synthesis, conversion, degradation (the breaking down of these complex molecules into their component parts), resynthesis, reconversion, and so on.

In 1940, two Columbia University biochemists, Rudolf Schoenheimer, also a refugee from Nazi Germany, and David Rittenberg, had described the problem from their research-oriented perspective: "The experimenter," they wrote, "loses track of [the fats, proteins, and carbohydrates consumed] as soon as they pass the intestinal wall and mix with the same substances in blood and organs." By then, Schoenheimer and Rittenberg had also provided the solution: use deuterium, a heavy and faintly radioactive form of hydrogen that had been discovered by Rittenberg's Columbia mentor, Harold Urey, to tag or label these dietary macronutrients so that they could then be followed, as necessary, through the steps of the intermediate metabolism. Researchers could feed animals macronutrients in which deuterium atoms replaced hydrogen atoms and then sacrifice the animals hours, days, or even weeks or months later to look for the radioactivity that would indicate in which tissues or organs the deuterium, and so the macronutrients, had accumulated.

Among the revelations that emerged from Schoenheimer and Rittenberg's research was that the fat tissue is a site of remarkable physiological activity. Some fats consumed in the diet, they reported, are transported directly to the fat cells and stored as fat, and some aren't. Moreover, the types of fats being stored are being continuously converted, one into another. Carbohydrates, too, are clearly being converted into fat, at least in the animals studied, because these animals would accumulate fat in their fat tissue; they would get fatter, even when fed only carbohydrates in the diet. Looking solely at the fats stored in the fat tissue at any one point in time, Schoenheimer explained in a talk he gave at Harvard in 1941 shortly before his death, they are "indistinguishable as to their origin." The "normal animal's body fats," he said, "despite their qualitative and quantitative constanc[y], are in a state of rapid flux."

In 1948, Ernst Wertheimer and his student Benjamin Shapiro produced the first comprehensive review of this emerging science and the revelations that had emerged from the Columbia research. "Mobilization and deposition of fat go on continuously," they wrote, "without regard to the nutritional state of the animal." And they added that "the classical theory, that fat is deposited in the adipose tissue only when given in excess of the caloric requirement"—what Newburgh had declared unconditionally and the medical establishment was now embracing—"has been finally disproved." Rather, fat accumulates in the adipose tissue when these forces of deposition exceed those of mobilization, and "the lowering of the fat content of the tissue during hunger is the result of mobilization exceeding deposition."

What was still unknown in 1948, though, was the specific physiological factors that determine this balance of fat deposition and mobilization. The fat cell couldn't be passively releasing fat into the bloodstream, Wertheimer and Shapiro explained, just as it couldn't be passively receiving fat from it. When the concentration of fat in the blood went up, for instance, as it did so obviously in the lipemia (hypertriglyceridemia) of uncontrolled diabetes, that lipemia was typically accompanied by emaciation and a decrease of body fat, the opposite of what would be expected if this flow of fat into the fat cells was a passive process. The balance of fat deposition and mobilization in fat cells, Wertheimer and Shapiro concluded, echoing Bauer and von Bergmann's argument, "must be controlled by a factor acting directly on the fat cell."

That controlling factor would turn out to be the central nervous system and the endocrine system and, particularly, insulin. Over the next fifteen years researchers in Europe, Israel, and the United States would report a series of revelations about insulin's role in stimulating both the conversion of carbohydrates into fat and then the storage of fat in the fat tissue. In 1954, for instance, researchers at Jefferson Medical College in Philadelphia, led by Franz Hausberger, reported that insulin, in laboratory animals, was more influential in stimulating the process of *de novo lipogenesis*—the creation of fat from carbohydrates—than it was in the use of blood sugar for fuel. This was one way insulin could work to maintain a healthy level of blood sugar, by signaling cells in the liver and fat tissue to take up some of the glucose in the circulation and convert it into fat for storage. In these animals, most of the carbohydrates consumed were first converted into fat and stored as such in adipose tissue before being mobilized into the circulation and used by cells for energy. As Denis McGarry would point out half a century later, diabetes may have been a disorder of carbohydrate intolerance—insulin, after all, is secreted primarily in response to the carbohydrates in the diet—but the relative absence of insulin or its excess was affecting the conversion of carbohydrates into fat and the storage and metabolism of fats as much as anything having to do directly with blood sugar.

Two more revelations were necessary before researchers could fully understand the influence of insulin on fat storage and metabolism, and both depended on the development of laboratory assays that could quantify what had to be measured. The first was a way to measure the amount of fat in the bloodstream, specifically in the form known as *free fatty acids*. These can be thought of as the component molecules of fat itself. Fat is stored in the body and consumed in the diet in the form of triglycerides, but triglycerides are made from three fatty acids (the "tri") bonded together by a glycerol molecule. When the fatty acids were found circulating in the bloodstream by themselves, they were called free fatty acids, and researchers couldn't understand the flow of fat into and out of fat cells and, hence, such conditions as diabetic lipemia, without being able to follow and quantify these free fatty acids in the circulation.

In 1956, researchers from three institutions—the National Institutes of Health, Rockefeller Institute in New York City, and the University of Lund in Sweden—reported independently that they had been able to accurately measure free fatty acids* circulating in the bloodstream and determine how the concentration changed with time and with diet. All three papers concluded that fatty acids are a likely form in which fat is made available to cells to meet energy demands, hence the form in which cells burn fat for fuel. Further, the availability of these fatty acids in the circulation seemed to be determined largely by the blood sugar levels and by hormones—insulin, specifically. When subjects in these experiments ate fat, the amount of free fatty acids in the circulation increased slightly, but it also increased slightly when the subjects were fasted and ate nothing at all. When they ate carbohydrates or were given injections of glucose, the fatty acid level plunged. The same happened with an injection of insulin itself, even when the dose was minuscule.

The implication of the three reports was that fatty acids are a critically important fuel source in the human body—a material of "great metabolic activity," as the National Institutes of Health (NIH) researchers Robert Gordon and Amelia Cherkes phrased it in their 1956 publication—stored in the fat tissue in the form of triglycerides when not needed, and then liberated into the circulation as free fatty acids to be used for energy by the cells when they are needed. If carbohydrates are available for fuel—if blood sugar is rising and insulin is being secreted in response—the fat tissue takes up these fatty acids and inhibits their release. When blood sugar levels are stable and insulin low, the fatty acids are mobilized into the circulation so that they can be used for fuel instead. While insulin served to promote the use of carbohydrates (glucose) for fuel and lock up fatty acids in the adipose tissue in the form of triglycerides, other hormones clearly worked to liberate the fatty acids.

---

* The technically correct term, as the NIH researchers pointed out, is "non-esterified" or "unesterified" fatty acids, because fatty acids can also be bound to proteins and so not actually always free. Esterification is the process by which three fatty acids are bound together with a glycerol molecule to store them as a triglyceride. "Non-esterified" or "unesterified" means only that the fatty acids are not bound to other fatty acids or the glycerol molecule.

Injections of adrenaline (epinephrine), for instance, caused fatty acids to flood the circulation. Because this hormone is naturally released by the adrenal glands as an integral part of the flight-fight response, the researchers suggested that the concentration of fatty acids rises and falls not just in relationship to the body's immediate needs for fuel, but "the anticipated need."

The remaining obstacle to understanding insulin's role in this physiological regulation of fat storage and metabolism was a profound one: how diabetologists conceived of the very nature of diabetes. Through the 1950s, diabetologists had continued to assume that diabetes was a disease of insulin deficiency, just as it was in pancreatectomized dogs. If that were true, if diabetes was always an insulin deficiency disorder, how could insulin promote fat accumulation when so many individuals with diabetes were obese? If, as Falta had said, a functioning pancreas was "necessary for fattening" and diabetes was caused by a dysfunctional pancreas, then shouldn't all patients with diabetes be thin? This was the paradox that confronted diabetologists as long as they believed that their diabetic patients were always or even mostly insulin deficient.

In the early 1930s, Falta and a young British diabetologist, Harold Himsworth (later Sir Harold and head of Britain's Medical Research Council, similar to the NIH in the United States), had reported independently that many of their diabetic patients did not respond as expected to insulin therapy. It was Falta who first suggested the terms "insulin sensitive" and "insulin resistant" to describe two apparently different diabetic states.* The insulin-resistant patients, "encountered in rather large numbers," as Russell Wilder wrote in 1935 discussing Falta's work, were the ones who require relatively large doses of insulin to control their blood sugar. And when they're given more insulin than necessary, they suffer little if any distress, manifesting little or no evidence of hypoglyce-

---

* The German diabetologist Ferdinand Bertram made a similar observation in the same era, classifying his diabetic patients according to their response to injected insulin: either *Insulin-mangeldiabetes* (insulin-lack diabetes) or *Gegen-regulations diabetes* (counterregulation diabetes), because the blood sugar was resistant to the insulin action.

mia or insulin shock. The insulin-sensitive patients, on the other hand, like pancreatectomized dogs, can get by on small doses of insulin and, when given too much, are "thrown into hypoglycemic shock."

Falta and Himsworth had made one of the most important observations in all of diabetes science. It would ultimately define how we see the two types of diabetes. Once again, though, diabetologists paid little attention. Whether a patient was insulin sensitive or insulin resistant seemed to have little relevance to treatment. Patients would use as much insulin as necessary to control their blood sugar. When Joslin's textbook discussed insulin resistance in the 1940 edition, it said they considered patients resistant to insulin only if they required more than 200 units of insulin daily for long periods of time. His clinic had seen only "6 and possibly 7" such cases among the nineteen thousand patients Joslin and his clinic had treated in four decades. ("Joslin, like all other investigators of the time," as the British diabetologist Edwin Gale wrote in 2013, "was locked into the assumption that diabetes was an insulin deficiency disorder.") But the difference between the insulin sensitive and the insulin resistant was indeed profound.

To understand that difference required the second of the era's two breakthroughs in laboratory technology: this was a means of quantifying how much insulin was circulating in the blood at any point in time. Insulin is a very small protein; it exists in the blood in concentrations that are nearly infinitesimal compared to those of cholesterol and the lipoprotein particles—low-density lipoproteins, for instance, known as LDL—that transport fat and cholesterol. When insulin levels are low, between meals or in the morning before breakfast, insulin might constitute one part per billion of the weight of the blood plasma. That might go up to five parts per billion when insulin is being secreted during a meal.

While these numbers were a testament to "the almost incredible potency of the hormone," as the New York Medical College diabetologist Rachmiel Levine described it in 1967, no method existed through the 1950s to measure such a scarce quantity in the circulation with any accuracy. Researchers trying to quantify the insulin in human blood relied on a variety of arcane tests that in turn depended on the ability of insulin to prompt the absorption of glucose by the tissues of rabbits or laboratory rodents—just a millionth of a gram of insulin was sufficient to send a mouse into hypoglycemic shock—or the oxidation of glucose in fat cells taken from these animals, with the awareness that fat was the tissue most sensitive to insulin. The best that could be said for any of these procedures is that they provided "interesting information," in the words of the two researchers, Rosalyn Yalow and Solomon Berson, who would resolve the problem. These assays could determine the presence of "insulin-like activity." They could not actually determine if it *was* insulin and, if so, how much.

In July 1960, *The Journal of Clinical Investigation* published an article by Yalow and Berson describing a method they had developed—the aforementioned radioimmunoassay—that could reliably and accurately measure in a sample of blood "as little as a fraction of a microunit of human insulin." It could and did work on other hormones just as well. In 1977, when Yalow was awarded the Nobel Prize for the discovery (as I said earlier, Berson died in 1972), the Nobel Foundation described the radioimmunoassay as bringing about "a revolution in biological and medical research," which it had.

Its immediate implications, from Yalow and Berson's very first papers, applied to diabetes and, with it, obesity and the physiological mechanisms by which we get fatter. Not only did patients with type 2 diabetes—"early or mild maturity-onset disease," as Yalow and Berson called it—tend to have high levels of insulin in their circulation, their insulin levels would keep rising after consuming carbohydrates until they "frequently exceeded the levels reached" in healthy subjects. What had been considered a disease of insulin deficiency was clearly not. These older diabetic patients had excessive insulin circulating through their blood, a condition known as hyperinsulinemia. Young patients with the acute form of the disease, on the other hand, showed the expected insulin deficiency.

This was the insulin resistance and insulin sensitivity that Falta and Himsworth had first observed. The coexistence of high levels of insulin with high blood sugar meant that the tissues of these older, heavier diabetic patients, those with type 2 diabetes, were somehow resistant to the action of the insulin. They had plenty of insulin in their circulation. The hormone was failing, though, to control blood sugar. The same would tend to be true for obese subjects as well. Before diabetes associations in the 1990s embraced the terminology "type 1" and "type 2" diabetes, the former would come to be known as insulin-dependent diabetes—because these patients required insulin therapy to be healthy—and the latter non-insulin-dependent diabetes, because these patients did not. The dependent form is a disease of insulin deficiency. Minkowski's dogs had been an appropriate model for this disease. The other, chronic form that associates with age and obesity and is non-insulin-dependent, is a disease of insulin resistance. For this form, the dog models—Minkowski's and then Allen's and his contemporaries'—had been profoundly misleading.

By 1965, when the American Physiological Society published its eight-hundred-page *Handbook of Physiology* on adipose tissue, the paradox that had confronted diabetologists for most of four decades had been resolved by Yalow and Berson's observation. The ever-accumulating evidence that insulin promotes fat accumulation and inhibits its mobilization (*lipolysis*, in the technical terminology) and its use for energy could now be resolved with the observed intimate association of obesity and diabetes: individuals with obesity and those with type 2 diabetes, and those, of course, with both, had high circulating levels of insulin. They were not insulin deficient; they were *hyperinsulinemic*; they had excessive amounts of insulin in their circulation, at least until their pancreas eventually failed to keep up with the necessary production, and they, too, became insulin deficient. If insulin was a fattening hormone, then we would expect individuals with a surfeit of insulin to be obese. And they were, whether with diabetes or not. In the latter stages of the disease, as the pancreatic exhaustion set in, these patients would lose weight, just as Joslin's mother had, for instance, late in her life, his famous Case No. 8.

In 1965, Solomon Berson gave the Banting Memorial Lecture at the annual meeting of the American Diabetes Association. Speaking for Yalow as well, he discussed another assumption of the diabetes community: that the diabetic condition—non-insulin-dependent or type 2—was caused or promoted by obesity. Joslin's textbook had stated this unequivocally in edition after edition: diabetes "is largely a penalty of the obesity." But Yalow and Berson's research now suggested that Joslin and generations of diabetologists might have confused cause for effect, and vice versa. "We generally accept that obesity predisposes to diabetes," Berson said, "but does not mild diabetes predispose to obesity? Since insulin is a most potent lipogenic [fat-forming] agent, chronic hyperinsulinism would favor the accumulation of body fat."

Six years later, in 1971, on the fiftieth anniversary of Banting and Best's discovery, Harvard's George Cahill gave the ADA's Banting Memorial Lecture. Cahill had done seminal research on insulin and its influence on both protein and fat metabolism, working originally at Joslin's Diabetes Center, collaborating with Albert Renold, a Swiss physiologist. Cahill and Renold together had edited the 1965 *Handbook of Physiology: Adipose Tissue*. Cahill's 1971 Banting Lecture was entitled "Physiology of Insulin in Man." In it, he summed up what had been learned about the actions of insulin on protein and fat metabolism since Banting and Best's original discovery. "Simply speaking," Cahill said, "the concentration of circulating insulin serves to coordinate fuel storage and fuel mobilization into and out of the various depots with the needs of the organism, and with the availability or lack of availability of fuel in the environment."

When insulin is secreted, Cahill explained, the organism (or human) knows it has been fed. It synthesizes fat out of the carbohydrates consumed and it stores that fat and the fat from the diet in the fat cells. "Low levels" of insulin, he said, herald "the 'fasted' state." These low levels signal the fat cells to mobilize the calories they have stored and make them available for fuel. Berson, in his Banting Lecture six years earlier, had referred to this as "the negative stimulus of insulin deficiency." While fat is mobilized from fat cells and insulin is low, the liver turns on its synthesis of ketone bodies, using the fat to generate ketones. These molecules then provide much of the fuel for the brain and the protein can be "spared" to stay in muscle and organs, where it's needed. Low levels of insulin—the "negative stimulus of insulin deficiency"—is why patients with type 1 diabetes become emaciated without insulin therapy. Their fat cells cannot hold on to the fat that has been stored. Wilhelm Falta had been right all along.

"A high insulin level signals fat synthesis and storage in adipose tissue," Cahill explained, "and a low level its release as free fatty acid back into the circulation." When the two processes are exactly balanced—what Cahill described as the "null point"—as much fat is leaving the fat tissue every day as is being deposited in it. This is the point, as Wertheimer and Shapiro had described it in 1948, at which the forces of mobilization and deposition on the fat cells cancel each other out. On one side of the null point, insulin is low, mobilization of fat from the fat tissue dominates, and the individual is getting leaner. On the other side, when insulin is high, deposition of fat dominates, and the individual gets fatter. "The average individual spends several hours a day to the right of the 'null' point," Cahill explained, with insulin in the circulation assuring that deposition is greater than mobilization, "and the remainder to the left," where the negative stimulus of insulin deficiency makes sure that mobilization dominates and the fat is used for fuel. "If lipid storage matches lipid mobilization, [the individual] is no fatter or thinner than the previous day."

Although Cahill did not say so explicitly, the implication should have been clear. As insulin levels responded to the carbohydrate content of the diet, the more time spent with insulin levels high and the forces of deposition overwhelming the forces of mobilization, the fatter the individual would be. When Cahill discussed what he called "aberrations in the 'null' point," one of them was insulin resistance, the condition in which insulin levels are unnaturally elevated, which would explain why type 2 diabetics tend to be fat; another was obesity, an insulin-resistant state, as Yalow and Berson had demonstrated.

Cahill also failed to discuss the relevance of this science to insulin therapy. The higher the dose of insulin therapy given to diabetics—as required by carbohydrate-rich diets—the more fat would tend to accumulate.

What researchers like Cahill had learned about insulin and fat storage since the 1930s suggested that it might be virtually impossible for some significant proportion of all patients with diabetes to maintain a healthy weight on carbohydrate-rich diets controlled by high doses of insulin. The diabetologists themselves had assumed that the weight gain could always be countered by eating less, as many still do, but the science of fat metabolism suggested that was not true. Like lean individuals when semi-starved, they might be able to lose weight in the short term, but they'd always be fighting hunger and a hormonal milieu in their bodies that was actively working to maintain significant body fat or accumulate it.

In the 1970s, the great debate in the diabetes community would still be whether or not diabetic complications could be reduced by tighter blood sugar control. Cahill was among the prominent voices arguing that it was. While clinical trials had yet to establish this unequivocally, considerable evidence suggested that individuals who kept their blood sugar under tight control delayed the microvascular complications—retinopathy, neuropathy, kidney damage. But better control required the use of more insulin, not less. And macrovascular complications—heart disease, in particular—had always associated with weight gain. That's why maintaining a healthy weight was invariably deemed as critical to those with diabetes as maintaining a healthy blood sugar.

From the 1970s onward, as diabetologists took to running clinical trials to assess the value of tighter control of blood sugar—hence, the use of higher and more frequent doses of insulin—they reported repeatedly that the more insulin was used, the fatter the diabetic subjects in these trials became. In a single year, patients might gain ten pounds more with what these trials referred to as "intensive insulin therapy" compared to the weight gain on the more standard insulin regimens. In patients with type 1 diabetes, the weight gain was often as much muscle—lean tissue—as fat. But in patients with type 2 diabetes, it was mostly fat and mostly around the waist.

When the diabetes community set out in the late 1990s, in three major clinical trials, to test the effects of intensive blood sugar control on the macrovascular complications of type 2 diabetes—heart disease, specifically—they saw the same effect. The results from the three trials were published in 2008 and early 2009: the ACCORD trial (for Action to Control Cardiovascular Risk in Diabetes), a twenty-country study called ADVANCE, and the Veterans Affairs Diabetes Trial (VADT). All three reported that intensive blood sugar control, much to the surprise of the diabetologists, had no apparent benefit regarding macrovascular complications, and may have done more harm than good in at least one of these trials. Moreover, two of the three reported that intensive insulin therapy caused significant weight gain. When the American Diabetes Association, the American Heart Association, and the American College of Cardiology published a joint position statement discussing the lessons to be learned from the three trials, the authors concluded that intensive blood sugar control might have failed to demonstrate a benefit because of "counterbalancing consequences" for heart disease, including low blood sugar (hypoglycemia) and weight gain.

Little had changed since the debates of the 1930s. As Robert Tattersall had written in his 2009 book, *Diabetes: The Biography*, diabetologists in the 1950s had been reluctant to start insulin therapy in their overweight, middle-aged patients "because it caused weight gain," but the diabetologists would still not accept the implications. Weight gain on insulin therapy is given three paragraphs in the thousand-plus-page textbook of the thirteenth edition of *Joslin's Diabetes Mellitus* in 1994, and three paragraphs again in the fourteenth and latest edition, published in 2005. In both editions, the authors discuss weight gain and obesity in the context of a complication of insulin therapy. In both editions, they discuss it happening at least in part because of the "direct lipogenic effects of insulin on adipose tissue, independent of food intake." In both editions, they discuss it as a vicious cycle, which implies it will only get worse with time: "The result of weight gain in insulin-treated patients is the often-cited vicious cycle of increased insulin resistance, leading to the need for more exogenous insulin [higher doses of insulin], to further weight gain, which increases the insulin resistance even more," as Robert Rosenzweig of Joslin's Diabetes Center and Harvard Medical School phrased it in the thirteenth edition, or "The result of weight gain in insulin-treated patients is further insulin resistance, leading to the need for more insulin and a potentially greater weight gain," as Alice Cheng and Bernard Zinman of the University of Toronto phrased it in the fourteenth.

In both editions, the authors assume that the cycle can be broken with weight loss, and then assume that this can be accomplished by dietary therapy, as diabetologists always have. "Diet therapy and weight loss are extremely important in reversing this process," says Rosenzweig. But the chapters on obesity in both editions acknowledge that the medical and nutrition communities did not actually know how to make this happen, how to induce overweight or obese patients to achieve sustained weight loss through dietary therapy. "Successful treatment of obesity," as Harvard's Jeffrey Flier phrased it in the thirteenth edition, "is rarely achievable in clinical practice." That was the long-term complication of Newburgh's energy balance thinking, and obesity researchers had never solved it.

In the 1970s, influential diabetologists had still been willing to accept the possibility that the lipogenic actions of insulin presented a challenge to their therapeutic thinking that might (barring a pharmaceutical miracle) be insurmountable. One of the few who clearly did was the University of Miami diabetologist Jay Skyler, who would become a president of the American Diabetes Association and the founding editor of its journal, *Diabetes Care*. Skyler was skeptical about the idea that hyperinsulinemia is the cause of the obesity that associates with type 2 diabetes, but he accepted the possibility. If it is true, he wrote in a chapter on nutritional therapy in a 1978 diabetes textbook, the implication was clear: "The treatment of obesity should entail disproportionate carbohydrate restriction, in an attempt to both diminish hyperinsulinemia and achieve normoglycemia [healthy blood sugar levels] by reducing carbohydrate load."

The catch was that a diet that disproportionately reduced the carbohydrate load, as Skyler had phrased it, would be disproportionately high in fat, the kind of diet that Rollo and a long line of physicians up to Newburgh, Marsh, and Petrén in the early years of the insulin era had proposed for diabetes, and the diabetes world had now left behind. By the late 1970s, diabetologists would not consider such a therapy as a serious alternative. The demonization of dietary fat that had started in the diabetes world with Allen and Joslin in the pre-insulin years, and that Joslin could not relinquish, had now spread throughout medicine and public health. The determination of an ideal diet for diabetes would be established by cardiologists, nutritionists, and public health authorities. The diabetologists themselves would renounce that obligation.

8

Good Science/Bad Science, Part II

Briefly summed up, the evolution is this: a premature explanation passes into a tentative theory, then into an adopted theory, and then into a ruling theory.

When the last stage has been reached, unless the theory happens, perchance, to be the true one, all hope of the best results is gone. To be sure, truth may be brought forth by an investigator dominated by a false ruling idea. His very errors may indeed stimulate investigation on the part of others. But the condition is an unfortunate one. Dust and chaff are mingled with the grain in what should be a winnowing process.

—T. C. CHAMBERLIN,
"The Method of Multiple Working Hypotheses," 1890

Why the delay? Why the acceptance? These are key questions, and the simple answer is that what we believe determines what we see.

—EDWIN GALE,
"The Hedgehog and the Fox: Sir Harold Himsworth (1905–93)," 2013

In the spring of 1950, Elliott Joslin looked back on a half century's experience in diabetes and summarized what he had learned. By then, over thirty thousand patients had been diagnosed as diabetic in his Boston clinic, and almost a third of them had died, many at a reasonably ripe old age. Joslin was justly proud that the average duration of life for his patients had tripled over the years, mostly because of insulin therapy and its near eradication of diabetic coma and acidosis as causes of death.

For those who had been stricken with diabetes in their youth, however, even the continuing advances in insulin therapy had not been enough to prevent them from dying prematurely. Children diagnosed before the age of ten would be expected to die of the complications of the disease well before their thirtieth birthday. For teenagers, the prognosis was slightly worse—they could expect to live less than sixteen years after diagnosis. Joslin attributed that to a natural tendency of teenagers to bridle at the rigor of the treatment needed to keep their disease under control.

These young men and women, though freed from the specter of diabetic coma, were now living long enough to have their veins and arteries ravaged by their disease. In the most recent survey from his clinic, Joslin wrote, "arteriosclerosis and cardiovascular-renal [kidney] diseases" had accounted for 70 percent of the deaths. Arteriosclerosis is the technical term for the thickening and hardening of arteries, when the lining of these vessels—known as the *endothelium*—is damaged. When plaques build up, this is known as atherosclerosis, from the Greek word *atheroma*, which means "porridge" and is what these plaques may look like when viewed in an autopsy. When the atheroma (fatty deposits) are in the coronary arteries, supplying blood and oxygen to the heart, the disease is known as coronary artery disease or cardiovascular disease, and this can be a precursor to a heart attack and all the troubles that follow.*

In 1948, Joslin's colleagues Ernest Millard and Howard Root had documented what they had learned from autopsies of 110 diabetic patients, all of whom had died relatively young. "One of the most striking facts," Millard and Root wrote, was the severity of the vascular pathology seen in the patients who had lived with diabetes the longest. Those few who had died only five or six years after their diagnosis—"of coma or an infection"—had arteries and veins that looked little different from those of healthy individuals of similar

---

* These words—arteriosclerosis, atherosclerosis, cardiovascular disease, coronary artery disease, and ischemic heart disease (for heart disease caused by narrowing of the arteries)—would often be used by authors interchangeably in the medical literature.

age. This observation suggested that diabetes, with insulin therapy, could be a relatively benign disease for the first half-dozen or so years. But the blood vessels of those who had been diagnosed before the age of thirty and then lived with the disease for fifteen years or longer were pathologically narrow and stiff. Almost every one of these diabetic men and women, despite having still died relatively young, suffered from arteriosclerosis, whether that was the immediate cause of their death or not. Severe coronary arteriosclerosis—coronary artery disease—was present in more than two-thirds of them.

Some factor or factors associated with the diabetic condition "markedly accelerates degenerative processes" in the larger blood vessels and in the smaller capillaries and arterioles throughout the body, Millard and Root suggested. This in turn was causing what diabetologists now refer to as the macrovascular and microvascular complications of the disease: not just heart disease and stroke (the macrovascular or large vessel complications), but neuropathy, retinopathy, and kidney (renal) disease, which are the microvascular (or small vessel) complications.

The critical questions for diabetes therapy from the 1930s onward would be: what is it about the diabetic condition that causes this severe vascular pathology and how best to treat and prevent it? High blood sugar was always the most likely culprit. This is why Joslin, Root, and their colleagues had been tireless advocates for tight control of blood sugar. They believed it to be the obvious way to prevent or at least delay the manifestation of vascular disease. The better job their patients did of keeping their blood sugar in a healthy range, whether by diet and/or by insulin, the fewer or less severe the complications. Whether this was a causal phenomenon, though, whether poor glycemic control and so high blood sugar caused vascular disease and diabetic complications, and good control prevented it, could not be established from the kind of observations these physicians were making. This is how Root and Joslin phrased the critical question in the 1940 edition of Joslin's textbook: "To a non-diabetic carbohydrate is benign," they explained. "But with the diabetic to use carbohydrate introduces the problem of hyperglycemia which develops when the carbohydrate is not burned. Is a persistent hyperglycemia a cause of arteriosclerosis in diabetes? It is an abnormal condition and any abnormal state would tend to wear out the machine."

One possibility was that patients with a relatively benign form of the disease were the ones who found it easy to maintain glycemic control. Sansum and Tolstoi and the physicians advocating for free diets argued that diabetic complications were inevitable in some patients. Joslin's patients got them, too, after all. So maybe the diet made little difference. These diabetologists published articles suggesting that over the course of a decade their patients eating as they liked and covering their carbohydrate-rich meals with large doses of insulin seemed to do no worse than Joslin's.

Physicians would argue, as Joslin and R. D. Lawrence did, that the benefits of maintaining healthy blood sugar levels could scarcely be doubted, while others would argue that those benefits, if any, came with risks as well—hypoglycemia, in particular—and were not worth the burden placed on their patients. Eventually diabetologists would launch numerous large and expensive clinical trials, as I've discussed and will again, to test the proposition that the better the glycemic control (by drug therapy), the fewer and the later in life the complications of the disease.

These discussions and debates, though, would rarely address what Joslin came to consider an obvious possibility: if those with diabetes developed atherosclerosis and arteriosclerosis more rapidly than those without, and if they developed it far younger than otherwise healthy individuals, maybe there was something about the diabetic condition that was the mechanism underlying *all* heart disease. Maybe the same pathological forces are at work in both those with diabetes and those free of the disease, though they are vastly accelerated in diabetic patients. And this would explain why those with diabetes get heart disease and atherosclerosis and all the vascular complications early in life, and those without diabetes get them later. If so, what diabetologists could learn about the pathological phenomena causing heart disease and arteriosclerosis in their diabetic patients might help everyone prevent or delay onset of the disease.*

---

* "These statistics," Joslin wrote, about the near universal prevalence of arteriosclerosis in patients with diabetes of sufficiently long duration, "indicate that research in the

Physicians and those clinical researchers who took to studying heart disease in the United States in the post–World War II decades, though, were concerned with what they thought was a very different heart disease epidemic, one that appeared to be killing healthy men in their prime. They would come to blame dietary fat for heart disease in the general population—a link between saturated fats, specifically, and the cholesterol levels in the blood, or *serum cholesterol*, as it's known. The diabetologists would then assume, with all the best of intentions, that those conclusions and their implications held true for patients with diabetes. The result would be yet another shift in the medical consensus on the nature of the foods a diabetic patient should or should not eat.

Beginning in 1971, the American Diabetes Association began prescribing carbohydrate-rich/low-fat diets for diabetic patients, largely because this is what the American Heart Association was suggesting for effectively *all* Americans. For the first time in history, patients with diabetes were being counseled en masse that as much as 60 percent of all their calories should come from the one macronutrient that their bodies could not safely metabolize without pharmaceutical assistance. The evidence did not exist at the time to know whether such a diet lessened the long-term complications of the disease or made them worse. But the authorities enlisted by the ADA to give advice believed implicitly that diabetic patients could sacrifice control of blood sugar by diet for the supposed benefits of a carbohydrate-rich/low-fat diet and those patients would come out ahead. The premise of this book, of course, is that they were wrong.

In these critical mid-twentieth-century decades, researchers generated competing hypotheses to explain the patterns of chronic diseases—heart disease and diabetes, specifically—that they were observing. One began as an attempt to explain the seemingly high prevalence of heart disease in the United States compared to other

---

prevention of arteriosclerosis should be carried out primarily in diabetics with a premature tendency to it, and especially in those in whom diabetes began before the age of 25, so that they can be followed up until they reach that period in life when arteriosclerosis begins to be routine in the general population."

nations; the other tried to explain the appearance of diabetes in populations worldwide as those populations transitioned—either through immigration, urbanization, or acculturation—to western diets and lifestyles, and the high prevalence of heart disease in those with diabetes.

The researchers involved had created competing paradigms, constructs for thinking about these diseases. These paradigms in turn determined not just the supposed causal factors driving the appearance of these diseases, but how they should be prevented and treated, and even what observations, experiments, or clinical trials should be made to shed light on the key questions, and how the evidence generated in those endeavors should best be interpreted.

The diabetes-centric hypotheses emerged from observations made worldwide by physicians who were witnessing diabetes epidemics as they appeared. These physicians then naturally speculated on the likely causes: what had recently changed, for instance, in the diets of those populations that might have been responsible. The link to heart disease emerged from the study of diabetic patients themselves and of patients whom physicians would come to think of as prediabetic and today we would say suffer from *metabolic syndrome*, a cluster of metabolic abnormalities that includes poor blood sugar control, high blood pressure, and excess weight. Together they implicated the quality of the carbohydrates consumed in the diet—specifically, processed flour and sugar—as the causes of diabetes, or at least of these diabetes epidemics. They implicated high blood sugar *and* insulin, the physiological responses to these highly processed carbohydrates, as causal factors in the genesis of heart disease in *everyone*, whether suffering from diabetes or not. Together they implied that the best dietary therapy for prevention and treatment of the diabetic condition is one in which carbohydrates are restricted and insulin use is minimized. This proposition differed only in its treatment of dietary fat from how Joslin conceptualized diet therapy in the early years of the insulin era.

The other paradigm, however, the heart disease–centric hypothesis, is the one that has been the basis of public health recommendations for over half a century. As a result, it is what most of us have grown up believing. It emerged out of observations of populations at large, those with a high prevalence of heart disease compared to those without—populations, say, in Mediterranean countries or Japan, with low rates of heart disease, compared to those in the United States. There was no consideration of comparing individuals with diabetes, as Joslin was suggesting, versus individuals without, or populations prior to the appearance of diabetes as a common disease and after.

Those investigating heart disease quickly came to focus on a single component of the blood, cholesterol. As this research evolved, the focus narrowed further to only the cholesterol in low-density lipoproteins, or LDL cholesterol (known as the "bad cholesterol"), elevated by a single component of the diet—saturated fat—and that was seen as the link between diet and atherosclerosis. The very definition of a healthy diet came to be a way of eating that would prevent cardiovascular disease, and so a diet low in fat and, specifically, saturated fat. The assumption was that eating this way is vital for anyone at high risk of heart disease, which includes all those with diabetes. And because a diet that restricts the calories from fat is a diet that is carbohydrate rich, this was the logic that led the American Diabetes Association in 1971 to begin advocating carbohydrate-rich/low-fat diets for all diabetic patients.

Rather than see these two hypotheses or paradigms as fundamentally different explanations for the same phenomenon and exploring them both, the medical and public health authorities embraced the saturated-fat/cholesterol hypothesis, despite the repeated failure of clinical trials to confirm that the reduction of saturated fat and cholesterol intake would actually prevent heart disease and, more important, lengthen lives. As a result, the supposedly ideal dietary prescription for diabetes came to depend on a conception of the dietary cause of heart disease that has remained controversial since its very inception and may very well be wrong.* The diabetes-centric thinking that competed with this conception, and with the thinking that the American Heart Association and the American Diabetes Association still disseminate widely (although not

---

* For a more comprehensive review of the evidence and the history of this science, see my book *Good Calories, Bad Calories* and *The Big Fat Surprise*, by Nina Teicholz.

quite so unconditionally as they once did), was considered of rela-tively minor significance. For this reason, it's important to review the history of both concepts, and to do so in the context of how science works to establish reliable knowledge, and how it doesn't. We have to understand how and why mistakes of this magnitude can be made and perpetuated afterward, despite the best inten-tions of the physicians, researchers, and public health authorities involved.

At the core of scientific controversies in medicine and public health, as I suggested earlier, is a fundamental conflict between the requirements of a science that is capable of establishing reli-able knowledge and the inherent constraints of medical practice and public health programs. Successful science requires a meticu-lous, dispassionate approach to the evidence and the patience to do whatever tests (experiments) are necessary to unambiguously rule out alternative explanations (hypotheses) for the phenomena being studied. "The method of science," as the philosopher of sci-ence Karl Popper described it, "is the method of bold conjectures and ingenious and severe attempts to refute them."

The process of science, in this sense, can be thought of as a highly evolved and institutionalized form of skeptical thinking. Variations on Richard Feynman's definition of the first principle of science, that "you must not fool yourself—and you are the easiest person to fool," can be found throughout the writings of philoso-phers of science and the scientists themselves. "Most people are concerned that someone might cheat them," as the Nobel laureate physicist Luis Alvarez wrote in his memoirs; "the scientist is even more concerned that he might cheat himself." Alvarez encapsulated this thinking as "only trust what you can prove." When the sociolo-gist of science Robert Merton discussed this perspective in 1968, in his analysis of the "Behavior Patterns of Scientists," he described the community of science as amplifying "that famous opening line of Aristotle's *Metaphysics*: 'All men by nature desire to know,'" and then added, "Perhaps, but men of science by culture desire to know that what they know is really so."

Scientists must manifest a fierce determination to establish the truth welded to a cold-blooded skepticism of their own work and thinking when progress appears to be made. "False conclusions are drawn all the time, but they are drawn tentatively," as the Cornell University astronomer and public communicator of science Carl Sagan explained in *The Demon-Haunted World: Science as a Candle in the Dark*. "Hypotheses are framed so they are capable of being disproved. A succession of alternative hypotheses is confronted by experiment and observation. Science gropes and staggers toward improved understanding." This process takes both time and the acceptance that reliable conclusions require unequivocal evidence to be believed. They are those hypotheses that have survived the gauntlet of ingenious and severe attempts to refute them. Until that evidence exists, conclusions must remain tentative.

This is where a conflict arises with the seemingly unavoidable limitations of medicine and public health. If lives are being lost, then actions have to be taken regardless of the state of the evidence at the time. This requires acting on judgments about what is most likely to be true. The critical assessment is whether or not assumptions are sufficiently likely or unlikely that action can be taken based on these probabilistic analyses. Can the dice be rolled with the confidence—a threshold of evidence that will always differ for all involved—that the benefits of the action or the medical or public health intervention will outweigh the harms? This applies even if the decision is to refrain from acting until better evidence is generated.

This conflict was often discussed openly by the American Heart Association and American Diabetes Association in publications communicating their dietary recommendations. The experts writing these guidelines knew that any uncertainty about whether they were right could not stop them from offering what they considered the guidance most likely to be right. Here's the Nutrition Committee of the American Heart Association doing so in 1978, eighteen years after first recommending that men at high risk of heart disease should eat significantly less fat; fourteen years after first extending that advice to all Americans old enough to walk; seven years after the American Diabetes Association revised its dietary guidelines based on this thinking and so the prescribed diets for patients with diabetes as well:

Within a framework of suggestive, but not unequivocal, scientific proof that dietary modification will ameliorate or prevent coronary heart disease, CHD, physicians must share the burden of uncertainty, and examine the issues where facts are incomplete and opinions differ. . . . Since our degrees of assurance regarding effectiveness of dietary modification on CHD are at varying levels of probability, research might focus on feasibility and safety of dietary modification. . . . Physicians often must make decisions in the absence of absolutely conclusive scientific proof. A reasoned resolution of the controversy is not currently possible.

With an "absence of absolutely conclusive scientific proof," physicians and public health authorities will inevitably say that they base their actions on the "best sense of the data," but their tendency will be to define that as the evidence that is consistent with their beliefs. They might acknowledge (at least when challenged, as was the case in 1978) that the evidence is equivocal and that their conclusions may be wrong, but they will also insist that they are sufficiently certain that they can act on them. They will convince themselves that they are not deluding themselves, despite a history of science and a dense literature on the philosophy of science suggesting that such confidence is rarely warranted. As Francis Bacon observed four hundred years ago, this is human nature. Neither scientists nor physicians are exempt.

The dangers of this thinking, though, go far beyond immediate harm. They are, in effect, precisely what the institutionalized skepticism of science evolved to prevent: the creation of a community of researchers and practitioners who embrace not just erroneous beliefs but a methodology for the acquisition of knowledge and a standard for assessing the reliability of that knowledge that cannot be trusted.

Once a position is taken on a subject based on premature, circumstantial, or equivocal evidence—a hypothesis embraced as so likely to be true that the authorities can bet human lives, quite literally, on it—the natural tendency is to accumulate evidence to support it, to justify the belief, and with it the value of circumstantial evidence, rather than to test it. This kind of thinking is known in cognitive behavioral sciences as "confirmation bias." Researchers come to consider evidence particularly meaningful when it agrees with their preconceptions, and to find reasons to reject the validity of the evidence when it doesn't. Francis Bacon, in 1620, described it this way:

> The human understanding, once it has adopted opinions, either because they were already accepted and believed, or because it likes them, draws everything else to support and agree with them. And though it may meet a greater number and weight of contrary instances, it will, with great and harmful prejudice, ignore or condemn or exclude them by introducing some distinction, in order that the authority of those earlier assumptions may remain intact and unharmed.*

Since the 1960s, nutrition and chronic disease researchers promoting the belief that heart disease is caused by saturated fat elevating serum cholesterol have often, as the AHA did, acknowledged that they engaged in this kind of confirmation bias. They did not, however, see it as a major problem or apparently understand the risks involved. (We have the benefit of hindsight; they did not.) The evidence that would become important to them would be the evidence supporting their beliefs, not the evidence that suggested they were wrong—what Bacon called "contrary instances." As early as 1957, for example, the University of Minnesota physiologist Ancel Keys, then in the process of becoming the most influential nutrition researcher in the world, wrote that "each new research adds detail, reduces areas of uncertainty, and, so far, provides further reason to believe" that something very much like his dietary fat–cholesterol hypothesis of heart disease was true. The caveat, though, was that "each new research" also added reasons to think it wasn't true.

---

* In her 2010 book *Being Wrong*, the journalist (and future Pulitzer Prize winner) Kathryn Schulz captured Bacon's thinking and that of the more recent half century of research on cognitive delusions this way: "As soon as we think we are right about something we narrow our focus, attending only to details that support our belief, or ceasing to listen altogether."

That's the essential nature of scientific investigations. Keys could imagine reasons himself to "ignore or condemn or exclude" that evidence, and so he did.

To understand the implications of this kind of situation, imagine a criminal trial in which a jury charged with establishing guilt or innocence hears only one side of the argument. That argument will be compelling by design, and it will go unchallenged. This is a likely scenario in any scientific endeavor in which the researchers come to believe they know the truth prior to acquiring definitive, unequivocal evidence. As the celebrated geologist T. C. Chamberlin explained in 1890, in such scenarios, we have to hope that the researchers got lucky, because we can expect that they had already renounced the skepticism necessary to establish reliable knowledge once they decided prematurely that they knew what was true.

In nutrition and chronic disease prevention, the diseases that are killing people prematurely do so slowly, routinely taking decades to manifest their harm. This should give more time for reasoned decisions to be made, for the science to grope and stagger toward understanding, as Sagan phrased it, which means time for ever more rigorous (ingenious and severe) experimental tests of the hypotheses. But those experiments—clinical trials in medicine and public health—also take a long time. The diseases that the researchers hope to prevent or delay in the subjects of the trial will take years or decades to manifest themselves. This means the trials will take that long to generate meaningful evidence. Tens of thousands of subjects will likely be needed, the costs will be exorbitant, and the trials unwieldy and exceedingly difficult to perform. New drug therapies cost in the neighborhood of a billion dollars to develop, in part because the clinical trials to test those drugs, to reliably establish both the benefits (efficacy) and the risks (safety)—to satisfy Food and Drug Administration (FDA) requirements—have to be large and long. Public health guidelines on nutrition have never been based on such rigorous evidence.

In 1961, the same year the American Heart Association began promoting low-fat diets for men at high risk of heart disease (those who are overweight, say, or have already survived a heart attack), the National Institutes of Health (NIH) began planning precisely the kind of trial necessary to test the hypothesis underlying that advice. As *The Wall Street Journal* reported at the time, such a trial would answer the "important question: can changes in the diet prevent heart attacks?" The researchers estimated that they might need perhaps fifty thousand subjects to consume a cholesterol-lowering diet for a decade, comparing their health to fifty thousand "controls" who ate as they always had. The NIH funded a pilot study to see if such a trial could be successfully carried out and determined that it could.

In 1969, a panel of experts assembled by the NIH to decide whether the nation should proceed with such a trial concluded that it had to be done precisely "because it is not known whether dietary manipulation has any effect whatsoever on coronary heart disease." Two years later, in 1971, the same year that the American Diabetes Association began promoting low-fat, carbohydrate-rich diets for all those with diabetes, the NIH published a four-hundred-page, two-volume report declaring again that "a definitive test" of the dietary-fat/cholesterol hypothesis of heart disease "is urgently needed." Because such a trial might cost as much as $1 billion, and might not produce a definitive answer (other billion-dollar, decade-long trials would still be necessary), the authors of the NIH report recommended that it not be conducted. And it never was.

As a result, decisions were made for decades based on circumstantial and equivocal evidence. Researchers had decided that the hypothesis was almost assuredly right, and so they argued that this evidence was compelling on the basis that it agreed with their preconceptions. A generation of nutrition and chronic disease researchers came to believe unconditionally that fats were villainous, just as Joslin had. The American Heart Association would continue to disseminate this belief, while simultaneously acknowledging (in the small print, effectively) that the experimental trials necessary to *rigorously* test this assumption had never been done.

Influential researchers would acknowledge in medical journals, as Thomas Dawber, founder of the famed Framingham Heart Study, did in *The New England Journal of Medicine* in 1978, that the idea that saturated fat caused heart disease was "an unproved hypothesis that needs much more investigation." But that would not stop the media, the AHA, and eventually the U.S. government in various guises from treating the hypothesis as though it were almost assuredly true. (Just two years later, Dawber insisted that his own study had provided "overwhelming evidence" that serum cholesterol "is a powerful factor" in the development of heart disease. "Yet," he noted, "many physicians and investigators of considerable renown still doubt the validity of the fat hypothesis.")

In 1984, when the National Institutes of Health took up this dietary fat hypothesis and promoted it widely, the NIH administrators had to do so based on the results of a drug trial, because all the diet trials done until then (and since) had failed to provide anything close to unequivocal evidence. Hence, the consensus of belief required a leap of faith, as it was later described by the NIH administrator, Basil Rifkind, who had orchestrated that consensus: if a drug that lowered cholesterol and seemed to keep middle-aged men with very high cholesterol alive a little longer than a placebo, as the trial results suggested, then a diet that lowered cholesterol— and did many other things as well, perhaps bad things—would also extend lives. "It's an imperfect world," Rifkind would say. "The data that would be definitive is ungettable, so you do your best with what is available."

By the late 1980s, the Surgeon General's Office and the National Academy of Sciences joined with the National Institutes of Health to put an official imprimatur on this dietary fat–cholesterol hypothesis of heart disease. With the American Diabetes Association embracing this thinking as well, the chapters on dietary therapy in diabetes textbooks would often be written not by diabetologists based on their clinical experience with patients or researchers based on their observations in randomized studies, but by registered dietitians disseminating this apparent consensus.*

Diabetologists would continue to discuss and debate the value

---

* In the 1971 edition of *Joslin's Diabetes Mellitus*, the eleventh, the chapter "General Plan of Treatment and of Diet Regulation" is written by two diabetologists, Leo Krall, an editor of the volume, and Elliott Joslin's son Allen. In the twelfth edition, published in 1985, the chapter on dietary management is written by four coauthors: three physicians, including Alexander Marble, president emeritus of the Joslin Diabetes Center, and a registered dietitian. In the thirteenth and fourteenth editions, the relevant chapters are authored only by a single dietitian.

of maintaining healthy blood sugar levels and how much insulin and other drugs should be used to do so. They would test their assumptions about drug therapy in a series of ever larger and more expensive clinical trials, but they would not do the same with diet. Rather, they assumed that the relationship between diet and heart disease had been established reliably by nutritionists and cardiologists. That the assumption might be wrong has been a minority position ever since, even as the evidence to support that possibility continues to accumulate.

Of all the reasons why the AHA, ADA, and other organizations wedded themselves so irrevocably to the dietary-fat/cholesterol thinking on heart disease, one observation stands out: this particular diet-heart hypothesis came first. For the better part of a century, no competing hypotheses existed, and that was more than enough time for the parties concerned to convince themselves they had the right answer.

It also helped that the hypothesis was based on a simple and straightforward observation, one that dated to the mid-nineteenth century, at least. The great German pathologist Rudolf Virchow, among others, saw that fat and cholesterol constitute a significant part of the atherosclerotic lesions, the plaques themselves, and that these two components of the blood were somehow invading the wall of the artery, the intima. In 1910, the German chemist (and later Nobel laureate) Adolf Windaus reported that plaques in the coronary arteries contained six or seven times as much cholesterol as could be found in the lining of a healthy artery. It was easy to imagine that an excess of cholesterol in the diet could lead to an excess in the blood, in serum cholesterol, which in turn could induce this accumulation of cholesterol in the plaques themselves. Three years later, a young Russian physiologist, Nikolai Anitschkow, demonstrated that he could induce atherosclerotic-like plaques in rabbits by adding cholesterol to their chow. "There can be no atherosclerosis without cholesterol," Anitschkow proclaimed, endowing the cholesterol hypothesis with the aura of experimental support. Now physicians began to consider it very likely right.

By 1924, when Ludwig Aschoff, the most influential German

pathologist of the post-Virchow era, toured the United States to lecture, he declared unconditionally that atherosclerosis was caused by unnaturally elevated levels of cholesterol and fats in the blood, responding to some aspect of the fat or cholesterol in the diet. "There is no doubt in my mind," Aschoff said. Joslin found Aschoff's reasoning sufficiently convincing that he quoted at length from Aschoff's lectures in his 1928 textbook. It "may or may not be the correct theory," Joslin wrote, "but it certainly applies to the conditions found in diabetes."

That cholesterol was the easiest of the blood fats to measure also provided an opportunity for physician researchers to buy into the hypothesis. By the mid-1930s, using a method developed by Rudolf Schoenheimer and his biochemist colleague Warren Sperry, physicians could test serum cholesterol levels in their patients with relative ease. Joslin and his colleagues used the level of serum cholesterol, which they could measure, as a surrogate for that of all blood fats, which they could not. Joslin justified this decision on the basis that the percentage of cholesterol "in the blood varies fairly consistently with the total fat in the blood."

When Joslin became aware in the late 1920s that his young diabetic patients, now saved by insulin from diabetic coma, were suffering prematurely from arteriosclerosis, he, too, focused naturally on cholesterol in the blood and in the diet as the most likely explanation. After all, until he had embraced Frederick Allen's thinking on the dangers of fat-rich diets; a fat-rich diet was what he had been advocating for his patients, including those who developed severe arteriosclerosis. Before insulin, those patients would often die of diabetic coma, which was associated with high levels of fat in the blood (diabetic lipemia). Joslin came to believe that both the lipemia and the coma were caused by the fat his patients were told to eat. His suspicions were reinforced by the close association of heart disease and diabetes with obesity in his older patients. Joslin found it easy to accept that these patients had accumulated too much body fat because they ate too much fat. Here, too, dietary fat could be implicated in the complications of the diabetic condition.

This path from cholesterol or fat in the diet to cholesterol or fat in the blood and then in the plaques themselves was another loose assemblage of assumptions and speculations, but the diabetologists of the era had no other credible hypotheses to consider. "Can it be that the prevalence of arteriosclerosis in diabetes is to be attributed to the high-fat diets we have prescribed and more especially to those diets having been rich in cholesterol?" asked Joslin in the 1928 edition of *The Treatment of Diabetes Mellitus*. "I suspect this may be the case."

By 1933, Joslin was acknowledging that the evidence was equivocal. Physicians had already documented patients dying of heart disease with low levels of serum cholesterol, and other patients thriving with high cholesterol. "Old arteriosclerotic diabetics as a rule do not exhibit a high blood cholesterol," Joslin acknowledged. "One cannot claim that the sclerotic plaques in the blood vessels have been deposited because of an increase of cholesterol in the blood. Indeed, in one study [it was] found that the cholesterol in the blood was lowest in those patients examined who showed the most marked arteriosclerosis." But he still couldn't put aside his suspicions that fat-rich diets caused heart disease. He could not connect the two "as closely as [he] had hoped," he wrote, but he assumed that given time and sufficient research the connection would be made.

The cholesterol hypothesis was never without its skeptics. Cholesterol is only found in foods of animal origin, those skeptics noted, and rabbits are herbivores—plant-eaters—that would never consume cholesterol-rich diets naturally, let alone the relatively large cholesterol doses that Anitschkow had given his rabbits. The disease produced in these rabbits was also distinctly different from that in man. "The condition produced in the animal was referred to, often contemptuously, as the 'cholesterol disease of rabbits,'" as the Harvard pathologist Timothy Leary* described it in 1935 (while still describing the disease in humans as "cholesterol overdosage"). Pathologists and biochemists—including Warren Sperry, co-inventor with Schoenheimer of the serum cholesterol assay—were examining human arteries removed at autopsy ("from

---

* Not to be confused with the Harvard *psychologist* Timothy Leary, famous for his promotion of LSD and hallucinogen research several decades later.

persons who had died suddenly, in most instances as a result of automobile accidents," as an editorial in *The Journal of the American Medical Association* explained) and reporting no correlation between cholesterol in the blood and the extent of atherosclerosis.

In 1936, when Russell Wilder of the Mayo Clinic assessed the evidence for his annual review of recent research in nutrition and metabolism, he noted that the arteriosclerosis produced in rabbits by feeding them cholesterol was always accompanied by cholesterol deposits in other organs and tissues. It suggested that the otherwise herbivorous rabbits, unaccustomed to dealing with cholesterol naturally, had a kind of storage disease of little relevance to humans. Moreover, Wilder noted, "a uniform failure has accompanied all attempts to produce arterial lesions by feeding cholesterol to animals the diets and cholesterol levels of which are more comparable to those of man: cats, dogs, foxes and monkeys. This negative evidence is more important for application to human conditions than is the positive evidence from the experience with rabbits." His conclusion: the existing evidence linking diet to heart disease via the cholesterol levels in the blood "would hardly influence an opinion not already prejudiced in favor of the idea."

In 1950, the cholesterol hypothesis came as close as it ever would to extinction. The University of Minnesota physiologist Ancel Keys reported that he had confirmed in men what Schoenheimer and Rittenberg had reported a dozen years earlier of laboratory animals: the blood level of cholesterol seemed to be mostly the product of the synthesis of cholesterol in the body, independent of how much cholesterol is actually consumed. Serum cholesterol levels could be lowered by cholesterol-free or virtually fat-free diets, as Keys and his colleagues now reported, but the serum cholesterol levels seemed to change little otherwise when subjects were fed the kinds of diets humans might actually eat. And, as Wilder had implied, Anitschkow's rabbit experiments had spawned a wave of research to see whether atherosclerotic lesions could be induced in other animals eating fat-rich or cholesterol-rich diets. The evidence continued to be uncompelling. Researchers would establish that they could induce human-like lesions in chickens and in (some strains of) pigeons (but not others), but those were among the very few species in which they succeeded. They also learned that such lesions appear naturally in pigs, cats, dogs, horses, cattle, marsupials, rats, sea lions, and a variety of primates. ("Since many herbivorous animals develop a significant degree of arteriosclerosis, diet apparently has little to do with the onset of the naturally occurring disease in most animals," as a comprehensive review of this evidence by University of California researchers asserted in 1963.)

Most relevant to the study of diabetes, even rabbits would not get atherosclerosis on cholesterol-rich diets if the rabbits were diabetic. This was first reported in 1949 by researchers at McGill University in Montreal led by the cardiologist Lyman Duff. In a series of publications over the next five years, Duff and his colleagues reported that they had induced diabetes in rabbits by injecting them with a chemical called alloxan that damages the insulin-secreting beta cells in the pancreas. Then they fed the rabbits cholesterol-rich diets, but to no noticeable effect. The diabetic rabbits developed atherosclerotic-like lesions only if they were given the high-cholesterol diets along with the insulin that they were otherwise lacking. Duff's experiments suggested that insulin was as critical to the formation of plaques—at least in rabbits—as the cholesterol-rich diets, but this observation would have little influence on how the science played out.

In the decade following the Second World War, social and political forces transformed both the focus and the context of the relevant research. The demographics of death and disease (mortality and morbidity) had changed dramatically in America over the first half of the twentieth century. Premature deaths from nutritional deficiencies and infectious diseases—tuberculosis, influenza, pneumonia, bronchitis—and parasitic infections had been largely eradicated by effective public health programs. Deaths in childbirth and infant mortality rates had been greatly reduced. As a result, more and more Americans were living long enough to die of the chronic diseases of old age—cancer and heart disease, most notably. In the postwar years, diabetes was still a relatively rare disease, afflicting, by Joslin's estimate, as few as one in every four hundred

Americans, while heart disease had emerged as the most common cause of death, killing one in every three to four. Heart disease had become the focus of attention of a public health establishment that was shifting its priorities in response.

This was the "newer public health," as described in 1953 by Ancel Keys in one of the articles that launched his rise to prominence. If a large proportion of the population was suffering and dying from chronic disease—heart disease and cancer, again, most notably—and there were good reasons to believe that situation could be remedied, Keys argued, then medical researchers and the government itself had an obligation to solve the problem and translate that solution into public health programs.

This was the rationale for much of the research that followed, as chronic disease research transitioned from physicians addressing the symptoms of these diseases to a public health perspective and national movements to prevent or delay their onset. As the focus shifted, funding became available to do the necessary scientific studies. At Harvard, for instance, the nutrition department was founded in 1942 by Fred Stare, who had no relevant training. "We were at war," Stare later recalled, "and we, like most of the other research labs, were interested and able to study problems with some relation to the war. . . . When the war was over, we wondered what the hell to do. We thought why don't we see if there's a relationship between nutrition and heart disease, and so we started working on that."

As politicians and health activists became aware of the rise of heart disease and cancer as major agents of premature death, they responded by advocating for federal programs dedicated to research, prevention, and treatment. Congress had taken the first step in 1937 when it authorized funding for a National Cancer Institute and a National Institute of Health. The latter became the National Institutes of Health in 1948 with congressional authorization for a National Heart Institute (which eventually evolved into the National Heart, Lung, and Blood Institute, the NHLBI, as it is today). The quarter of a century that followed were the "golden years" of NIH expansion, as an official history describes it, and NIH grants for medical research would jump from a few million dollars a year in 1947 to $100 million a decade later, and more than $1 billion by the mid-1970s.

The postwar years also saw the rise to prominence of the other institution that would play a critical role in how this science evolved: the American Heart Association. Founded in 1924 as a private organization for doctors, the AHA transformed itself after the war into a national health agency dedicated to raising money for heart disease research and disseminating information and guidance based on what this research revealed. In 1948, the AHA held its first nationwide fundraising campaign—"Listen to Your Heart"—aided by thousands of volunteers, including some of the most famous celebrities of the era. The AHA succeeded in widely disseminating the awareness that Americans were more likely to die of heart disease than any other illness.

With these institutions now funding the science, a new generation of researchers emerged in the United States to study heart disease, with the cholesterol hypothesis as the dominant, if not, for many of them, the only explanation that might link the American diet or lifestyle to heart attacks. The tools they would employ to do this research were new and of unknown value, particularly epidemiologic surveys, which had been used in the past only to help identify the causes of infectious diseases (cholera, most famously) and deficiency diseases like scurvy or pellagra. Those diseases were relatively rare and progressed noticeably over the course of weeks or months; the chronic diseases that would now be studied were all too common and manifested observable symptoms only over the course of years and decades. These researchers made up the protocols and methods as they went along. They had little idea of what they were doing, or how best to do it, and as their critics would point out in the decades that followed, they had little to no idea of all the many ways the accumulating evidence from these surveys might be misinterpreted.

They could justify their naïveté, as they did, by saying they had no apparent alternative. With almost a million Americans dying each year from heart disease, the researchers argued that the rigorous, meticulous experiments considered by investigators in harder sciences and even by some of their own colleagues to be an essential requirement for establishing reliable knowledge was a luxury that they could not afford. They had to establish their best sense of the science and act on it. As they did, the dietary-fat/cholesterol hypothesis took on a life of its own and became essentially impervious to any evidence against it.

Once again, a single individual—in this case, Ancel Keys—gets credit for promoting the hypothesis and working diligently to assure its acceptance. Unlike Joslin, Keys did not become the dominant influence in the field because he was the physician with the most clinical experience treating the disease (Keys had two doctorates in science but no medical degree) and the writer/editor of seminal textbooks. Rather, as with Louis Newburgh in the science of obesity, he seized upon the early days of a new field of research, with seemingly complete faith in his scientific intuition and the strength of will to forcefully defend his arguments against all who doubted him. He promoted his ideas neither tentatively nor cautiously, as the best scientists would argue was necessary. Both Keys's allies and his critics, of which he had many, would describe him as uncompromising and relentless, the kind of individual who is said not to suffer fools gladly, but who will also define foolishness as the quality of holding opinions in conflict with his own. Keys, in fact, would be wrong about numerous important issues, including the role of lipoprotein fractions (LDL and HDL) and of saturated and unsaturated fats in influencing serum cholesterol levels.

Eventually Keys would seem to question whether his dietary-fat/cholesterol hypothesis was right, or at least how relevant it was to human health—"I've come to think that cholesterol is not as important as we used to think it was," *The New York Times* quoted him saying in 1987. But by that time the U.S. government was in the midst of its public health campaign to lower the cholesterol levels of the nation and, with it, the diabetic diet had also changed irrevocably.

Keys was a remarkable man with an extraordinary breadth of life experience. He had been recognized as intellectually gifted in his youth—he was among those children famously studied by the University of California, Berkeley, psychologist Lewis Terman in his Genetic Studies of Genius project—and during his school years had shoveled guano in a bat cave, worked in a coal mine and a lumber camp, as a clerk at Woolworth's, and, between his undergraduate years at the University of California, Berkeley, on an oil tanker that took him to China and back. His first PhD, also from UC Berkeley, was in oceanography and biology, and his early research and publications were almost entirely on the physiology of fish. Only after he traveled to Europe in the early 1930s, studied with the Nobel laureate August Krogh, and then obtained a second doctorate at Cambridge University in physiology, did he begin studying humans. In 1937, while a Harvard researcher, he organized and carried out an expedition to the Chilean Andes, spending three months working at elevations between 10,000 and 20,000 feet studying the ability of humans to adapt to high altitudes.

After the Andes expedition, Keys moved to Minnesota: first the Mayo Clinic, which he declared "awfully provincial," and then the University of Minnesota, where he founded the Laboratory of Physiological Hygiene and developed the emergency rations for army troops in the field known as K-rations (the K apparently stood for "Keys"), which would be used widely in the coming war. During the war years, he carried out a celebrated study of human starvation using conscientious objectors as his subjects and publishing the results, along with much of the world's accumulated knowledge on famine and starvation, in what became the seminal publication in the field: *The Biology of Human Starvation,* two volumes, more than 1,300 pages long. By 1961, when Keys was on the cover of *Time* as the face of nutrition science in America, he was quoted saying that he considered it his scientific obligation "to find out why people got sick before they got sick."

His focus on heart disease, as with virtually all of the influential researchers of the era, emerged only in the postwar years. "The news in the American public press was no longer about the war and its political and economic aftermath," Keys recalled in a short memoir written half a century later. "Among reports of new events were increasing notices of executives dying from heart attacks. . . . Middle-aged men, seemingly healthy, were dropping dead. There was talk of the stress of being an executive. But what was the difference between those attacked and those who stayed well?"

In 1951, Keys took a sabbatical year to study at Oxford University. While attending an international conference in Italy, as he later told this story, he realized that his European colleagues considered heart disease of little concern in their populations. When a professor at the University of Naples invited Keys to come visit, Keys and his wife, Margaret, loaded their car "with apparatus for measuring serum cholesterol and headed for Naples." Margaret Keys was a medical technician who had become adept at such measurements, and that's what they would measure. In Naples and then in Madrid on the same trip, they assessed the prevalence of heart disease by speaking with local physicians and visiting hospital wards, and Margaret Keys took blood samples from members of the working class and the wealthy and measured the cholesterol content. Then they linked what they observed to any noticeable differences in diet. In both cities, they saw the same pattern: heart disease, high serum cholesterol, and fatty diets among the wealthy, but none of the three among the working class. Keys would acknowledge the preliminary nature of the research—amateurish by today's standards—but he nonetheless found the associations they had observed to be compelling.

He first publicly aired his hypothesis at a 1952 conference in Amsterdam, later describing his proposed chain of causality as "fatty diet, raised serum cholesterol, atherosclerosis, myocardial infarction [heart attack]." Few attendees at the conference, Keys recalled, took him seriously. He knew he was speculating—"direct evidence on the effect of the diet on human atherosclerosis is very little," he would admit in one of these early presentations, "and is likely to remain unsatisfactory for a long time"—but the lack of direct evidence did not deter him. His articles (and the lectures on which they were often based) ended with a plea for further research of the kind he had just done, epidemiological investigations into possible associations between diet and heart disease, but also the assertion that "public health programs must take cognizance of the information already at hand." Then he would suggest that Americans would benefit from avoiding much to most of the fat they consumed.

Through the 1950s, Keys's research took him around the world, often accompanied by the Harvard cardiologist Paul Dudley

White, the most influential heart disease specialist of his generation.* "All doors of medical schools and hospitals were open to him," Keys wrote, and so he and Margaret "trail[ed] along taking measurements, including serum cholesterol, on samples of men in the populations." Their travels took them to Africa, back to Italy, to Scandinavia, Japan, Hawaii, and Los Angeles. At each stop, Keys felt he had found at least some evidence that the fat-cholesterol relationship was integral to the genesis of heart disease. In Uganda, for instance, a pathologist showed them a display of 200 hearts removed at autopsy, 198 of which had clean arteries; two had "extensive atheroma." The pathologist told them that those two "were the hearts of butchers whose pay for their work was the offal, the entrails, which formed a part of their diets." In eastern Finland, Keys and his party visited a remote logging camp and were told that these hardworking lumberjacks were plagued by heart disease. They shared a snack with the loggers: "slabs of cheese the size of a slice of bread on which they smeared butter," Keys wrote; "they washed it down with beer. It was an object lesson for the coronary problem." Keys would later suggest that "the habit of smoking cigarettes was the major risk factor for cardiovascular and all-cause deaths" in this population, but that, too, was decades after his cholesterol/dietary-fat hypothesis had been widely embraced.

As Keys published new evidence for his hypothesis, or variations on the same evidence while making his argument in different journals, he, too, as I said, generated his share of formidable critics, questioning both the quality of his evidence and his interpretation. In 1957, the UC Berkeley biostatistician Jacob Yerushalmy, writing with New York State health commissioner Herman Hilleboe, accused Keys of dealing "uncritically or even superficially" with data—specifically, associations in vital statistics between mortality from heart disease and fat consumption—that might ultimately be "worse than useless."

That same year, the American Heart Association published a sixteen-page assessment of the evidence, coauthored by five of

---

* White would be called in to consult when President Dwight Eisenhower had his first heart attack in 1955.

the leading cardiologists and nutritionists in the country. They described researchers—presumably including Keys, if not Keys specifically—as taking "uncompromising stands based on evidence that does not stand up under critical examination." Their conclusion was that the evidence linking dietary fat to heart disease was still equivocal: "difficult to disentangle . . . from caloric balance, exercise, changes in body weight and other metabolic and dietary factors that may be involved." As such, they argued, no public health recommendations could or should be given, although they could understand why physicians might want to discuss with individual patients their best sense of the data.

Just four years later, in 1961, the American Heart Association reversed itself and embraced the position it would take from then on. Rather than assume that a medical association should first do no harm, should not disseminate public health advice based on "evidence that does not stand up under critical examination," as the AHA's authorities had argued in 1957, the AHA did just that, now publishing a two-page report promoting Keys's philosophy. The authors were a six-member ad hoc committee that included Keys and the Chicago cardiologist Jeremiah Stamler, who would become, perhaps even more so than Keys, the most outspoken proponent of the dietary-fat/cholesterol hypothesis. The underlying science had remained unchanged in the intervening years and no new evidence had been collected, but now Keys, Stamler, and their coauthors argued that "the best scientific information available at the present time" strongly suggested that Americans would reduce their risk of heart disease by reducing the fat in their diets, or replacing the saturated fats with polyunsaturated fats.

This was the first time the American Heart Association would invest itself publicly in the hypothesis that heart disease is caused by the fat we eat, the first public health recommendation by the organization about how Americans should eat to prevent heart disease. By taking sides publicly on a very active scientific controversy, the AHA was committing itself to a position, despite the complete absence at the time of clinical trials that might confirm it was right.

In the decade that followed, journalists came to see the AHA as a credible and unbiased source for information on heart disease and diet—a *New York Times* article on the AHA's 1961 report described Keys's ad hoc committee as "the highest medical and scientific body of the world's most authoritative institution in the field of heart disease." Influential newspapers and magazines published articles on heart disease or on the ongoing research and correctly reported that insufficient evidence existed to conclude that dietary fat and serum cholesterol were the drivers of atherosclerosis, but they would assume, as apparently did the researchers quoted (and just as Joslin had thirty years earlier), that given time and sufficient effort that evidence would surely be found, an assumption diametrically opposed to the skepticism that is institutionalized in the practice of science itself.

As the media took to reporting that the hypothesis was almost assuredly true, the AHA guidance to eat low-fat diets became ever more unconditional. By 1970, the AHA was advocating low-fat, high-carbohydrate diets not just for those high-risk men who had already had heart attacks or smoked cigarettes, but to virtually everyone in the nation, "including infants, children, adolescents, pregnant and lactating women and older persons." As the AHA guidelines and the media became ever more assertive, government organizations—most notably the U.S. Congress and the U.S. Department of Agriculture—in turn felt a need to pass on the same guidance.

Clinical trials had still not confirmed even that men at imminent risk of a heart attack would extend their lives by eating this way. But the hypothesis had effectively been transformed into dogma, nonetheless. Those who believed the hypothesis to be accurate continued to assess the evidence as Keys had, embracing those results that justified their belief, explaining away the results and observations that suggested otherwise.

In the late 1950s, physicians in the United States and Europe had begun conducting and reporting on increasingly elaborate and expensive clinical trials designed to do what Keys himself never would, which was to put his hypothesis to experimental tests: specifically, that a diet that lowered serum cholesterol levels would prevent heart disease and, more important, extend the lives of those who ate it.

To interpret the significance of these trials requires understanding the distinction between a low-fat diet and a cholesterol-lowering diet. The American Heart Association was recommending Americans eat a low-fat diet, which replaced a proportion of the fat with carbohydrates. This is what Keys had been arguing for beginning in 1952, the AHA in 1961, and which the American Diabetes Association would embrace in 1971. The National Institutes of Health began disseminating the same low-fat-diet prescription in 1984. They did this despite the knowledge that cholesterol levels in the blood respond not to the total amount of fat in the diet but to the chemical nature of these dietary fats. Saturated fats—the primary fat in butter, for instance—elevate cholesterol levels (compared to carbohydrates) while unsaturated fats, as in vegetable oils, tend to lower them. Cholesterol-lowering diets replace saturated fat with unsaturated fats. Low-fat diets, which replace saturated and unsaturated fats with carbohydrates, would not have the same effect.

The experts writing the guidelines for these public health organizations were aware of these distinctions but worried that many Americans would be unable to grasp them. Conveying the idea that we should all eat less fat, and so low-fat foods, was thought to be a simple message that the public could understand. The authorities writing the guidelines also assumed that Americans would lose weight if they ate less fat (based on the idea that there are nine calories in a gram of fat, compared to only four in carbohydrates and protein) and if Americans could be induced to avoid fatty foods, they would be eating fewer saturated fats in the process. Since the AHA inaugurated the practice of disseminating dietary guidelines in 1961, these guidelines have focused primarily on reducing total fat consumed rather than entering into the weeds of types of fat.

By 1984, though, when the NIH declared a consensus of opinion that the fat/cholesterol hypothesis of heart disease was correct and prescribed a *low-fat* diet for all Americans, only two small clinical trials had actually tested that specific prescription. A Hungarian trial, published (in Hungarian) in the journal *Therapia Hungarica*, suggested that men who had already had a heart attack might delay a recurrence by eating a very-low-fat diet. A British study, published in 1965 in *The Lancet,* a British journal, suggested the opposite: "A low-fat diet has no place in the treatment of myocardial infarction [heart attack]," the authors reported.

As for trials of *cholesterol-lowering* diets, the results from half a dozen would be published over the years providing evidence that could charitably be called ambiguous: two studies (the Anti-Coronary Club Trial in New York City and the Los Angeles Veterans Administration Trial) concluded that subjects eating a cholesterol-lowering diet had less heart disease than those eating their usual fare, but also observed that they would die just as soon if not sooner by doing so. In the Veterans Administration Trial, eating the cholesterol-lowering diet was associated with a decrease in deaths from heart disease, but an increase in cancer deaths and no difference in mortality. Indeed, when the trial results were published, the principal investigator, Seymour Dayton, worried publicly about diets that fed significant amounts of unsaturated fats—seed oils, for instance. "Was it not possible," Dayton asked, "that a diet high in unsaturated fat . . . might have noxious effects when consumed over a period of many years? Such diets are, after all, rarities among the self-selected diets of human population groups."

A Finnish trial conducted in two Helsinki mental hospitals and published in 1972 concluded that men who ate the cholesterol-lowering diet had fewer heart attacks and lived slightly longer than they otherwise might have; for women, the diet had no consistent benefit. A trial with a similar design conducted in Minnesota mental hospitals and rehabilitation homes, and on which Keys was originally a co-investigator, found that the cholesterol-lowering diet may have actually shortened the lives of the men who consumed it. Results from the Minnesota trial were available by 1975 but the details published only sixteen years later, after the principal investigator, Ivan Frantz, Jr., retired and without Keys's participation as an author. The existing evidence linking diet to heart disease via the cholesterol levels in the blood, as Russell Wilder might have put it, would still "hardly influence an opinion not already prejudiced in favor of the idea."

When the NIH had decided in 1970–71 not to fund a $1 billion National Diet-Heart Study, despite its experts declaring the urgent necessity to do so, it had proceeded instead with two clinical trials that could test hypotheses related to the fat-cholesterol question. The two trials together would cost more than a quarter of a billion dollars and would take a decade to carry out.

The first to be completed was the Multiple Risk Factor Intervention Trial, known as MRFIT (pronounced "Mister Fit"), which cost $115 million. Its goal was to test whether heart disease in men could be prevented if the three primary risk factors for heart disease that had been identified by the famous Framingham Heart Study—serum cholesterol, blood pressure, and cigarette smoking—were all addressed simultaneously. Researchers would later describe the trial as "throwing the kitchen sink" at heart disease. Twelve thousand men were recruited, all at high risk of having a heart attack—cigarette smokers, for instance, with high blood pressure and serum cholesterol levels in the top 3 percent of the population. Six thousand were prescribed multiple interventions: a cholesterol-lowering diet, a blood pressure medication for hypertension, and smoking cessation classes if they smoked cigarettes. The researchers tracked the health of these men for an average of seven years and compared them to another six thousand men at equally high risk who had received no such advice or guidance.

The MRFIT researchers published their results in 1982. *The Wall Street Journal* headline above its story on the study captured the conclusion succinctly: "Heart Attacks: A Test Collapses." As it turned out, those men who had received no guidance experienced fewer deaths than the men who had been in the multiple intervention group. The researchers considered three possible explanations for why the results were so disheartening and concluded that "some aspect of the intervention program has a deleterious effect on mortality," most likely the medication prescribed for high blood pressure. The results of the MRFIT trial could then be considered irrelevant, as they were, to whether or not the fat-cholesterol hypothesis itself was true.

The second trial, known as the Lipid Research Clinics (LRC) Coronary Primary Prevention Trial, cost $150 million, by far the most expensive clinical trial conducted until then. It tested the proposition that a cholesterol-lowering drug known as cholestyramine given to men considered very likely to suffer a heart attack—

again, those with cholesterol levels in the top 3 percent of the population—would not only prevent heart disease but also extend their lives. The researchers followed their 3,800 subjects for eight years. They published their results in January 1984: the 1,900 men who had been randomly assigned to take cholestyramine experienced fewer heart attacks than the 1,900 men who had not, and had fewer heart disease deaths as well. The researchers published the trial results as a ringing endorsement of the fat-cholesterol hypothesis and suggested that its implications should be extended widely. The results, they wrote, left "little doubt of the benefit" of cholesterol-lowering therapy.

The LRC study was a "landmark," Antonio Gotto, president of the AHA, told *The New York Times*, and the results had been "anxiously awaited." Now, as the NIH administrator Basil Rifkind would later describe it, he and his colleagues at the National Institutes of Health would take their leap of faith, using the results of the LRC trial to assert that the fat-cholesterol hypothesis had been right all along, and that Americans, from age two on up, should be urged to eat fat-restricted diets.

"It is now indisputable," Rifkind said in *Time* when the LRC results were released in January 1984, "that lowering cholesterol with diet and drugs can actually cut the risk of developing heart disease and having a heart attack." The *Time* article reporting on the LRC trial was headlined "Sorry, It's True. Cholesterol Really Is a Killer," and the article about a *drug* trial began, "No whole milk. No butter. No fatty meats. Fewer eggs." Three months later, *Time* ran a follow-up cover story—"Hold the Eggs and Butter"—quoting Rifkind saying that the trial results "strongly indicate that the more you lower cholesterol and fat in your diet, the more you reduce your risk of heart disease." The article also quoted Gotto saying that if everyone went along with the goal of properly lowering their cholesterol, "we could look forward to the time when atherosclerosis is conquered."

These assertions, though—Rifkind's leap of faith—came with multiple untested assumptions, the most obvious being that the cholesterol-lowering benefits of a drug applied to diet as well. And if they did, that there would be no unforeseen consequences—no risks or long-term complications—from a diet that attempts to achieve the same physiological effect. Another obvious question was whether the benefits observed in a single study of middle-aged men with extremely high cholesterol levels "could and should be extended to other age groups and women and . . . other more modest elevations of cholesterol levels," as the LRC investigators asserted.

The skeptics, of whom there were now many, pointed out that the cholestyramine taken by the men in the study may have helped prevent heart disease and even deaths from heart disease, but not the likelihood of actually dying prematurely, which would be death from any or all causes: sixty-eight of the men taking the cholesterol-lowering drug had died during the study, only three fewer than those in the control group. In other words, the use of a cholesterol-lowering drug for eight years by men with *very* high cholesterol had reduced their risk of dying prematurely by less than two-tenths of a percent, "no whisper of benefit," as described by the University of Chicago biostatistician Paul Meier. Calling the LRC results "conclusive," as Rifkind had, Meier said, constituted "a substantial misuse of the term." Thomas Chalmers, dean of the Mount Sinai School of Medicine and among the most respected authorities on the statistical analysis of randomized controlled trials, told the journal *Science* that the LRC investigators had perpetrated "an unconscionable exaggeration" of the data. Both Chalmers and Meier were members of the NIH data monitoring committee, but their critical comments had little effect. The NIH now declared that a consensus existed that dietary fat caused heart disease, and that Americans—men, women, and children—should avoid it.

With publication of the results from the cholestyramine trial, the NIH launched what Robert Levy, a former director of the National Heart, Lung, and Blood Institute, had called a "massive health campaign" to continue the job of convincing the public of the benefits of eating less fat and lowering serum cholesterol levels. A series of reports were released—from a newly established National Cholesterol Education Program at the NIH, from the Surgeon General's Office and the National Academy of Sciences—diligently reiterating the evidence that supported the fat-cholesterol hypothesis and, as had become customary, mostly omitting the evidence that did not. The surgeon general's report, published in 1988, blamed two out of every three premature deaths in America on fatty diets. A cover letter by Surgeon General C. Everett Koop, who had been so active in warning against the health risks of smoking, asserted that "the depth of the science base . . . is even more impressive than that for tobacco and health."

Regrettably it was not. Since the mid-1980s, the accumulating evidence has continued to challenge the validity of the fat-cholesterol hypothesis. While cardiologists have fully embraced cholesterol-lowering drugs (a class of drugs known as statins, specifically) to prevent heart disease, researchers have been unable to generate anything close to unambiguous evidence that saturated fat or even serum cholesterol levels are a cause of heart disease, and not for lack of trying. By 1991, for instance, epidemiologic surveys in populations had revealed that high cholesterol was *not* associated with heart disease or premature death in women. Rather, the higher the cholesterol in women, the longer they lived, a finding that was so consistent across populations and surveys that it prompted an editorial in the American Heart Association journal, *Circulation*: "We are coming to realize," the three authors, led by UC San Francisco epidemiologist Stephen Hulley, wrote, "that the results of cardiovascular research in men, which represents the great majority of the effort thus far, may not apply to women."

The sex bias of the research extended far beyond heart disease—virtually all meaningful clinical trials of chronic disease prevention in the United States had been done in men. That prompted the NIH in the early 1990s to launch what remains the largest clinical trial of diet ever done, the Women's Health Initiative (WHI) Dietary Modification Trial, with only women as subjects. The researchers hoped to determine whether a low-fat, fiber-rich diet would prevent breast cancer. The trial cost at least half a billion dollars. (Different sources provide different estimates.) The researchers randomized twenty thousand women to consume a diet with only half the fat content of what Americans typically ate, replacing the fat with fruits, vegetables, whole grains, and fiber. Another twenty-nine thousand women were given no dietary counseling, and the two groups of women were then followed for seven years to assess the risks and benefits. The results were published in 2006. Not only did the low-fat, high-fiber diet fail to prevent breast cancer, it also had no beneficial effect on heart disease or, for that fact, body fat, diabetes, or any other disease or disorder.

Once again, though, the results of the trial failed to influence the multitude of health organizations and government organizations that had embraced the fat-cholesterol hypothesis and were disseminating its dietary implications as public health guidelines. These authorities took the position that the trial had simply failed to get the correct answer. The World Health Organization, for instance, released a statement in response to the trial with the headline "The World Health Organization notes the Women's Health Initiative Diet [*sic*] Modification Trial, but reaffirms that the fat content of your diet does matter." The trial, of course, had suggested the opposite.

The latest reviews of the evidence continue to reiterate how equivocal it remains. When the American Heart Association assembled a dozen authorities to publish a 2017 review—an "AHA Presidential Advisory on Dietary Fats and Cardiovascular Disease"—the authors, many of them long-time proponents of the fat-cholesterol hypothesis, were no longer addressing the issue of whether avoiding fat in the diet would make anyone live longer, but only whether or not it might prevent heart disease. They acknowledged that only four clinical trials had ever satisfied their criteria for adequately testing the hypothesis, all dating to the 1960s and early 1970s, and had produced at least some evidence that the hypothesis was right.* When they treated the data from those four trials together—

---

* These included the Los Angeles Veterans Administration trial; a British trial published in 1968 in men who had already had heart attacks and replaced the saturated fat in their diets with soybean oil; the Finnish mental hospital study; and a study in Oslo carried out by a physician who counseled his subjects to avoid saturated fat and eat more unsaturated fats (from fish, for instance). The Norwegian findings were compromised by the fact that the physician's patients reduced their sugar consumption by almost half, which could have been, among many other possibilities, why the diet seemed to prevent heart disease.

a *meta-analysis*—they concluded that diets that replaced saturated fat with unsaturated fats could reduce heart disease risk, albeit not those that replaced saturated fat with carbohydrates, which was the essence of the AHA (and U.S. government) dietary guidelines for all those years. The AHA authorities also found reasons to exclude from their analysis all those trial results, including those of the Women's Health Initiative, that did not align with their thinking. "Readers may wonder," they added, "why at least 1 definitive clinical trial has not been completed since then [the 1960s]," and they reiterated all the reasons why, but did not mention that the absence of such a definitive trial should still leave their interpretation open to question. Three years later, a collaboration of a dozen international authorities published a "State-of-the-Art" review in the *Journal of the American College of Cardiology* proposing a halt to public health recommendations calling for the restriction of saturated-fat-rich foods. "The totality of available evidence does not support further limiting the intake of such foods," they concluded.

Since the late 1990s, the most authoritative reviews on the benefits and risks of medical interventions such as low-fat or cholesterol-lowering diets have been published under the imprimatur of an international organization known as the Cochrane Collaboration, which was established to provide a standardized methodology that would minimize the effect of bias. Implicit in the Cochrane Collaboration's existence is the assumption that a considerable number of medical authorities believe analyses like those of the AHA are unreliable, that authors will create a rationale for choosing the trials and results they consider meaningful, consciously or unconsciously, based on whether or not the trials resulted in outcomes that align with the authors' beliefs.

Since 2000, researchers from the University of Bristol in the United Kingdom have been using the Cochrane methodology to review the evidence for the benefits of low-fat or cholesterol-lowering diets. They have updated the reviews every half-dozen years, but their conclusions have changed little because no new meaningful trials have been conducted since the Women's Health Initiative published its results in 2006. In the 2020 edition of their review, the Bristol researchers suggested in their summary statement that the existing evidence is still sufficient to recommend low-fat diets for the population. But they also concluded that they could find "little or no effect of reducing saturated fat on all-cause mortality"—i.e., the chance of dying prematurely—or even on cardiovascular disease mortality, which would be the chance of dying prematurely of a heart attack. "The effects on total (fatal or non-fatal) myocardial infarction, stroke and CHD events (fatal or non-fatal) were all unclear," they added. The authors also stated that the data "are too limited to be able to answer [the] question" of whether or not the public health pronouncements to avoid dietary fat and specifically saturated fat have been beneficial.

Such an assessment is worrisome. The acceptance as true of such a critically important hypothesis for more than half a century, based on relentlessly equivocal evidence, had resulted in a reversal of direction in what the authorities consider the burden of proof. In the 1950s and 1960s, the absence of definitive or compelling evidence was cited as reason to avoid disseminating the dietary implications as public health advice. Harm could be done. By the 2010s, the same absence of definitive evidence was used to argue the opposite: that public health guidelines based on the hypothesis should continue to be given because now it was unknown whether ceasing to do so might cause harm.

One additional important clinical trial, testing the fat-cholesterol hypothesis, would never be included in these systematic reviews of the benefits of cholesterol-lowering diets because its subjects were overweight or obese patients with type 2 diabetes, and one goal of the trial was weight loss. These researchers will not include trials of weight-loss diets in their meta-analyses of the benefits of low-fat diets because they believe that if the participants in the trial do lose weight, that might confound the results. Any reduction observed in heart disease risk might be a result of the weight loss, not the fat content of the diet. In this case, however, the subjects in the trial both achieved and maintained weight loss *and* failed to show any benefit, suggesting that the trial is very relevant to this controversy.

This was the largest and most ambitious diabetes lifestyle trial ever done, the $200 million Look AHEAD (Action for Health in Diabetes) trial. Beginning in 2001, investigators from eighteen cites nationwide recruited more than 5,100 overweight or obese patients who were also suffering from type 2 diabetes. Half of them were assigned to what was then the standard of care for patients with diabetes, and the other half to an "intensive lifestyle intervention . . . which promoted weight loss through decreased calorie intake and increased physical activity." Those in the latter group were encouraged to work out five days a week, for three hours total. They were counseled to eat at most 1,500 calories if they weighed less than 250 pounds or 1,800 calories if they weighed more, and specifically to avoid eating fat (less than 30 percent of calories) and saturated fat. Those who still struggled to lose weight were given additional counseling and, if necessary, weight-loss drugs and exercise equipment and enrolled in supervised exercise programs and even cooking classes. This level and duration of intervention was unprecedented for a clinical trial, and certainly nothing like it had ever been conducted on patients with type 2 diabetes.

The results were published in July 2013 in *The New England Journal of Medicine*. The investigators had clearly demonstrated that they could get the men and women in their trial to lose what they considered significant weight and sustain the weight loss—almost 9 percent of their body weight on average at the end of the first year and 6 percent after eight years (although that would only be 16 and 12 pounds respectively for someone who weighed 200 pounds when entering the trial). The successful weight loss suggested, as well, that the investigators had succeeded in getting their patients to eat the diet they were counseled to eat. "Many important benefits" had resulted, the investigators reported, including better glycemic control, physical functioning, quality of life, and even modest reductions in sleep apnea, urinary incontinence, and symptoms of depression.

What it clearly did not do, though, despite the dietary restrictions, is reduce the risk of heart disease or strokes. The trial was stopped prematurely for "futility" after the patients, on average, had been following their intensive lifestyle intervention and maintaining their weight loss for a few months shy of ten years. Those patients suffered just as many heart attacks and strokes as those who had done little more than visit their physicians regularly and take their prescribed medications. Even more surprising, the patients randomized to ten years on the low-fat, low-saturated-fat, calorie-restricted diet had slightly *higher* LDL cholesterol at the end of the study than those who ate whatever they wanted, providing still more evidence against the benefits of a low-fat, low-saturated-fat diet, even when severely calorie restricted, and so evidence against the fat-cholesterol hypothesis of heart disease itself.

A simple explanation for what happened in the decades since Ancel Keys proposed the first iteration of the fat-cholesterol hypothesis is that public health authorities came to believe it was true because it filled a vacuum in our understanding of why Americans seemed so very likely to die from heart disease. It supplied a viable and seemingly reasonable answer to the question of what aspect of diet was responsible. But clinical trials testing the hypothesis failed so often and so consistently to confirm it because it is *not* the correct hypothesis.

Researchers approaching the problem from a different perspective, as we will see, would suggest alternative explanations that better explain the patterns of disease that were being witnessed, but once the American Heart Association and the National Institutes of Health had declared the fat-cholesterol thinking to be almost assuredly true, it would have to be dislodged—proved, in effect, not true—before these authorities would take any alternative explanations seriously. That was a far more difficult job than filling the knowledge vacuum in the first place, more so since the clinical trials necessary to test these competing hypotheses were not even being considered, let alone funded.

Ultimately, the fat-cholesterol hypothesis has been kept alive not just by this institutional inertia—researchers and public health authorities preferring to believe what they had always believed—but by an observation that is only tangentially related to diet: the seeming success of statin drugs in reducing morbidity and mortality from heart disease in high-risk patients. Since these drugs reduce levels of LDL cholesterol, public health and medical authorities continue to assume that fat-rich foods that raise LDL cholesterol (and do many other things as well, many of them known to be beneficial) are the foods that cause or increase our risk of heart disease and should be avoided.

Many reasons can be found to remain skeptical of this logic, foremost being the ever-dwindling evidence that saturated fat plays a meaningful role in this disease process. A related reason requires a reminder of the kind of observations that the fat-cholesterol hypothesis set out to explain. Among those that Ancel Keys found to be compelling evidence for the fat-cholesterol hypothesis was a dramatic decrease in the prevalence of heart disease reported in Europe during the Second World War, specifically in those populations that had suffered lengthy food shortages. Because Keys suspected that dietary fat caused heart disease, he interpreted this evidence to support his belief: the reduced availability of meat, eggs, and dairy products, and so the saturated fats they contain, was responsible. "A major lesson gained from World War II," Keys wrote in a 1975 review, "is the proof that in a very few years the incidence of CHD [coronary heart disease] could drop to a level of the order of one-fourth the preceding rate."* But the clinical trials, even those four that the AHA in 2017 had considered meaningful, had shown no such precipitous drop, nothing even close, over the course of similar time periods, "a very few years." Rather they had shown, at most, very subtle effects on the prevalence of heart disease and no effects on heart disease mortality.

While Keys and other proponents of the fat-cholesterol hypothesis had shown little interest in other possible explanations for the disease trends observed in the world wars, there were many. As other investigators noted, World War I and II had profound effects on the diets and lifestyles of the populations affected (gasoline shortages, for instance, left people to get around by foot or bicycle and so they were more physically active). Moreover, the prevalence of other chronic diseases had also dropped precipitously during the

* The German pathologist Ludwig Aschoff had made the identical observation about the decrease in heart disease prevalence in Germany toward the end of the First World War and in the postwar period. The accumulation of fat in the arteries, he suggested, "would entirely disappear in malnutrition, especially when there is a deficiency of lipoids [fats] in the diet."

conflicts—diabetes, most conspicuously. This happened even in the United States, far from the war zones, where people had plenty of food to eat and rationing was limited only to specific commodities. If the relative scarcity of dietary fats could not explain why the prevalence of these other diseases would also diminish with the wars, a seemingly obvious question to ask is, What could?

9

Good Science/Bad Science, Part III

How then can one with certainty determine which of two concurrent phenomena is cause and which effect, and whether either is in fact cause and both are not simultaneous effects of a third factor or, indeed, that each is not the effect of two quite distinct causes?

—RUDOLF VIRCHOW,
"On the Standpoints in Scientific Medicine," 1847

In any scientific endeavor, the questions that researchers ask when confronted with a problem or an observation will determine the answers they get. The technology available to quantify and observe a phenomenon will determine what questions can be asked, but the intuition of the researchers and their preconceptions often play the critical role.

In the early 1950s, Ancel Keys had set out initially to answer the question of why Americans seemed to be experiencing an epidemic of heart disease. To answer that question, he compared heart disease rates in populations in the United States, Europe, and Asia. And, as we've seen, because Americans ate more saturated fat and less unsaturated fat than the populations in Japan and the Mediterranean, both relatively free of heart disease, he thought he had his answer. The fat-cholesterol hypothesis then became the singular focus of the relevant medical research community in the United States from the late 1950s onward, and the *only one* that would be tested in any meaningful way. It would fail the tests, but that failure would be dismissed as a failure in the tests themselves rather than a failure of the hypothesis.

Physicians outside the United States, however, had been speculating since the late nineteenth century on a closely related observation: the inevitable appearance of a cluster of common chronic disorders in populations throughout the world as they became, in a word, westernized. This concept would later become known as a *nutrition transition*, although what is observed first is a disease transition, leading to the assumption that the diseases appearing are related to changes in diet and lifestyle that precede their appearance or are occurring simultaneously.

As populations throughout the world came into contact with Europeans and their descendants, and began to embrace European foods and lifestyle, they manifested the same chronic diseases and disorders that seemed to be increasingly common in Europe and those of European ancestry elsewhere—dental caries, appendicitis, obesity, diabetes, hypertension, heart disease, and even cancer as well as a cluster of bowel-related disorders from the seemingly benign (constipation and hemorrhoids) to fatal malignancies. This transition from traditional lifestyles and diseases to these western lifestyles and diseases accelerated worldwide in the nineteenth and early twentieth centuries, as did attempts to explain it.

British researchers had taken the lead on this question, perhaps because the British Empire had missionary and colonial hospitals scattered throughout the world, and this disease transition was observed in many of them. Physicians working in outposts far from England would report the relative absence in the local native populations of the common chronic diseases that afflicted Europeans, and even those Europeans living in these same outposts. As the local peoples, though, took to "living more and more after the manner of the whites," in the words of the physician and 1952 Nobel Peace Prize winner Albert Schweitzer, who had witnessed this disease/nutrition transition at his missionary hospital in equatorial West Africa, they began to suffer from the very same disorders.

Among the more dramatic observations documenting this kind of disease/nutrition transition was the difference in the diseases afflicting the Black populations in Africa in the early to mid-twentieth century and those of Black Americans sharing the same genetic ancestry but living many thousands of miles away in a dramatically different food environment. In the early 1950s, for instance, George Campbell, director of a diabetes clinic at the King Edward VIII Hospital in Durban, South Africa, noted that his patients fell into two distinct categories of diseases. The local whites suffered from diabetes, coronary heart disease, hypertension, appendicitis, gall bladder disease, and other chronic disorders. The rural Blacks (Zulus) did not. In 1956, Campbell spent a year working at a hospital in Philadelphia. He later wrote that he was "absolutely staggered by the difference in disease spectrum" between the Black population there and the Black population of South Africa. Among the Blacks of Philadelphia he saw the same chronic disorders that characterized his white patients in Durban.

Denis Burkitt, a missionary physician working as a cancer epidemiologist with Britain's Medical Research Council, made a similar observation in 1970 in a *Lancet* article entitled "Relationship as a Clue to Causation." Burkitt had worked at Mulago Hospital in Kampala, Uganda, from 1948 to 1966. He had then returned to London, where he established a network of hospitals in Africa that would report monthly to him on the patients they were admitting and the diseases from which they suffered. "Although atherosclerosis is still exceedingly rare in Africa," Burkitt wrote, "it is very common in the American coloured community. A recent visit to the Charity Hospital in New Orleans and to the Cook County Hospital in Chicago emphasized for me the great frequency of obesity, diabetes, and atherosclerosis in the Negro population. Amputation for diabetic gangrene is one of the commonest surgical operations performed in the general surgical unit at the Charity Hospital, where 80% of the patients are Negro. I have never seen a case in an African."

To the physicians paying attention to this phenomenon worldwide, the fact that middle-aged men in America suffered disproportionately from heart disease, and that heart disease and diabetes mortality plummeted during the scarcity conditions of wartime, were important observations. A possible explanation was that the western foods that caused these chronic diseases when populations went through a nutrition transition and became westernized were the same foods that were unavailable or rationed during wartime and also the foods responsible for heart disease being the number one killer in America.

The excessive consumption of saturated or animal fats was not a viable explanation for these disease transitions because populations that consumed significant animal fats prior to western contact and acculturation—the pastoral populations of East Africa, for instance, or the Native Americans of the Great Plains and the First Nations people in Canada, or the Inuit living near the Arctic Circle—experienced the same disease transitions after they began consuming western foods. In those cases, these populations consumed, if anything, less fat (and even less saturated fat in the case of the African pastoral populations) after westernization, not more.

Of all these chronic diseases, diabetes would be among the most conspicuous and the one that attracted the most attention because the disease was so dire and its diagnosis in later stages—the thirst, hunger, urination, and weight loss—so unmistakable.* The emergence of diabetes as a common chronic disease in the United States itself is a well-documented example of how this occurred in coincidence with such a nutrition transition, and how it prompted physicians and public health authorities who were observing this transition as it happened to suggest what they thought was an obvious explanation. Here Elliott Joslin, once again, plays a critical role.

In 1898, the same year that Joslin opened his diabetes clinic, he also published an analysis with the Harvard pathologist Reginald Fitz documenting all they had learned from a painstaking review of the medical records of Massachusetts General Hospital, Boston's major urban hospital. Out of 48,000 patient reports going back three-quarters of a century, they had identified a total of 172 patients who had been diagnosed with diabetes and treated at Mass General. These patients represented just 0.3 percent of all those admit-

---

* In 1912, Richard Cabot, a physician and medical educator at Massachusetts General Hospital in Boston, published a review in *The Journal of the American Medical Association* of the accuracy of diagnoses based on three thousand autopsies: he concluded that diabetes was the disease most often diagnosed accurately. In 95 percent of the cases, the attending physicians had apparently gotten it right.

ted to the hospital, but nonetheless Joslin and Fitz could discern a clear trend. From 1824 through 1850, the hospital often went an entire calendar year without admitting a single diabetic patient. But as the century progressed, the number and proportion of diabetic patients steadily increased. In the thirteen years between 1885 and 1898, as many diabetic patients were admitted to Mass General—eighty-six—as in the prior sixty-one years.

In 1921, when Joslin reviewed the state of affairs for *The Journal of the American Medical Association*, he recognized this ongoing trend for what it very likely was: the beginnings of the diabetes epidemic, one that now afflicts one patient in every four, for example, in hospitals of the Veterans Affairs. By the early 1920s, the fact "that diabetes has increased rapidly in recent years," as it was described by Louis Dublin, vice president and statistician of the Metropolitan Life Insurance Company, had become the fodder of newspaper and magazine articles.

In 1924, two Columbia University researchers, Louise Larimore and Haven Emerson, a former New York City health commissioner, assessed the extent of the epidemic in a lengthy analysis published in *Archives of Internal Medicine*. Diabetes mortality had increased steadily throughout major urban U.S. centers, they reported, with numbers that were already staggering. From 1900 to 1920, the number of deaths per year attributed to diabetes had increased in some cities many times over. In New York City, for which Emerson and Larimore had the best data, the annual deaths from diabetes (per one hundred thousand residents) had increased fifteenfold since the end of the Civil War. This represented an almost fiftyfold increase in the *percentage* of all deaths that were attributed to the disease: from 1 in every 2,500 to 1 in every 50.

Some of this increase, Emerson and Larimore acknowledged, could be explained by the aging of the population, a consequence of the success that public health initiatives were having on childhood mortality and infectious diseases like tuberculosis, the most prominent causes of premature death in the nineteenth century. Some of the increase could be explained by physicians being more aware of diabetes as a disease, possibly fatal, which meant a greater likelihood they would attribute a death to diabetes on a death certificate. But with little doubt, the disease was steadily afflicting more and more Americans.

Because the nineteenth century had also experienced a conspicuous explosion in the amount of sugar Americans were eating, Emerson and Larimore considered this to be a likely explanation for the epidemic they were describing. At the beginning of the century, Americans had been eating and drinking a few pounds of sugar, on average, per person per year (less than a teaspoon of sugar a day). By the end of the century, that number had increased to seventy pounds (more than 20 teaspoons of sugar a day), and sugar represented a significant portion of all the calories Americans consumed. By the early 1920s, the sugar industry was selling, and Americans were apparently ingesting, one hundred pounds of sugar per capita each year.*

The Industrial Revolution and the creation of the beet sugar industry had transformed sugar from an expensive luxury to a staple of the diet, the least expensive source of calories. Children, in particular, had been targeted by this dietary shift. The chocolate, ice cream, and candy industries had all been born in the 1840s, followed by the soft-drink industry in the 1870s and 1880s with the appearance of Dr Pepper and then Coca-Cola and Pepsi-Cola. By 1895, Coca-Cola was available in every state in the country.

Diabetes mortality steadily climbed as Americans consumed more and more sugar, and it fell on those few occasions—most notably, the rationing during the First Word War—when sugar was hard to come by. To Emerson and Larimore, the changes during the period of the world war were "particularly striking." The critical question for them was whether the increased consumption of sugar was merely a proxy for eating more calories, causing both obesity and diabetes, or was there something unique about the sugar that triggers the diabetic condition?

---

* All of these numbers are at best informed guesses based on assessments of sugar production per state and imports. In the 1917 edition of Joslin's textbook, he provides statistics on sugar consumption in the United States from 1800 to 1917, although without referencing his source. By Joslin's accounting, sugar consumption increases from eleven pounds per capita yearly average in the first decade of the 1800s to seventy-three pounds per capita yearly between 1910 and 1917. In the 1840s, for reasons unspecified, the average per capita consumption is only two pounds per year.

This wasn't the first time physicians and public health authorities had considered increased consumption of sugar as a likely explanation for the appearance of diabetes in a population. In 1907, the British Medical Association had held a meeting on "diabetes in the tropics," where physicians working in the Indian subcontinent and elsewhere in Asia and Africa had also discussed the increasing prevalence of diabetes among their patient populations. "There is not the slightest shadow of a doubt that with the progress of civilization, of high education, and increased wealth and prosperity of the people under the British rule, the number of diabetic cases has enormously increased," the University of Calcutta physician Rai Koilas Chunder Bose noted, estimating that perhaps one out of every ten of his "well-to-do" male patients suffered from diabetes. The physicians and public health authorities attending the British meeting had considered a range of possible explanations, everything from mental strain to indolence—both considered common features of affluent western lifestyles—but the increase in sugar consumption that came with prosperity was consistently on their list of likely suspects. As early as the sixth century BCE, Hindu physicians in India had diagnosed diabetes in the rich of that era, Bose noted, and attributed it to "their overindulgence in rice, flour and sugar."

Frederick Allen had also considered the idea seriously and had discussed it at length in his 1913 monograph, *Studies Concerning Glycosuria and Diabetes*. "The consumption of sugar is undoubtedly increasing," he had written. "It is generally recognized that diabetes is increasing, and to a considerable extent, its incidence is greatest among the races and the classes of society that consume [the] most sugar." Allen had surveyed the writings of European diabetologists and divided these authorities into three categories: those who dismissed the idea (Carl von Noorden, most notably), those who considered the evidence ambiguous (as Bernhard Naunyn had), and those (like the French authority Raphaël Lépine) who were convinced it was true and cited as evidence that diabetes had become common among the laborers in sugar refineries. Even those who rejected the idea, as Allen noted, still counseled against allowing their diabetic patients to eat sugar. This led Allen to conclude that the general attitude of diabetologists might be "doubtful or negative" toward the notion that sugar causes diabetes, but "the practice of the medical profession is wholly affirmative."* Allen imagined a scenario in which individuals who appeared healthy but were predisposed to become diabetic, whose pancreatic function was weakened or weaker than it should be, were then pushed over a threshold into the diabetic condition—sugar (glucose) appearing in their urine—by chronic consumption of sweets, pastries, and sugary beverages.

In this case, Joslin disagreed with Allen's assessment, and Joslin, once again, would be the dominant influence determining what diabetologists thought. The early editions of Joslin's textbooks discussed the sugar hypothesis and the evidence for and against it in two pages, dismissing it as unlikely; later editions would afford it ever less credibility. By the 1971 edition, as a new generation of physicians and nutritionists was challenging Keys's notion that dietary fat was the problem with modern diets and suggesting, mostly unaware of this history, that it was sugar instead, Joslin's textbook no longer considered it worthy of discussion. The idea that chronic sugar consumption might precipitate the diabetic condition was afforded not even a sentence in the nine-hundred-page volume.

Joslin's bias, as we've discussed, inherited largely from Allen, was that dietary fat either caused diabetes or made the complications worse. Joslin also thought the goal of insulin therapy was to increase the carbohydrate tolerance of diabetic patients and that the more carbohydrates that could be consumed without sugar (glucose) appearing in their urine, the healthier the patients would be. Although biochemists knew better, Joslin had seemingly little or no awareness that dietary sugar (sucrose, to be precise) might be more harmful than other carbohydrate-rich foods.

As early as 1917, Joslin had argued against the sugar hypothesis, citing the experience of the Japanese as conclusive: here was a population living "upon a diet consisting largely of rice and barley,"

---

* This would remain the case. "Sugar and candies do not cause diabetes," wrote Garfield Duncan, medical director of Pennsylvania Hospital in Philadelphia, in 1935, "but contribute to the burden on the pancreas and so should be used sparingly."

he wrote, "yet so far as statistics show, the disease is not only rare but mild in that country." He acknowledged the upsurge in sugar consumption in the United States and Europe in the nineteenth century coincident with the rising tide of diabetes mortality, but he insisted the Japanese experience argued against making such a connection: "Fortunately, the dietary habits and the statistics upon diabetes of Japan would seem to save us from this error."

In the 1928 edition of his textbook, Joslin insisted sugar was benign because the two relevant trends—sugar consumption and diabetes prevalence—while both increasing, were doing so at different rates: sugar consumption had risen by 35 percent between 1900 and 1925, he wrote, and diabetes mortality by 75 percent. Moreover, some populations—Australians, most notably—consumed as much sugar as Americans but still reported a lower rate of diabetes mortality. (Joslin did think, though, that it would be worth examining the prevalence of diabetes "in the employees of candy factories and candy shops" to see if "acute dietary excesses" might precipitate the disease.)

And so it went. By 1934, Joslin was citing a three-page analysis published by Clarence Mills, a physician associated with the University of Cincinnati, as sufficient reason to dismiss the sugar-diabetes hypothesis. Mills had obtained a League of Nations report on per capita sugar consumption in three dozen nations between 1923 and 1928 and he compared it against a similar ranking of nations by diabetes mortality rates over the same years. "Of the thirteen countries highest in consumption of sugar," Mills had written, "eleven are found among the thirteen highest in death rate from diabetes." That might have been considered evidence for the sugar hypothesis, but Mills found the individual variations in the data more telling. In Norway, for instance, diabetes mortality had trended downward after 1922, while sugar consumption had continued to go up. In Australia, sugar consumption had been high and diabetes mortality had remained constant. And in the United States, sugar consumption had remained relatively stable but deaths from diabetes steadily increased. The lack of uniformity, Mills wrote, "leaves little doubt in my mind that consumption of sugar as such is not related in a causal way to the death rate from diabetes." Other investigators,

Joslin among them, had noted that the arrival of insulin therapy in 1923 could easily explain the lack of uniformity in the data and, particularly, why diabetes mortality might have decreased in the post-insulin era—insulin therapy was saving lives—even as sugar consumption was rising.

In post–World War II editions of his textbook, Joslin cited research published by the British diabetologist Harold Himsworth as reason enough to dismiss the sugar hypothesis, even as Himsworth would eventually disavow his own conclusions. As a young diabetologist in the early 1930s, Himsworth had shared Joslin's biases both against dietary fat and for the role of carbohydrates in the diet of diabetic patients. "Sugar is what must be given, and enough insulin to enable the sugar" to resolve diabetic ketoacidosis, Himsworth wrote, while also believing the goal of insulin therapy was to build up a patient's carbohydrate tolerance, implying that the more carbohydrates in the diet, the better.

In 1935, Himsworth published two lengthy analyses supporting his—and Joslin's—beliefs. The first reported on a dietary survey he had conducted of 135 diabetic patients at the University College Hospital in London—young patients with the acute form of the disease and older, heavier patients whose disease was more chronic. Working with E. M. Marshall, a dietitian, he had questioned these patients to determine what they recalled eating prior to their diagnosis. He then compared what his diabetic patients told him with what he learned from surveying several hundred healthy individuals.

Himsworth acknowledged that such dietary recollections are fraught with errors, but he was nonetheless struck by how many of the diabetic patients had recalled an unusual fondness for fatty foods. "A considerable number of patients confessed to their weakness for eating butter in spoonfuls, without consuming bread at the same time," Himsworth and Marshall wrote.

Researchers in the discipline of physiological psychology—led by Curt Richter at Johns Hopkins, a seminal figure in the field—would later report the same behavior from laboratory rodents rendered diabetic, suggesting that the preference for fatty food was the result of the animals' inability to metabolize carbohydrates, the best way, in effect, that they could provide their bodies with the necessary energy. Himsworth concluded that it was this craving for fat that had likely caused the diabetes. He then published a second lengthy review of dietary surveys, acknowledging that diabetes mortality was increasing with prosperity in populations throughout the world, not because of the sugar consumed, but because the wealthier the populations the more fat they consumed.

With Himsworth and Joslin in agreement, the two diabetologists then created the scientific equivalent of an echo chamber to make the argument that dietary fat triggered diabetes and sugar was benign. Joslin would cite Himsworth's "painstakingly accumulated" data implicating dietary fat as the cause, while Himsworth would cite Joslin's dismissal of the sugar hypothesis, along with Mills, as reason enough to eliminate that possibility.

In fact, both Himsworth and Joslin made critical errors in their analyses. In coming to his conclusions that fat-rich diets caused diabetes, for instance, Himsworth had to explain why pastoral populations like the Maasai warriors in Africa and the Inuit living near the Arctic Circle, eating exceedingly fat-rich diets, had vanishingly low rates of diabetes. (They, too, would later experience epidemics of diabetes with westernization.) He did so by noting that "the evidence as to the presence or absence of diabetes" in the Maasai "is so scanty that no opinion can be expressed," and by misinterpreting two papers written about immigrant populations in northern Canada to conclude that the Inuit really did eat a carbohydrate-rich diet, despite all other observations to the contrary. "The popular assumption that [the Inuit] take a very high-fat low-carbohydrate diet is false," he wrote, citing a study of "fisherfolk" in Labrador and not apparently noticing that these were not Inuit but the descendants of Scottish immigrants eating what Samuel Hutton, a missionary physician in Labrador at the turn of the century, called the "settler dietary," i.e., western foods.*

---

* The second study was a 1927 article entitled "Health of a Carnivorous Race: A Study of the Eskimo," in which the author, William Thomas, does note that the Inuit "contrary to general opinion . . . eats relatively little fat or blubber." Those pure fat sources are used for fuel. Himsworth interpreted this as meaning the Inuit did not eat a high-fat diet, but they clearly did. As Thomas notes, "There is no edible vegetation," and so the Inuit consumed meat, fowl, and fish from the northern latitudes, all of which would have been fat rich.

Joslin's lack of recognition of the unique biochemical properties of sucrose—i.e., sugar itself—haunted his diabetes research. Grains, like the rice and barley staples of the Japanese diet, and starchy vegetables like potatoes, are composed primarily of long chains of glucose molecules that are broken down upon digestion and enter the circulation as glucose, hence their direct effect on blood sugar levels. The glucose is then metabolized by cells throughout the body. Sucrose is a molecule of glucose attached to a molecule of fructose, the sweetest of the carbohydrates.* Unlike glucose, the fructose that makes it through the small intestine is metabolized primarily in the liver. As such, it has fundamentally different effects in the human body than other carbohydrates, which means sucrose does as well. Neither Joslin nor Himsworth considered this relevant to diabetes. Nor did they ever publicly address whether the Japanese, with their low rates of diabetes, ate relatively little sugar, which they did. As late as the 1960s, the Japanese were consuming, per capita, roughly the same amount of sugar that the Americans and the British had been eating and drinking a century earlier, when diabetes in those nations had also been uncommon. The Japanese experience was not evidence against the sugar hypothesis, but evidence for it.

In time, Himsworth backtracked, publicly acknowledging that he had been wrong about dietary fat. While fat consumption did *tend* to track with diabetes incidence in populations, Himsworth said in a 1949 lecture to the British Royal College of Physicians (and in one of the last papers he authored before leaving clinical research behind to become director of Britain's Medical Research Council), research on laboratory animals had come to contradict his belief that dietary fat acted as a trigger of the disorder. "The consumption of fat has no deleterious influence on sugar tolerance," Himsworth admitted, "and fat diets actually reduce the susceptibility of animals to diabetogenic agents." Himsworth suggested that maybe "other,

---

* High fructose corn syrup, which came to replace roughly half of the sucrose in the U.S. diet beginning in the late 1970s, is most often found in the form of HFCS 55, which is 45 percent glucose and 55 percent fructose. As such, it is a subtle variation on the 50-50 fructose-glucose content of sucrose. The human body treats them, for the most part, identically.

more important, contingent variables" could be found that *associated* with fat consumption in western diets and might trigger the appearance of the disease. He speculated that it might be as simple as eating too much, because of the association between diabetes and obesity, and because "in the individual diet, though not necessarily in national food statistics, fat and calories tend to change together." He omitted mention of sugar, though, which is another variable that tracks with fat and calories in both national and food statistics *and* individual diets, and which could have explained, by its relative absence, the low rates then of diabetes in the Inuit and the Maasai, just as it could in the Japanese.

The sugar-diabetes hypothesis might have died with Joslin and Himsworth's dismissal, but the observations on which it was based—the emergence of diabetes as a common disease in populations newly exposed to modern western diets and lifestyles—continued to be made in the post–World War II decades. In this research, Israel was one of several natural laboratories, as Jewish immigrants flooded into the country postwar from disparate populations throughout Europe and the Middle East. By the mid-1950s, Israeli physicians were documenting significantly higher rates of heart disease in populations that had come from Eastern and Western Europe (Ashkenazi) than in those that had emigrated from the Middle East.

In 1961, Aharon Cohen, a physician at the Hebrew University–Hadassah School of Medicine in Jerusalem, reported a similar disparity in the incidence of diabetes, in this case between two different generations of immigrants from Yemen on the Arabian Peninsula. The Yemenites who had been living in Israel since the 1930s, Cohen reported, had a prevalence of diabetes similar to other Israeli populations and not dissimilar to that of major urban centers in Europe and the United States. But nearly fifty thousand Yemenite Jews had arrived in Israel in 1949 and 1950 in a legendary airlift known as "Operation Magic Carpet," and in these very recent arrivals diabetes was still an exceedingly rare disease, one-fiftieth as common as in the Jewish immigrants who had arrived two decades before.

That people of the same ancestry would have such vastly different rates of diabetes suggested that the environment in Israel—food or lifestyle—was the key factor in triggering the appearance of the disease. When Cohen and his colleagues surveyed the Yemenite immigrants about how their diets had changed with immigration, the singular difference was that "the quantity of sugar used in the Yemen had been negligible," as Cohen reported in *The Lancet* in 1961. "In Israel there is a striking increase in sugar consumption, though little increase in total carbohydrates."

A similar observation was made by the South African diabetologist George Campbell, who focused his research on a population of immigrants who had arrived in South Africa from India in the late nineteenth century to work as laborers on the sugar plantations in Natal Province (now KwaZulu-Natal), around Durban. Campbell estimated that one in three middle-aged men in this population suffered from diabetes. "A veritable explosion of diabetes is taking place in these people," Campbell wrote in 1966.

The prevalence of diabetes in Natal was far higher than the one in one hundred estimated at the time for all of India itself. If a predisposition existed, it, too, had to be triggered by the local environment. But the fat content of the diet in South Africa, Campbell reported, was little different than it was in India. The amount of sugar consumed, though, clearly was: in India, yearly sugar consumption per capita averaged 12 pounds, compared to almost 110 pounds for those of Indian ancestry in Natal. In the early 1950s, according to Campbell, the urban Zulus, with an increasingly high prevalence of diabetes, were eating on average 85 pounds of sugar each year, while the rural Zulus, in which diabetes was still rare, consumed 6 pounds.

Campbell's observations led him to two conclusions about the sugar-diabetes relationship: First, the available sugar consumption data from around the world suggested that populations could tolerate as much as 70 pounds of sugar per capita yearly—roughly what Americans and the British had been consuming in the 1870s—before diabetes prevalence would begin the kind of epidemic increase that Cohen had documented in Israel and Campbell was observing among the Natal Indian and urban Zulu populations in South Africa. Second, diabetes might have an incubation period similar to the roughly two decades before lung cancer appears in cigarette smokers. From the patient histories he had taken in his clinic, Campbell noted "a remarkably constant period in years of exposure to town life"—eighteen to twenty-two years—before diabetes appeared.

Through the 1970s, the discussions of the emergence of diabetes as a common disorder in populations newly exposed to western diets and lifestyles were shaped primarily by the thinking of four British physician-researchers—Thomas "Peter" Cleave, John Yudkin, and then, ultimately, Hugh Trowell and Denis Burkitt. They took a hypothesis that began as a direct challenge to the dietary-fat/cholesterol hypothesis of heart disease—that the dietary cause of *both* heart disease and diabetes was the types of carbohydrates consumed—and converted it over the course of twenty years into a hypothesis that could be reconciled with what the AHA was treating as dogma: that heart disease was still caused by saturated fat raising cholesterol, but that the absence of fiber in modern diets (a result of the process of refining or processing carbohydrates) was a cause of the other related common chronic diseases. The direct link to diabetes effectively vanished from the science, as did the possibility that sugar and processed grains, in particular, might be responsible.

Cleave, a surgeon captain in the British Royal Navy, established the ultimate context of the discussion. Cleave, like Keys, had been an academic prodigy, graduating from medical school in 1927 at twenty-one. He had then joined the British navy and served at naval hospitals in the United Kingdom, Malta, and Hong Kong. His travels gave him an appreciation for how the pattern of chronic disease prevalence could differ profoundly between populations, particularly so with the infiltration of western influences. In the postwar years, Cleave became director of medical research at the British Institute of Naval Medicine (retiring in 1962) and corresponded with hundreds of physicians around the world, asking about disease prevalence in their patient populations, most notably peptic ulcer. His 1962 book on ulcers contained pages of testimony from these physicians reporting the relative absence of ulcers in isolated or traditional populations that had little access to sugar, white flour, and white rice, and the frequency of the disease in the communities that ate these foods regularly.

As Cleave would argue, the primary change in traditional diets worldwide with economic development and westernization was the addition of refined, carbohydrate-rich foods—sugar, white flour, and white rice. These, therefore, were the very likely causes of the chronic diseases that appeared shortly thereafter. Add these foods to any traditional diet, no matter how otherwise healthy, how rich or poor in fat, and they will cause these common chronic diseases. This could explain why the same diseases—diabetes, obesity, and heart disease, most notably, but also peptic ulcers, dental caries, varicose veins, and a host of others—appeared after westernization in populations that had lived almost exclusively on animal foods, as well as in primarily agrarian populations like the Hunza in the Himalayas or those in Asia (the Japanese, for instance) whose traditional diets consisted of significant whole grains and starchy vegetables. In effect, all the common chronic diseases that are associated with western diets and lifestyles, Cleave argued, were manifestations of a single primary disorder that he named "the saccharine disease" (although he acknowledged that "refined-carbohydrate disease" might have been more to the point).

As a student of evolutionary biology, Cleave invoked a Darwinian hypothesis—"the Law of Adaptation," he called it—to explain this nutrition/disease transition and his saccharine disease hypothesis: species require "an adequate period of time for adaptation to take place to any unnatural (i.e., new) feature in the environment, so that any danger in the feature should be assessed by how long it has been there." As Cleave argued and others would then echo, the refining of sugar and white flour and the dramatic increase in the consumption of these processed carbohydrates since the mid-nineteenth century were the most significant changes in human nutrition since the introduction of agriculture roughly ten thousand years before. "Such processes," Cleave wrote, "have been in existence little more than a century for the ordinary man and from an evolutionary point of view this counts as nothing at all."

In the immigrant populations of the kind that Campbell and

Cohen were studying or the traditional populations in which other physicians were observing the explosive emergence of diabetes as a common disease—Native American tribes, for instance, throughout the United States, as documented by the University of Oklahoma diabetologist and epidemiologist Kelly West—the increases in the consumption of sugar and white flour that whites in the United States and Europe had experienced over a century or more were occurring over just a few decades. Their response to these foods, Cleave suggested, should be all that much more dramatic—everything from greater tooth decay to higher levels of obesity and diabetes. If researchers studied a population of Native Americans or South Pacific Islanders, or a population of Natal Indians, as Campbell had, consuming seventy or a hundred pounds of sugar per person yearly, and compared them to a population of European ancestry consuming the same amount, the latter would exhibit a lesser prevalence of obesity and diabetes because they would have had more time to adapt to these foods.*

Cleave first published his ideas at length in 1956 in *The Journal of the Royal Naval Medical Service*, in an article entitled "The Neglect of Natural Principles in Current Medical Practice." In 1961, he read a letter George Campbell had written to the *British Medical Journal* arguing (as Campbell later described it) that "radical dietary change" was the primary cause of diabetes and reached out to suggest they work together. Their resulting collaboration led to the 1966 publication of *Diabetes, Coronary Thrombosis, and the Saccharine Disease*, which laid out the evidence and their thinking. The book's foreword was written by Sir Richard Doll, then the director of the Statistical Research Unit of Britain's Medical Research Council and already celebrated (and knighted) for his seminal research linking cigarettes to lung cancer. If only a small part of what Cleave and Campbell argued turned out to be correct, Doll wrote, the book represented "a bigger contribution to medicine than most Univer-

---

* As early as 1940, Cleave was writing in British medical journals that animals and humans could trust their tastes not to mislead them on a healthy diet so long as the foods they were consuming were in their natural form and "responsible for the evolution of those very instincts. . . . By the natural law, the more a food is altered from its natural state, the more dangerous to health it becomes."

sity departments or medical research units make in the course of a generation."

The reception of *Diabetes, Coronary Thrombosis, and the Saccharine Disease* was decidedly mixed. In the *British Medical Journal*, the reviewer was Sir Derrick Dunlop, whose work on diabetes we've discussed, and who found the book both infuriating and thought-provoking. He said he could see why Doll might think it was so potentially important, but then dismissed it, perhaps ironically, as saying nothing new: health authorities already "generally agreed," he wrote, that "the excessive consumption of refined carbohydrates is the main nutritional error in this country and in most others." "Many of us," he added, "think it far more deleterious than the excessive consumption of animal fats."

While Cleave and Campbell were outsiders in the British nutrition world, with limited influence beyond publication of their book, John Yudkin was not. Yudkin, too, was a physician, but he had also earned a PhD in biochemistry from Cambridge University. His doctoral research had been instrumental to the work that later won a Nobel Prize for the French chemist Jacques Monod. Yudkin's experience in World War II, serving in the medical corps in West Africa, had motivated his interest in nutrition, identifying a vitamin deficiency as the cause of a skin disease among the local soldiers. After returning to England postwar, Yudkin founded the first dedicated academic nutrition department in Europe at what was then Queen Elizabeth College and would later become part of the University of London. His first foray into the chronic disease literature had been provoked by Keys and his dietary fat–cholesterol theory. Yudkin had been among the researchers in 1957 who publicly criticized Keys's interpretation of the evidence, writing in *The Lancet* that heart disease (like diabetes) was associated with prosperity in populations throughout the world, but that could have been explained by any number of dietary and lifestyle factors, of which dietary fat was only one.

By 1963, Yudkin was echoing Cleave's argument that species adapt over generations—"anatomically, physiologically, and biochemically"—to a particular diet and combination of foods. "If we are looking for the dietary cause of some of the ills of civilization," Yudkin wrote in *The Lancet*, "we should look at the most significant changes in man's diet." That was not changes in the types of fats consumed or the total amount of fat, Yudkin argued, but the enormous increase in sugar consumption that had come about with the Industrial Revolution, just as Frederick Allen had suggested about diabetes half a century earlier and Cleave, Campbell, and others were now arguing.

Yudkin focused on heart disease and would eventually bolster his argument with a series of findings coming from American biochemists and biophysicists—at the University of California, Rockefeller University in New York City, and Yale University—suggesting that coronary heart disease is associated with a clustering of numerous metabolic and hormonal disruptions, not just the elevation of cholesterol or LDL cholesterol. All of these metabolic abnormalities implicated the carbohydrate content of the diet in the disease, in turn implying a common pathology underlying obesity, heart disease, and diabetes. As with Cleave, Yudkin had come to conclude that dietary fat had little or nothing to do with heart disease. Unlike Cleave, he focused exclusively on sugar as the agent of disease, not the refining of carbohydrates in general.

Over the next decade, before he, too, retired, Yudkin conducted experiments feeding sugar to college students (often the subjects of academic health research because of their availability) and a range of laboratory animals including mice and pigs, reporting that it could trigger most to all of the metabolic and hormonal abnormalities that associated with obesity, diabetes, and heart disease. Other researchers, including Aharon Cohen and his collaborators in Israel, were doing the same.

By the spring of 1973, Yudkin's work and the publication of his book *Pure, White, and Deadly* had generated sufficient attention that a U.S. congressional committee led by the South Dakota Democrat and former presidential candidate George McGovern held a hearing titled "Sugar in Diet, Diabetes, and Heart Disease." Cohen, Campbell, Cleave, and Yudkin testified, as did Peter Bennett of the National Institutes of Health. Bennett had established an NIH laboratory in Phoenix, Arizona, to study a local Native American tribe, the Pima, in which, as Bennett and his collaborators had documented, diabetes had quickly come to afflict one in every two adult members of the tribe. "The only question that I would have," Bennett told the assembled representatives, "is whether we can implicate sugar specifically or whether the important factor is not calories in general which in fact turns out to be really excessive amounts of carbohydrate." Although he then added a caveat to that statement: "Of course, one of the more important components of carbohydrate intake is sugar." The U.S. Department of Agriculture also had two witnesses testifying: Walter Mertz, head of the agency's Nutrition Institute, and his colleague Carolyn Berdanier. In experiments with laboratory rats, the two USDA researchers told the committee, sugar raised the level of both glucose and fats in the bloodstream and caused the rats to become diabetic. "They die at a very early age," Berdanier said.

The congressmen were receptive to the argument, but the cognitive dissonance they experienced foreshadowed its eventual downfall. They could believe that sugar caused diabetes, but not heart disease. Here was a collection of researchers, mostly foreign—even Bennett hailed from the United Kingdom originally, and Mertz from Germany—arguing that what the American Heart Association, the U.S. media, and American doctors and cardiologists had now so fully embraced was simply not true. Both Cleave and Yudkin dismissed the fat-cholesterol hypothesis out of hand. Mankind had been eating saturated fats for hundreds of thousands of years, Cleave said. "For a modern disease to be related to an old-fashioned food is one of the most ludicrous things I have ever heard in my life. . . . But when it comes to the dreadful sweet things that are served up . . . that is a very different proposition." Yudkin was equally adamant. When the British nutritionist testified that "high blood cholesterol in itself has nothing whatever to do with heart disease," suggesting that he would advise a patient with high cholesterol to eat less sugar, rather than less fat, McGovern responded simply, "That is exactly opposite what my doctor told me."

After the hearings concluded, the sugar industry went on the offensive. It hired pollsters to survey physicians throughout the United States, only to learn that most of them now thought sugar, at the very least, accelerated the onset of diabetes. Even the Harvard nutritionist Jean Mayer, then as influential as Keys, was now writing that sugar "plays an etiological role in those individuals who are genetically susceptible to the disease" (although that raised the question of whether anyone ever gets diabetes who isn't genetically susceptible, with the exception of those rare individuals with pancreatic injuries or tumors). In 1974 and 1975, the sugar industry hosted a pair of conferences to address whether "the risk of becoming diabetic [is] affected by sugar consumption"—the first in Washington, D.C., and the second in Montreal—inviting only those researchers who had been outwardly skeptical of the sugar-diabetes connection.

Even these skeptics, however, took seriously the possibility that some proportion of the public might be particularly sensitive to the sugar content of the diet and that this proportion had to be quantified and understood. After the 1975 Montreal meeting, the sugar industry's International Sugar Research Foundation circulated a memo to its members in which it quoted a University of Toronto diabetologist, Errol Marliss, saying the industry had an ethical obligation to "establish definitively what contribution sucrose can and does make to the course of diabetes—and other diseases . . . and that this would require the support of well-designed research programs."

Neither the sugar industry, however, nor the National Institutes of Health acted on the recommendation. By 1975, the NIH was in the process of spending more than $260 million on the MRFIT and Lipid Research Clinics trials that would test facets of the cholesterol-fat hypothesis. If the NIH or the American Heart Association had the resources to fund tests of an alternative hypothesis, they lacked the institutional interest. As for the sugar industry, rather than fund clinical trials, it put its resources into a public relations campaign to exonerate sugar as a cause of any chronic diseases (other than, perhaps, tooth decay) on the basis that researchers had already established that dietary fat was responsible.

The industry's lobbying and public relations arm in the United States—the Sugar Association—produced and disseminated twenty-five thousand copies of a lengthy report, "Sugar in the Diet of Man," countering the research cited by Yudkin, Campbell,

Cohen, Cleave, and other "enemies of sugar," as the industry's internal documents were now describing them. Each chapter focused on a different chronic disorder. The chapter on heart disease, written by Keys's University of Minnesota colleague Francisco Grande, argued against Yudkin's assessment of the evidence and in support of Keys and the cholesterol-fat hypothesis. The chapter on diabetes was coauthored by Edwin Bierman of the University of Washington, who was responsible for revising the American Diabetes Association guidelines to mirror the AHA position. "The causes of primary diabetes mellitus in man remain unknown," Bierman wrote with Ralph Nelson of the Mayo Clinic, but "there is no evidence that excessive consumption of sugar causes diabetes."

While the sugar industry was conducting its public relations campaign to absolve sugar of any harmful effects on health, the scientific controversy, albeit not the science itself, was resolved by the British missionary physicians Denis Burkitt and Hugh Trowell. They gave the medical and public health communities a way to continue assuming the fat-cholesterol hypothesis was true, while reconciling it with the observations from around the world that had generated the sugar–refined carbohydrate hypothesis: the association of chronic diseases with western diets and lifestyles. The problem with refined or processed carbohydrates and sugar, they said, was simply that they lacked fiber—the bulky, indigestible roughage in vegetables, starches, legumes, and grains that was removed in the refining process.

Burkitt and Trowell's thinking made it possible to believe that these refined carbohydrates, sugar and white flour, could be benign, while still accepting that fat, when it came to heart disease, was not. Burkitt took to condemning modern diets equally for their high fat content ("We eat three times more fat than communities with a minimum prevalence of [western] diseases," Burkitt would say. "We must reduce our fat") and for their fiber deficiency, which he considered "the biggest nutritional catastrophe in [the United Kingdom] in the past 100 years," the cause of almost all other common chronic disorders.

Nutritionists were skeptical of Burkitt and Trowell's fiber hypothesis, but it nonetheless caught on almost immediately. While the sugar/refined-carbohydrate hypothesis was being rejected on the basis that the epidemiologic evidence was inconclusive and no clinical trials had rigorously tested it, nutrition authorities were willing to accept that fiber-rich foods were required in a healthy diet based on, essentially, the identical evidence. They did so because it fit their preconceptions.

Burkitt played the dominant role in the dissemination of the fiber hypothesis. He was the only one of these four British researchers who had not yet retired in the 1970s, and he had earned the reputation necessary to have his ideas taken seriously. He had worked with African troops during the Second World War and moved back to Africa in 1946 to serve as a missionary surgeon. In 1948, he joined the staff of Mulago Hospital in Kampala, Uganda. By the early 1960s, he had become famous—"one of the world's best-known medical detectives," as *The Washington Post* would call him—for his identification of a fatal childhood cancer, the first human cancer ever linked to a viral cause. It would be known ever after as Burkitt's lymphoma. That investigation had taken him on a ten-thousand-mile journey across the African continent, an experience that taught him to appreciate the lessons that could be learned by attending to the geographical distribution of disease.

Burkitt left Uganda and the missionary service in 1966, returning to England to work as a cancer epidemiologist for the Medical Research Council. There he met Sir Richard Doll and Harold Himsworth and persuaded them to let him establish cancer registries in undeveloped countries from which he could collect the kind of incidence data that had led him to his lymphoma breakthrough. Doll then introduced Burkitt to Cleave and suggested he read Cleave and Campbell's *Diabetes, Coronary Thrombosis, and the Saccharine Disease*. Burkitt later described Cleave as possessing "perceptive genius," and the logic of the book as "irrefutable."

"I knew from my experience in Africa that he was perfectly right," Burkitt wrote. "But the profession wouldn't listen to him." Burkitt now requested that the reports he was receiving monthly from his cancer registry—a network of 150 African hospitals—

include information on the chronic diseases that Cleave had suggested were diseases of westernization: heart disease, diabetes,
gallstones, appendicitis, diverticular disease, and others. He also
sent questionnaires to hospitals worldwide; "near to a thousand,"
Burkitt said, eventually responded. What he learned confirmed
Cleave's assertion that western foods/lifestyles were closely associated with the prevalence of these chronic diseases.

In 1969, Burkitt began publishing in prominent medical journals on both sides of the Atlantic, supporting and expounding on
Cleave's work, placing it firmly in the context of epidemiological
methodology: how the geographical and chronological relationship between diseases provided a clue to their cause. If disparate
diseases or disorders were associated in their geographical distribution and also tended to cluster together in patients, Burkitt
argued, it was reasonable to suspect a common or related cause,
no matter how different the diseases themselves might seem. He
invoked syphilis as a metaphor to explain his thinking: "Before
the cause of syphilis became known," he wrote in 1970 in *The Lancet*, "it was observed that certain skin rashes, characteristic penile
lesions, perforation of the palate . . . and aortic aneurysms tended
to develop in the same groups of people. This association led to the
realization that all these conditions were manifestations of a single
disease." Hence, as Cleave had proposed, obesity, diabetes, heart
disease, and all the diseases that associated with them were likely
to have the same underlying cause.

What Cleave and Campbell had called "saccharine diseases" and
Yudkin "diseases of civilization," Burkitt now called "western diseases." The greater "the extent to which Western customs, especially
dietary habits, have been adapted," the greater the prevalence of the
diseases. "Changes made in carbohydrate food may of course be
only one of many etiological factors," Burkitt wrote in 1971 in the
Journal of the National Cancer Institute, "but in some instances they
would appear to be the major one."

But Burkitt was already beginning to shift his focus. Among the
diseases that clearly fit the pattern of western diseases were those
related to the bowel or the gastrointestinal tract: constipation, hemorrhoids, appendicitis, diverticulosis, polyps of the large bowel,

ulcerative colitis, Crohn's disease, and colorectal cancer. While Cleave had focused on dental caries as an indisputable example of a western disease, one that tied all of them to the refinement of the carbohydrates—"the greatest cause of dental caries seems undoubtedly to be the production of acid from the bacterial fermentation of starch and sugar," Cleave had written*—Burkitt focused instead on constipation, a condition that could be alleviated by eating fiber-rich foods and so, as physicians had often argued in the past, might be related to the absence of fiber in modern diets. In 1968, Burkitt had met Neil Painter, an Oxford surgeon who had published studies suggesting that a "bulky diet" could prevent diverticulosis. In 1969, Burkitt met Alec Walker, a Scottish biochemist working in South Africa, who had been studying the effects of bran and fiber-rich foods on constipation and stool transit time. Walker's research, too, seemed to confirm the basics of Burkitt's hypothesis.

As a cancer epidemiologist, Burkitt focused his attention on colorectal cancer, and the network of hospitals he had created in Africa confirmed its association with westernization. Burkitt would later credit Harold Himsworth for guiding him away from the presence of refined sugar and flour as toxic elements of the diet to fiber deficiency instead, paying attention to those factors that were absent in searching for the cause of disease, rather than those that were present. "Denis," Burkitt recalled Himsworth saying to him, "do you remember the story in Sherlock Holmes when Holmes said to Watson: 'The whole clue, as I see it, to this case lies in the behavior of the dog.' And Watson said: 'But, sir, the dog did nothing at all.' 'That,' said Holmes, 'is the whole point.'" Since fiber was conspicuously removed in the process of refining carbohydrates, it satisfied the criteria. Since it was clearly linked to constipation, that made it Burkitt's prime suspect.

By the early 1970s, Burkitt was suggesting a chain of causality that led directly from the fiber deficiency of western diets to constipation—removing the fiber from starches and grains slowed

---

* Cleave had read *Nutrition and Physical Degeneration*, first published in 1939 by the Cleveland dentist Weston Price (and still in print), reporting on Price's travels around the world and his prolific research documenting the appearance of dental caries and jaw malformations in traditional populations with their transition to western diets.

the transit time of the stool through the colon—and from there to all the bowel-related diseases, including colorectal cancer. Supporting his argument were measurements of transit time he had made in more than a thousand subjects throughout the United Kingdom (including his family members) and urban and rural Uganda. As Burkitt described it, constipation, hemorrhoids, and appendicitis would appear within just a few years of a population transitioning to a fiber-deficient diet, while the constipation allowed carcinogens present in the stool more time to do their damage and cause malignancies in the rectum and colon. Cleave and Campbell's refined carbohydrate hypothesis, which had already morphed into Yudkin's sugar hypothesis, was beginning its transformation into Burkitt's fiber hypothesis.

Burkitt's fellow missionary Hugh Trowell finished the job, expanding Burkitt's thinking on fiber deficiency so that the proposition might also explain the etiology of obesity, diabetes, and heart disease. Trowell had arrived in Africa in 1929, working first in Kenya and then moving to Mulago Hospital in Kampala, where he had first met Burkitt upon his arrival in 1948. As Trowell later described it, the experience of British doctors working for the colonial service and missionary hospitals in the Kenya highlands in the 1930s had been uniquely informative: watching the native population of "three million men, women and children . . . emerge from pre-industrial tribal life and undergo rapid westernization." Diseases that were nonexistent in this population in the 1930s began appearing as the twentieth century progressed, just as diabetes had in the United States a century before. In 1956, Trowell reported what he believed was the first clinical diagnosis of coronary heart disease in a native of East Africa—an obese high court judge who had lived in England and had been eating a western diet for twenty years. After moving back to England in 1959, Trowell wrote and published *Non-Infective Disease in Africa*, a methodical attempt to catalog the spectrum of diseases afflicting the native population of sub-Saharan Africa. The common chronic disorders that associated with westernization—obesity, diabetes, heart disease, hypertension, hemorrhoids, diverticulosis, colorectal cancer, caries, and others—had been conspicuous by their absence.

When Trowell returned to Uganda in 1970, he found the disease transition in just a decade to be remarkable: "the towns were full of obese Africans and there was a large diabetic clinic in every city. The twin diseases had been born about the same time and are now growing together." In Kampala, Trowell reconnected with Burkitt, listened to him lecture, and took up the fiber hypothesis himself.

Over the next half-dozen years, Trowell reassessed the epidemiologic studies and the existing animal and human experiments to make the argument that the evidence supporting Cleave's refined-carbohydrate idea—his saccharine disease hypothesis—could just as well be used to argue that the critical variable was the fiber deficiency.* In 1972, he began publishing articles suggesting that all the western diseases associated with the bowel-related disorders would have been prevented in traditional diets by the fiber that was present in the whole grains, the starches, vegetables, fruits, and legumes. Heart disease would be rare because dietary fiber sequesters bile acids, increases the rate that cholesterol and blood fats are excreted in the stool, and so helps keep serum cholesterol levels under control. Because the presence of fiber in unprocessed foods reduces the available energy (by 4 percent, Trowell calculated), it is much less likely that a person would overeat and become obese. Diabetes would be prevented, because the fiber present in whole foods slowed the absorption of carbohydrates into the bloodstream, reducing insulin requirements.

Burkitt and Trowell's fiber hypothesis still failed to explain the relative absence of chronic western diseases in populations that had traditionally consumed little fiber—the pastoral populations like the Maasai in East Africa, for instance, where both Burkitt and Trowell had worked as missionaries—and the appearance of these diseases with westernization. These were the same observations that Keys's fat-cholesterol hypothesis had failed to explain, as had Himsworth's fat-diabetes hypothesis, because the low-fiber diets

---

* It was Trowell who suggested that they add the subtitle *Some Implications of Dietary Fibre* to the first book he and Burkitt edited on the subject in 1975, *Refined Carbohydrate Foods and Disease*, which otherwise might have been perceived as more supportive of Cleave's thinking than their own.

traditionally consumed by these populations were also relatively, if not extremely, fat rich. But these could always be explained away, as Keys had, by assuming these populations were sufficiently unique that their experience was not relevant to that of others.

By the mid-1970s, Burkitt and Trowell's fiber hypothesis had been widely embraced. Fiber was "the tonic of our time," as *The Washington Post* wrote in 1974, reporting on a review of the science by Burkitt, Painter, and Walker in *JAMA*. Even those skeptical of what the Harvard nutritionist Jean Mayer called the "furor over fiber" (which included the now yearly doubling in sales of fiber-rich products) were arguing, as Mayer did, that dietary fiber was a necessary part of a healthy diet. "A good diet," Mayer wrote, "high in fruits and vegetables and with a reasonable amount of undermilled cereals—will give all you need of useful fiber." The assumption that such a diet would lead to long life and good health, however, let alone prevent or delay the onset of diabetes, obesity, and heart disease or of colorectal cancer and other bowel-related disorders, was still based on faith, not science.

Motivated by Burkitt and Trowell's speculations, nutrition researchers worldwide began publishing studies looking at the effects of dietary fiber on cholesterol and blood fats, on the colon and stool, on body weight and blood sugar, in whatever laboratory animals seemed appropriate and human subjects they could recruit. As was common for the era, the trials would last a week to a few months, often including no more than a dozen subjects. The researchers inevitably traded off fiber-rich carbohydrates for fiber-deficient, refined carbohydrates, and so fruits, vegetables, starches, and whole meal breads replaced sugar and white flour, which meant any effects they might observe could still be explained by either the presence of the fiber or the absence of the sugar and refined flour. The results were inconclusive by definition, the trials too small and too short to draw any inferences about chronic disease states and the validity of either Burkitt and Trowell's fiber hypothesis, Cleave's saccharine disease thinking, or Yudkin's sugar hypothesis.

In the mid-1980s, as health agencies took to promoting low-fat diets for all Americans, the U.S. Department of Agriculture, the

National Institutes of Health, the Surgeon General's Office, and the National Academy of Sciences were all willing to embrace the fiber hypothesis, assuming, as the surgeon general did, that fat-rich foods are eaten "at the expense of foods high in complex carbohydrates and fibers . . . that *may* [my italics] be more conducive to health."

Sugar restriction was counseled only for those "particularly vulnerable to dental caries" and maybe those who had to limit their caloric intake for weight control. The USDA Dietary Guidelines cautioned Americans to "avoid eating too much sugar" in 1980 and then claimed unconditionally five years later that "sugar does not cause diabetes," without any research being published that could have established that as a fact. In 1986, an FDA report declared that sugar could be "generally recognized as safe," allowing its continued almost ubiquitous use as a food additive, on the basis that "no conclusive evidence" proved otherwise.

Only the National Research Council, in its 1989 report *Diet and Health: Implications for Reducing Chronic Disease Risk*, acknowledged that the "foods and dietary components that alter the risk of chronic diseases and elucidation of their mechanisms of action" had yet to be established. It also acknowledged that virtually everything having to do with the purported benefits of dietary fiber remained to be proven.

In the United Kingdom, the general state of nutritional confusion and cognitive dissonance could be seen clearly when the government's Committee on Medical Aspects of Food Policy released a report in 1989 titled *Dietary Sugars and Human Disease*. Harry Keen, among the United Kingdom's most renowned diabetologists, chaired the report, and his committee members included a dozen equally celebrated nutritionists, physiologists, and biochemists. They declared unconditionally that the consumption of sugar plays "no direct causal role" in the development of either heart disease or diabetes. But they also added the caveat that anyone who had "special medical problems such as diabetes and hypertriglyceridaemia [high triglycerides]" should limit their consumption to forty pounds a year, the equivalent of the amount of sugar consumed in England per capita in the early years of the Victorian era.

While health agencies have yet to fund a major clinical trial testing either Yudkin or Cleave's hypothesis, targeting the restriction
of sugar and/or refined carbohydrates to see if that would prevent
the onset of diabetes or heart disease, this was not the case with
Burkitt and Trowell's fiber hypothesis. When it was tested—in two
major trials in the 1990s, one of which was run by the National
Cancer Institute and cost $30 million—it failed the tests. The
larger and more rigorous the trials, the more consistently negative
the results.

Reviewers working under the auspices of the Cochrane Collaboration reviewed the relevant evidence twice, in 2002 and in 2017,
and their conclusion was the same. As stated in the 2017 review,
the evidence "does not support that increased dietary fibre intake
reduces the risk of [colorectal cancer]." But the authors also said
that the reliability of the data was "questionable," which it invariably is, and that "we have no reliable evidence to refute the use of
dietary fibre."

As for Trowell's expansion of Burkitt's hypothesis implicating
fiber-deficient diets in obesity, diabetes, and heart disease, that was
essentially tested by the Women's Health Initiative Dietary Modification Trial, which we discussed. That trial had concluded that
a low-fat, relatively fiber-rich diet provided no apparent benefits.

Once again, though, the institutional acceptance of the hypothesis rendered the outcomes of the actual tests mostly irrelevant.
It must be true, as this thinking went: otherwise, why would we
believe it and be spending so much money testing it? When the
trials failed to support the benefits of fiber in the diet, they merely
prompted researchers and the media to suggest all the reasons why
the trials themselves might have failed, and then to suggest, as *The
New York Times* did in 2002, that there were still "plenty of reasons
to say, 'please pass the fiber.'" Those reasons, however, had either
never been seriously tested, or had failed to be confirmed by the
trials that did test them.

Yet once again, this left a vitally important and familiar question unanswered. If fiber deficiency was not the explanation for
the epidemic emergence of major chronic diseases—diabetes, most
notably—then what was? In 2003, I asked Sir Richard Doll that

question. "Burkitt's hypothesis got accepted pretty well worldwide, quite quickly, but it has gradually been disproved," Doll said. "As far as a major factor in the common diseases of the developed world, no, fiber is not the answer." He then suggested that maybe Cleave had been right, that it was the quality of the carbohydrates consumed in modern diets—the sugar and white flour, specifically—that was the key.

10

The End of
Carbohydrate Restriction

*In the end the principal trouble with the trial was that it produced
results the world did not want to hear and when that happens the
assumption is that there is something wrong with you and your trial
because, surely, the world cannot be wrong. . . . The reality is that much
of what is extolled in medicine for preventing or delaying disease is
based on supposition rather than fact.*

—CURT MEINERT,
*The Trials and Tribulations
of the University Group Diabetes Program,* 2015

Diabetologists had entered the 1950s still debating the benefits
of tight blood sugar control. Their focus would become drug
therapy, not diet, as it has remained.

In the early years of World War II, French physicians working
with compounds developed by German chemists had stumbled
serendipitously on a drug that could be taken orally and lower
blood sugar. This had been a holy grail of diabetes research since
the earliest days of insulin therapy, when it was understood that
insulin was rendered inactive by the digestive tract. An oral hypoglycemic agent could free those with diabetes, as Charles Best had
said, "from the tyranny of the insulin syringe." For many patients
with mild to moderate cases of the adult-onset disease, these drugs
would do just that.

The first such drugs were known as sulfa drugs, a class of compounds that had been developed for their antimicrobial properties, primarily to treat infections, but had occasionally been toxic. By the end of the war, Auguste Loubatières, a diabetologist at the University of Montpellier, showed that these compounds could stimulate insulin secretion in those diabetic patients who still had functioning beta cells in their pancreas. This suggested they might work to control blood sugar in patients with either adult-onset diabetes—type 2—or mild, early cases of the juvenile form. By the mid-1950s, German authorities had approved a sulfa drug known as BZ-55 for just this use.

Physicians still worried about its possible adverse effects. While sulfa drugs had a history of use for microbial infections, the course of treatment in those cases was typically no more than a few weeks. Even so, the compounds had caused allergic reactions and, in extreme cases, liver and bone-marrow damage and death. Now physicians were considering the use of similar drugs to be taken for a lifetime to control blood sugar. When Joslin's clinic tested BZ-55 on their patients, it produced unacceptable side effects: toxicity in 5 percent of patients.

By then, German chemists had another drug in the pipeline, tolbutamide. In the United States, it would be tested and marketed by the Upjohn Company. Just as Eli Lilly had done with insulin in 1922, Upjohn distributed tolbutamide to influential diabetologists in the United States and Canada who would then cautiously experiment with the drug on their patients and report their experience. Throughout 1956, the results were presented at conferences and published in the journals. Tolbutamide kept blood sugar under control without insulin in most diabetic patients, specifically, as expected, in those older patients with mild cases of the disease. Side effects were minor—skin rashes typically—and appeared in a very small percentage of patients, fewer than one patient in a hundred at Joslin's clinic. "With the advent of these new chemicals," one Upjohn physician told *The Saturday Evening Post*, "we may be upon the threshold of a new era of therapy in diabetes, for they give promise that a significant portion of our diabetic population may be unshackled from syringe and needle."

Beyond its clinical promise, the availability of tolbutamide clarified much of what diabetologists did *not* know about the long-term consequences of glycemic control, whether by diet, oral hypoglycemic agent, or even insulin. "The consumption, day after day, year after year, of a substance which is foreign to the body is not considered to be good practice from a medical point of view; yet this is what the use of oral hypoglycemic agents for the treatment of *diabetes mellitus* entails," wrote three University of Toronto diabetologists in 1962. "Can the patient's diabetes be brought into equally satisfactory control by other more physiological means? If not, do the injurious side-effects of the oral agent outweigh injuries resulting from less well controlled blood sugar level if the oral agent is not used?"

Long-term, randomized controlled trials could answer these questions. The Food and Drug Administration did not yet require such trials for drug approval—these would be mandated only after the thalidomide* tragedy of the early 1960s—but physicians and public health authorities were coming to understand their necessity for establishing the long-term risks and benefits of medical interventions. In the case of tolbutamide, according to the Johns Hopkins Medical School epidemiologist Curt Meinert, the first randomized clinical trial involving diabetes was initiated only because the daughter of a congressman had been diagnosed with adult-onset diabetes and prescribed tolbutamide. The congressman reached out to Max Miller, an influential diabetologist at the University Hospitals of Cleveland, wanting "to know if blood sugar control was beneficial in reducing the complications of diabetes. Miller's answer was that no one knows because there have not been any trials to address that question." That response, Meinert added, "came as a shock to the Congressman."

Miller, Meinert, and a "small cadre" of colleagues organized a multicenter, placebo-controlled trial that would compare tolbutamide to insulin therapy and to diet alone.† The trial, known as

---

* Thalidomide was a sedative first marketed in 1956 and used by pregnant women worldwide to relieve nausea. It had never been tested for safety. By 1962, it was clear that the drug had caused severe, often fatal congenital malformations in thousands of children.

† Phenformin, another hypoglycemic agent, would be added as part of the trial beginning in 1962.

the University Group Diabetes Program (UGDP), received its first funding in 1960 and eventually cost $30 million. Patients were randomly allocated to one of the three treatments and monitored for a decade. The results, leaked first to *The Wall Street Journal* in May 1970, sparked a vitriolic controversy among diabetologists that reverberated for decades.

Not only were the subjects prescribed tolbutamide more likely to die prematurely than those getting a placebo, but the subjects on insulin therapy, surprisingly, did little or no better. An ad hoc committee of the American Diabetes Association then asked, "If insulin—the diabetic's medicinal remedy sine qua non—does not permit patients to live longer than does a diet, would not this class of patients, in respect to longevity, be just as well off with diet alone?" In *Diabetes: The Biography,* the British diabetologist Robert Tattersall described the outcome of the UGDP as suggesting that "most forms of treatment in adult-onset diabetes were a waste of time."

Many diabetologists concluded that the trial must have been flawed. The FDA proposed that warning labels should be put on the oral hypoglycemics. That notion was "assailed" by diabetologists, *The Wall Street Journal* reported, adding that "the single scientific study on which the FDA based its actions 'is scientifically unacceptable to many specialists in diabetes.'" The controversy grew increasingly heated, as physicians, for the most part, continued to prescribe tolbutamide and other, newer oral hypoglycemics in both the United States and Europe.

If it was true, though, that hypoglycemic drugs and insulin had little or no effect on diabetic complications, as the UGDP investigators had concluded, then diabetic patients who wanted to maximize their health would have to rely on their diet to do it. Clearly the patient with "severely symptomatic diabetes . . . deserves treatment with insulin," the Harvard diabetologist Ron Arky wrote in 1971 in *The New England Journal of Medicine*, but what about the great majority of all the rest? A reasonable assumption was that the benefits of any drug therapy for these patients with type 2 diabetes were at best minimal, and diet alone was the safest bet. "Renewed enthusiasm for the various dietary approaches in the treatment of diabetes has surfaced," Arky wrote. But which dietary approach should diabetic patients follow? Arky's editorial had been written in response to another *New England Journal* article reporting on a diet trial carried out by University of Washington physicians, who would then use the results to shift dietary thinking on diabetes ever after. The primary authors of the new study were John Brunzell, a young doctor reporting on his first foray into clinical trials, and Edwin Bierman, his mentor, who had been wrestling with the relationship between obesity and adult-onset diabetes for a decade.

Bierman had come to the University of Washington from New York City, where he'd graduated in 1955 from Cornell University Medical College (now the Weill Cornell Medical College). Bierman's Cornell classmates had included Arky and Robert Atkins, a cardiologist who would become infamous in medical circles for his 1972 best-selling diet book, *Dr. Atkins' Diet Revolution*. The three were friends, according to Arky. Nonetheless, in the late 1960s, as Atkins began to get media attention for his high-fat, very-low-carbohydrate ketogenic diet for weight loss and weight control, based in part on the logic that a diet that lowered insulin and blood sugar would minimize fat accumulation and heart disease risk, Bierman took to promoting a dietary philosophy for diabetes that took the opposite position: the healthiest diet for diabetic patients was carbohydrate-rich with minimal fat.

Bierman and Brunzell had recruited for their trial nine healthy subjects and thirteen with "mild diabetes." For ten days, they were all fed a liquid diet formula of 45 percent carbohydrates with 40 percent fat and then for another ten days a diet of 85 percent carbohydrates with no fat (the remaining 15 percent was protein). It may have been the first time clinicians had ever tested a diet completely absent fat on human subjects, let alone on patients with diabetes.* Brunzell and Bierman reported that both groups of subjects seemed to do a better job of controlling their blood sugar on the high-carbohydrate, fat-free diet. They suggested, and Arky echoed the thought in his editorial, that something about the diet might be working to make the cells of some of these patients with mild dia-

---

* When I interviewed Ron Arky in February 2004 he said, "In today's world, I'm not sure that paper would be accepted, because who's going to go on an 85 percent carbohydrate diet."

betes more sensitive to insulin and doing a better job of controlling glycemia. Whether that would be true of a diet that was consumed for longer than ten days couldn't be answered. Arky described the study in his *New England Journal* commentary as merely exposing "the enormous gaps in our knowledge about a disease that was allegedly 'cured' by the discoveries of Banting and Best almost five decades earlier."

Brunzell and Bierman suggested that the results of their trial "may have implications for the optimal diet for the patient with mild diabetes." While reducing calories was still of "paramount importance" for obese patients, they wrote, a carbohydrate-rich diet that seemed to improve blood sugar control "combined with the *possible advantage* [my italics] of reduced serum cholesterol levels, warrants consideration and further study over a longer period of this type of diet for diabetic management."

Within months of the *New England Journal* publication, Bierman put aside the caveat that such a diet requires study over a longer period of time and initiated the process of shifting both the philosophy and the guidelines of the American Diabetes Association to align with what Keys and the American Heart Association had been promoting for a decade. To justify his thinking, Bierman embraced one of Frederick Allen's conclusions from the pre-insulin era: any diet will benefit patients with diabetes if it prompts them to eat less and lose weight. This was "the common thread," Bierman proposed, running through every diet intervention for diabetes from the sugar and milk diets of the nineteenth century to von Noorden's potato diet, Allen's starvation therapy, and even Newburgh and Marsh's high-fat diet.

Bierman then put the results of his ten-day diet trial into the context of those physicians who had promoted liberal carbohydrate diets in the insulin era. A handful of physicians had continued to publish their experiences with liberal carbohydrate diets in the post–World War II decades. In these reports, some of the diabetic subjects required more insulin to keep their blood sugar under control, but others did not.* By increasing the proportion of car-

---

* The report cited most frequently was a 1963 publication by the Iowa physicians William Connor and Daniel Stone. They reported on thirty-one diabetic patients who had

bohydrates in the diet—typically fiber-rich vegetables and legumes, while still prohibiting sugar and restricting bread consumption—and reducing dietary fat to balance the calories, these diets suggested a reasonable, if untested, compromise between the growing belief among cardiologists and nutritionists that dietary fat caused heart disease and the need to control blood sugar in diabetic patients.

When the American Diabetes Association in 1967 had first published "principles of nutrition" for patients with diabetes, a one-page "special report," the authors—eight members of a food and nutrition committee led by Laurance Kinsell of the Institute for Metabolic Research at Highland-Alameda County Hospital in Oakland, California—were still focusing on glycemic control as the primary goal, balancing carbohydrates to insulin dose, and noting that "in some patients greater fat and protein and consequently lesser carbohydrate may be required for [blood sugar] stability." Kinsell and his coauthors had been wary of low-fat diets because of their ability to elevate circulating levels of triglycerides, the blood fats of diabetic lipemia. They had also been open to the possibility that replacing saturated fat with polyunsaturated fat, mostly from vegetable sources, could reduce the risk of heart disease in diabetic patients, too, by lowering cholesterol. (Kinsell had been the first to report that the critical factor influencing serum cholesterol levels was not the amount of fat in the diet or the proportion of fat to carbohydrates, but the nature of the fats consumed, whether animal based or vegetable.) Ultimately, Kinsell's committee had left the decision up to individual physicians, who should decide on a patient-by-patient basis "whether or not control of plasma lipids as well as plasma glucose is of sufficient importance to warrant special dietary instruction."

Bierman became chairman of the ADA's Food and Nutrition

---

eaten diets that restricted cholesterol and fat, as the AHA was counseling, and increased carbohydrate consumption to over 60 percent of calories. Connor and Stone compared the health status of these patients after a year to twenty-five patients eating the more traditional carbohydrate-restricted, higher-fat diet. Their experimental diet, they reported, lowered cholesterol without increasing insulin requirements. These were "the preliminary results of a prolonged prospective study," they wrote. "We shall report later after longer clinical observations." They never did.

Committee in 1971 and took the opportunity to move the ADA guidelines fully into alignment with the AHA. In a two-page special report, "Principles of Nutrition and Dietary Recommendations for Patients with Diabetes Mellitus," Bierman's committee set about permanently shifting the association's perspective on diabetes and diet:

> There no longer appears to be any need to restrict disproportionately the intake of carbohydrates in the diet of most diabetic patients. Increase of dietary carbohydrate, even to extremes, without increase of total calories, does not appear to increase insulin requirement in the insulin-treated diabetic patient. In the less severe typically obese diabetic, substitution of carbohydrate for fat does not appear to elevate fasting blood glucose or worsen glucose tolerance in response to standard glucose loads.

Bierman's guidelines are another conspicuous example of the challenge of disseminating medical advice when the evidence is still so clearly ambiguous. Bierman's committee was balancing on a tightrope between conveying the message that eating less fat might prevent heart disease in diabetic patients and so more carbohydrates might be consumed instead, and the proposition that some types of carbohydrate-rich foods might be deleterious as well. The recommendations still included the advice that diabetic patients avoid "ingestion of simple sugars . . . because of the associated hyperglycemic peak," which meant avoiding sugar and flour, a significant portion of all the carbohydrates in the modern American diet.

Among the members of Bierman's ADA committee were researchers whose own studies suggested that the advice they were now proposing could worsen the long-term complications of diabetes—atherosclerosis and heart disease, in particular. The two-page report acknowledged the state of the evidence—what was known and what was not—in a paragraph that captured all the uncertainty of the time. The following italics are mine, contrasting what Bierman and his ADA committee were now hoping was true and what they also knew had never been tested and was very much open to debate:

> It is not yet possible to obtain uniform agreement as to the optimal proportion of carbohydrate and fat calories recommended for the diabetic. *The long-range effect of extreme changes, such as a high carbohydrate diet, on diabetic complications in a population not adapted to such diets is not known.* On the other hand, a liberalized carbohydrate intake for the diabetic will necessarily be associated with a decrease in dietary fat and cholesterol. *At the present time, there is not enough evidence to determine to what extent restriction of dietary fat and cholesterol is desirable.* However, diabetic patients appear to respond to reduction of saturated fat and cholesterol intake with a lowering of circulating cholesterol, as do nondiabetic persons. *There is at present no firm evidence that this restriction will retard the development of diabetic complications, particularly atherosclerosis.* However, the epidemiologic and experimental evidence relating circulating lipids to atherosclerotic cardiovascular disease also appears to apply to the diabetic patient.

While recognizing that there remained "important gaps in our knowledge," Bierman's 1971 guidelines effectively put an end to the centuries-long history of carbohydrate restriction for patients with diabetes. "Medical Group, in a Major Change, Urges a Normal Carbohydrate Diet for Diabetics" was the headline in *The New York Times*. "The American Diabetes Association has recommended that physicians encourage their diabetic patients to eat the same amount of carbohydrate foods—sugars, starches and celluloses—as people who are unaffected by the disease," the article began. "Diets high in carbohydrates do not raise the blood sugar," the *Times* quoted Bierman saying. "That's the misconception that most physicians have had during the last 30 years."

Just as the American Heart Association had become increasingly assertive about the potential benefits of low-fat diets with each successive edition of its guidelines, the ADA now did the same. In

1979, when the ADA published its next edition of the guidelines, John Brunzell coauthored the report, along with Frank Nuttall, a diabetologist and nutritionist working at the Veterans Administration hospital in Minneapolis.* They acknowledged that "the field of nutrition is a dynamic science," hence as "new facts emerge and concepts change" the guidelines would "continue to undergo evolution and modification." Nonetheless, they now specified that carbohydrates "should usually account for 50–60% of total calorie intake."

In 1986, after the results of the Lipid Research Clinics trial had been published and the NIH was launching its public health campaign to induce Americans to eat less fat, the minimal carbohydrate content of the diet prescribed by the ADA was nudged upward again to synchronize the messages, "ideally up to 55–60% of the total calories." Although the recommendations also included the suggestion that diets might have to be individualized "dependent on the impact on blood glucose and lipid levels." As for fat and cholesterol in the diet, the message remained unequivocal: "intake should be restricted."

The science, of course, was anything but unequivocal. Diabetologists were finding it increasingly difficult to reconcile what they had come to believe about dietary fat and cholesterol with an accumulating body of laboratory and clinical evidence about other factors that were also clearly associated with heart disease risk. Just weeks before the 1986 ADA guidelines were published, the National Institutes of Health hosted a conference for diabetologists to establish a consensus about the role of diet and exercise in treatment and prevention of type 2 diabetes. It turned out to be a very difficult task. Those attending the conference and the organizers as well—George Cahill, who had moved from Harvard to the Howard Hughes Medical Institute and chaired the meeting, and Robert Silverman of the NIH, who organized it—acknowledged candidly that, despite the

* By then, Bierman had become head of the American Heart Association's nutrition guideline committee.

ADA guidelines, much of what they thought they knew about the nature of a healthy diet for a diabetic patient had been contradicted by at least some evidence presented at the conference. They agreed that weight control was still the most important goal of dietary therapy for type 2 diabetes, perhaps the only effective method of controlling blood sugar and blood lipids, but then they acknowledged that they had little idea how to get their diabetic patients to achieve a healthy weight and then maintain it. "The long-term effectiveness of any diet therapy is terrible," the NIH endocrinologist Clifton Bogardus said, "and will remain terrible until we learn why people become obese." Some attendees were skeptical about claims that exercise could benefit either weight control or glycemic control (both "were disputed by speakers at the conference," *Science* reported); about the safety, effectiveness, and palatability of fiber-rich diets ("frankly, from the data we've seen, we're not impressed," Silverman said); and about a relatively new concept called the *glycemic index*, which is a measure of how quickly carbohydrate-rich foods are absorbed into the circulation and raise blood sugar (Cahill's personal opinion of that evidence: "a bucket of fluff").

Then there were the ADA guidelines themselves. Cahill and his colleagues seemed willing to accept that lowering cholesterol by diet—reducing total fat, saturated fat, and cholesterol—was a good idea, if for no other reason than because it was being disseminated so widely and now so vigorously by the NIH itself and its newly inaugurated National Cholesterol Education Program. But a fat-reduced diet meant a carbohydrate-rich one, and now that, too, had been challenged directly.

Among the researchers who gave a presentation at the NIH conference was the Stanford University endocrinologist Gerald Reaven. His presentation was a review of decades of research, much of it his own, that tied together the concept of insulin resistance—the failure of insulin to efficiently lower blood sugar that Falta and Himsworth had first identified in the 1930s and then Yalow and Berson had confirmed in the early 1960s with their radioimmunoassay—with type 2 diabetes, hypertension, obesity, and heart disease. If Reaven was right, and "his data speak for themselves," Cahill told

Science, then the carbohydrate-rich diet that the ADA had been advocating for fifteen years would clearly do more harm than good. Indeed, if Reaven was right, the same was true for all the chronic disorders associated with insulin resistance. The diabetologists at the conference had been confronted with mutually exclusive theories of nutrition—competing paradigms—and it had left them at a loss. "High protein levels can be bad for the kidneys," Silverman said. "High fat is bad for your heart. Now Reaven is saying not to eat high carbohydrates. We have to eat something." And then he added, "Sometimes we wish it would go away, because nobody knows how to deal with it."*

---

* The existence of the conflict was readily apparent in the 1986 ADA nutrition guidelines, although the authors ultimately chose to ignore the implications. The report included five "conclusions [that] can be drawn from the literature," four of which implied that the dietary advice being disseminated with such confidence was the wrong advice for those with diabetes.

11

Diabetes and Heart Disease

Science is helplessly opportunistic; it can pursue only the paths opened by technique.

—HORACE FREELAND JUDSON,
The Eighth Day of Creation, 1979

When unproved hypotheses are enthusiastically proclaimed as facts, it is timely to reflect on the possibility that other explanations can be given for the phenomena observed.

—EDWARD AHRENS, JR., AND COAUTHORS,
"Dietary Control of Serum Lipids in Relation to Atherosclerosis," 1957

The problem, once again, was timing.

In the 1960s, the fat-cholesterol hypothesis filled a vacuum in our understanding of the relationship between diet and heart disease. It did so with evidence that could neither confirm it was right nor definitively prove that it wasn't, but health agencies and associations had embraced it nonetheless. They took, as the NIH's Basil Rifkind had implied, a leap of faith. Once that vacuum had been filled, though, any competing hypothesis had the far more difficult challenge of displacing the existing paradigm. It had to convince all those influential researchers, clinicians, and public health officials who had come to believe that the problem had been solved, and had acted on that belief, that they were wrong, perhaps tragically so. The health agencies and associations that were disseminating dietary guidelines based on the fat-cholesterol hypothesis would also have to accept that their advice had been ill-conceived, a difficult admission to make. By not doing so, they maintained their credibility, but sacrificed the underlying science in the process—the understanding of what might be the true mechanistic relationship of diet, diabetes, and heart disease.

It's precisely to prevent this kind of situation that good scientists promote the importance of advancing ideas cautiously and tentatively, to trust only what they can prove, as the Nobel laureate physicist Luis Alvarez memorably put it. That way, when an idea turns out to be misguided—as is most likely with any scientific hypothesis—those who promoted it can accept that reality, adjust their thinking as necessary, and move on. Also true is that scientists commonly ignore this wisdom, which is why one of the more famous lines about scientific progress is that it proceeds "funeral by funeral," a paraphrasing of an observation made by the great German physicist Max Planck that new theories are embraced not because the evidence and the logic convince those scientists who believed and endorsed the old theories, but because those proponents eventually die off.

For those researchers studying the confluence of obesity, diabetes, and heart disease, the conflict was between two competing explanations for the role of diet in these disorders, and, specifically, in atherosclerosis and heart disease. The conflict was accentuated for diabetologists because their patients were so very likely to develop atherosclerosis and to die from it.

The fat-cholesterol hypothesis presupposed two entirely different causes of heart disease depending on whether a patient had diabetes or did not. For those without diabetes, atherosclerosis was caused by eating saturated fat, which raised LDL cholesterol. Those who did have diabetes, though, typically had normal cholesterol levels. Even their LDL cholesterol was normal. So why did they suffer so young and so severely from atherosclerosis? Why did so many of them die from it?

While Joslin had assumed in the 1920s that it was because of the fat they had fed their diabetic patients pre-insulin, diabetologists, including those at Joslin's clinic, would assume implicitly that the accelerated atherosclerosis in their patients was caused by the metabolic abnormalities that were so specific to the diabetic condition itself—the high blood sugar and poor glycemic control. This was why heart disease was considered the most deadly of the macrovascular complications of the disease. And this was a primary reason why diabetologists considered glycemic control so important, because it might prevent or delay heart disease. For these diabetic patients, saturated fat in the diet would supposedly exacerbate their heart disease risk just as it would in healthier individuals. This is why diabetologists were now telling their patients to avoid it, but it was not the reason why these patients were at such high risk.

The alternative hypothesis or competing paradigm was what scientists and philosophers of science would describe as more parsimonious, a simpler hypothesis. This is what Reaven's work had implied, and what Joslin had in mind when he had suggested in 1950 that "research in the prevention of arteriosclerosis should be carried out primarily in diabetics with a premature tendency to it." This hypothesis proposes that atherosclerosis and heart disease in everyone proceeds through identical mechanisms, these same diabetes-related factors: not just high blood sugar and poor glycemic control, as this was beginning to be understood by the 1960s, but also insulin resistance and the hyperinsulinemia that inevitably goes with it. And because those with diabetes are *more* insulin resistant, *more* hyperinsulinemic, and have *poorer* glycemic control than those without—the diagnostic criteria, in effect, of diabetes—they suffer heart disease earlier and more severely.

By this thinking, the diet that best prevents heart disease in anyone, and *particularly* so in diabetes, is the diet that does the best job of controlling blood sugar, reversing the insulin resistance—if such a thing is possible—and minimizing circulating levels of insulin. In those with diabetes, such a diet would also serve to minimize the use of hypoglycemic medications, whether insulin or oral drugs like the sulfonylureas, just as Joslin would have advised in the first decade of insulin therapy.

That, in turn, led to the cognitive dissonance that manifested itself so conspicuously at the 1986 NIH consensus conference on diabetes: the diet that would do the best job of glycemic control and minimize circulating levels of insulin would very likely be the diet with the fewest carbohydrates and the most fat, different only in the details from what Newburgh, Marsh, and Petrén had been proposing in the earliest years of the insulin era. And this dietary prescription was the opposite of what the ADA had come to embrace. This alternative scenario resolves the cognitive dissonance by assuming that the association of saturated fat, LDL cholesterol, and heart disease had led down a blind alley—one of the many hypotheses in science that at one time seemed reasonable but never pan out.

Among the critical factors that determine the progress of science, as I said earlier, is what specific questions that researchers ask. But the questions they can ask are determined by the technologies available to observe and study the problem, as also noted. These technologies will determine what can be seen and measured. As the technologies evolve or new technologies come into play, more and more facets of the problem can be observed and quantified, and the hypotheses created to explain them will evolve as well. Often the researchers who make the significant breakthroughs are the ones who develop the necessary technology, as Yalow and Berson had in the understanding of type 2 diabetes as a condition of insulin resistance and hyperinsulinemia. The psychologist Daniel Kahneman, a Nobel Prize winner in economics in 2002, described the context of this technology-dependent scientific progress simply as "What You See Is All There Is" (WYSIATI, for short). By having very limited information of the kind available when scientific or medical disciplines are new and the technologies for studying the relevant problems are relatively primitive and limited, researchers can put together compelling stories about what they're studying. They can never know, however, if their story, their hypothesis, is correct because they don't know what they're not seeing, what new measurement technologies will tell them about the problem.

That this had happened with the cholesterol hypothesis of heart disease is clear. Physicians, cardiologists, and even diabetologists, as we saw earlier, came to believe elevated cholesterol levels in the blood cause heart disease because cholesterol was the component of the blood that they could measure and that seemed relevant to the disease. The test to do so was inexpensive. Because Ancel Keys had been aware of a cholesterol hypothesis and Margaret Keys had become adept at measuring cholesterol in blood samples, cholesterol is what they measured and it remained at the core of their thinking ever after.

As the technology improved, though, and new methods and devices became available to measure other constituents of the blood, allowing researchers to learn more about the pertinent physiology of the human body—how cholesterol is ferried around the circulation, in what particles, and with what other lipids (blood fats); how and why they might infiltrate the endothelial lining of the veins and arteries; the role of insulin and other hormones in facilitating these processes—new relationships could be observed between the components of the blood and heart disease itself. From these observations, new hypotheses could be generated about what might ultimately be cause and to what effect.

In 1950, when a team of Harvard biochemists published the details of a new analysis scheme for studying the components of human blood, their article began with a paragraph that captured the remarkable but nonetheless unsurprising biological complexity of the substance they were studying:

Human blood is a tissue of great complexity: of formed elements, erythrocytes, lymphocytes, granulocytes and platelets; of plasma proteins; serum albumins of more than one kind, with a sulfhydryl group and without, globulins of many kinds, some euglobulins, some pseudoglobulins, some lipoproteins, glycoproteins or mucoproteins; proteins concerned with blood coagulation; the metal-combining protein; the hormones which, by definition, are the chemical messengers transported in the blood, but which have but rarely been separated from the blood and characterized in this state of nature; diverse antibodies and complement components concerned with immunity; and the enzymes from which substrate proteins must be separated.

Any of these blood constituents could be playing causal roles in diseases of the arteries, in arteriosclerosis and atherosclerosis, specifically, as the researchers newly capable of quantifying them now argued. These researchers, however, were coming late to the game. The advocates of the cholesterol hypothesis were already declaring that they had successfully answered the relevant questions. They had to shift their thinking as necessary to adapt to new revelations that seemed indisputable—that saturated fat, for instance, raised cholesterol levels, not all dietary fats as Keys had been proposing; and that the cholesterol in low-density lipoproteins was associated with heart disease, not all cholesterol, as Keys had also insisted— but never straying from the cholesterol focus of the story.

It didn't help that the new measurement technology could be prohibitively expensive—specifically a device known as an ultracentrifuge that only became commercially available in the late 1940s and cost \$30,000 for a single machine (equivalent to more than \$350,000 in today's dollars). Only a very few laboratories could afford one; practicing physicians and their clinics could not. Other new methods of chemical analyses were considerably less expensive but required a level of biochemical and technical expertise that few physicians possessed.

Public health and medical authorities understandably saw little benefit to embracing new measurements, and so new risk factors for heart disease, that could not be performed by the physicians or clinics whose job it was to treat and, ideally, prevent the disorder. Hence, what the researchers would learn from these chemical assays and from their ultracentrifuges, vitally important information for understanding the mechanisms of the disease itself, would remain very much minority positions, with little impact. The messaging from health authorities remained the same.

While hundreds of millions of dollars would be allocated to testing the fat-cholesterol hypothesis (and failing to confirm it), virtually none would be spent on trials that could test the proposition that emerged, time and time again, from these parallel lines of research: that it's the quantity and particularly the quality of the carbohydrates we eat and the influence of these carbohydrates, specifically, on insulin signaling and glycemic control, that has the greatest influence on our health, not just for heart disease, but obesity, diabetes, and a host of other related chronic conditions.

The ultracentrifuge and advances in biochemical analysis initially developed during the war years had set off parallel revolutions in the understanding of what physiologists were calling intermediate metabolism—not just in our understanding of the mechanisms working to store fat in the adipose tissue, but also those working to synthesize fat in the liver, shuttle fat and cholesterol around the body, and distribute them to tissues and cells as necessary. By the early 1950s, this was yet another scientific discipline that was populating with researchers who had entered the field after the war and were putting these new assays and technologies to work to understand the pathology of heart disease and atherosclerosis.

Among the first revelations to emerge was that virtually all of the cholesterol in the blood is carried around in relatively giant molecules known as lipoproteins, two classes of which were differentiated early by their density and chemical characteristics: alpha and beta lipoproteins. The alpha lipoproteins are what we today call high-density lipoproteins or HDL. The very first reports in 1951, from a team of physicians at New York Hospital–Cornell Medical Center, documented abnormally low levels of these HDL particles in patients who had already had heart attacks as well as in diabetic patients with their accelerated atherosclerosis. This was the beginning of the awareness that the cholesterol in HDL particles was *inversely* or *negatively* associated with heart disease; the less cholesterol in these HDL particles, the greater the risk of heart disease. This is why the cholesterol in HDL eventually came to be known as "the good cholesterol." The more we have circulating in the bloodstream, the less likely we are to have a heart attack, although whether this is a cause or merely an association is still debated.

Nonetheless, the observations on HDL cholesterol were mostly ignored for the next twenty years because they seemed to contradict the proposition that high cholesterol itself was a bad thing. It "simply ran against the grain," as Tavia Gordon, biostatistician for the renowned Framingham Heart Study, later wrote. "It was easy to believe that too much cholesterol in the blood could 'overload' the system and hence increase the risk of disease . . . but how could 'too much' of one part of the total cholesterol reduce the risk of disease? To admit that fact challenged the whole way of thinking about the problem."

In 1977, the association between low HDL cholesterol and heart disease was confirmed definitively by Gordon and a collaboration of researchers, analyzing the data from five large population studies, including Framingham. The "striking" revelation, as the researchers reported in two articles on the analysis, was that the amount of cholesterol in HDL particles was far and away the *single* best predictor of heart disease risk of all those they had studied. Individuals with low HDL cholesterol were four times more likely to have a heart attack as those whose HDL was higher but whose LDL cholesterol (aka "the bad cholesterol") was also high. And this inverse association between HDL cholesterol and heart disease held true for every age group in these populations, in both men and women, and in every ethnic group. The HDL cholesterol was the only cholesterol-related risk factor that did. In 1977, the only three lifestyle interventions that were known to raise HDL cholesterol, as Gordon and his coauthors noted, were physical activity, weight loss, and eating a low-carbohydrate, and so high-fat, diet, the opposite of what the AHA and ADA were now recommending.

The beta lipoproteins would be subdivided into two fractions of their own: the low-density lipoproteins, familiar today as LDL particles, and very-low-density lipoproteins, or VLDL, which played little role in the *public* discussion of heart disease prevention but are critical to understanding the pathology of the disease. Most of the triglycerides (the fats) in the blood are found in these VLDL particles, while the LDL particles contain most of the cholesterol.

These lipoproteins are composed primarily of a fat (a lipid and, in this case, a type known as a *phospholipid*) and a protein. The phospholipid, along with some cholesterol, makes up the outer membrane; the protein (known today as apolipoprotein B, or apoB, for short, with the B standing for "beta") provides the structure that holds it together. Inside is more fat—in the form of triglycerides, essentially a droplet of oil—and more cholesterol. The size and the density are determined by the amount of triglycerides inside the membrane—how big the oil droplet, and how much cholesterol is present.

These beta lipoproteins are synthesized in the liver, which then secretes them into the circulatory system, the great majority as very-low-density lipoproteins. As they travel through the circulation, the triglycerides they're carrying are broken down into fatty acids by enzymes known as *lipoprotein lipases* on the membranes of cells. If the lipoprotein lipases are on fat cells, then the fatty acids will be stored; if on skeletal or cardiac muscles, they'll be used for fuel. In this sense, lipoproteins are delivery vehicles that the liver loads with fat—triglycerides—and then ships out into the circulation so that the fat can be deposited in fat cells for storage and used in other tissues for energy. As that happens, as the oil droplet in the lipoprotein gets ever smaller, the lipoprotein itself gets smaller and more dense, and a very-low-density lipoprotein (by this somewhat confusing terminology) transforms into a *low*-density lipoprotein. This is why LDL particles would come to be thought of as *remnants* of VLDL particles, and VLDL particles as *precursors* of LDL.

In the decade following the end of the Second World War, one of the very few laboratories in the world that had access to ultracentrifuges capable of studying these lipoproteins was at the University of California, Berkeley. It was run by John Gofman, a physician and physical chemist. Gofman had spent the war years working with the Manhattan Project developing a process to isolate plutonium for hydrogen bombs. After the war, he turned his attention and expertise back to medicine, heart disease specifically, using two ultracentrifuges, one to separate out the lipoproteins from the blood serum taken from patients, the other to determine how those lipoproteins differed by density.

In 1950, writing with his research colleagues at UC Berkeley, Gofman published an article in *Science* that was controversial at the time but would later be credited with inaugurating the modern era of cholesterol research. It described how the ultracentrifuge could be used to identify these different classes of lipoproteins—a process known as *fractionation*, resulting in the identification of these different lipoprotein *fractions*—and how the low-density lipoproteins, in particular, seemed to be abnormally elevated in patients who had heart attacks.

This was the beginning of the idea that the cholesterol in the LDL particle itself is the causal agent in atherosclerosis—hence, "the bad cholesterol"—but that was not the primary focus of Gofman's article. The more important observation, Gofman argued, was that the amount of cholesterol in the circulation (as measured by the kinds of test that Margaret Keys had done) did not reflect or correlate with the concentration of these low-density lipoproteins: "At a particular cholesterol level," Gofman wrote, "one person may show 25 percent of the total serum cholesterol in the form of [low-density lipoproteins], whereas another person may show essentially none in this form."

This could explain why patients with atherosclerosis often had relatively low cholesterol, and those without often had high. Maybe it wasn't the cholesterol carried in the lipoproteins that was causing the atherosclerosis—that was *atherogenic*—but the lipoproteins themselves. The fact that cholesterol happened to be easy to measure did not necessarily mean it was the agent of disease. Maybe the problem was too many of these lipoproteins (or too few, as with HDL particles), or maybe the very structure of these molecules was defective in those who developed atherosclerosis. This might be particularly relevant to diabetes, Gofman and his colleagues explained: since high cholesterol "does not account for the extraordinary susceptibility" of these diabetic patients to atherosclerosis, maybe some abnormality of these lipoproteins did.

By focusing on the lipoproteins that ferried cholesterol around the circulation rather than the cholesterol itself, the dietary implications of the research changed. In the late 1950s, Gofman and his colleagues were reporting that both LDL and VLDL particles were found in abnormal concentrations in patients with heart disease.* But these fractions responded quite differently to dietary modifications. The amount of LDL in the circulation could be elevated by eating saturated-fat-rich foods, Gofman confirmed, but the VLDL went up in response to eating carbohydrates. The VLDL concentra-

---

* They were not measuring HDL in these studies.

tion in the circulation could be reduced only by restricting carbo-hydrates, a fact that Gofman and his collaborators deemed critical to the dietary prevention of heart disease.

In 1958, two years before Keys and the AHA began institutional-izing the idea that men at high risk of heart disease should avoid the fatty foods in the diet and eat more carbohydrates instead—hence lowering total cholesterol—Gofman was calling the mea-surement of total cholesterol a "false and highly dangerous guide" and arguing that "generalizations such as 'we all eat too much fat,' or 'we all eat too much animal fat'" would do more harm than good for a significant portion of the population, raising their risk of heart disease if they followed the advice. Among those who would be harmed by eating a low-fat diet, Gofman's research implied, would be patients with diabetes or anyone overweight or obese, precisely those who are at highest risk of having a heart attack.

Other researchers were by then coming to similar conclusions, specifically those researchers who had the ability to measure and quantify the triglyceride content of the blood—the fats carried by the VLDL particles rather than the lipoprotein particles them-selves. At Rockefeller Medical Center in New York, Edward "Pete" Ahrens was studying the kind of lipemia—Joslin's "milky appear-ance of the serum"—that characterized the blood of patients with diabetes in the years before insulin therapy and with poorly con-trolled diabetes afterward. Ahrens would go on to become among the most influential and respected scientists in the field of lipid metabolism, although his influence would not be sufficient to suc-cessfully challenge the fat-cholesterol hypothesis. To make progress in his research, Ahrens needed a method to study fat and choles-terol compounds circulating in the blood, so he pioneered the use of a technique called gas-liquid chromatography to do so (helping in the process to "revolutionize the field of lipid chemistry," as the American Heart Association later said of his work).

Ahrens's first contribution to the heart disease research had been to establish definitively, building on the work of Laurance Kinsell in Oakland, that circulating cholesterol levels respond not to all the fat consumed, as Keys was then claiming, but to the saturation of the fats. Saturated fats, whether of plant or animal origin, tended to increase the total amount of cholesterol in the blood, while unsaturated fats, whether of plant or animal origin, lowered it. In the course of that research, feeding liquid diets to volunteers, Ahrens and his Rockefeller colleagues realized that while cholesterol was responding to the type of fat consumed, the triglycerides in the blood were responding to the carbohydrates; the lower the fat content of the diet and so the higher the carbohydrates, the greater the concentration of triglycerides in the blood.

In 1961, Ahrens reported that only a small minority of the patients he had seen in his medical practice with lipemia, the condition that Joslin had found so worrisome, had what Joslin would have assumed and Ahrens now called "fat-induced lipemia." In these patients, the lipemia was caused, apparently, by a genetic defect that left them unable to clear fat from their circulation after a meal.*

In the great majority of Ahrens's patients, the lipemia was caused by carbohydrate consumption, not by fat, what Ahrens now called "carbohydrate-induced lipemia." "This phenomenon of carbohydrate-induced lipemia is still not commonly appreciated," he wrote, while pointing out that diabetologists had clearly demonstrated in the early days of the insulin era that the lipemia in their patients could be resolved by feeding their patients high-fat, low-carbohydrate diets of the kind Newburgh and Marsh had used at the University of Michigan. As Ahrens and his Rockefeller colleagues now reported, these patients with carbohydrate-induced lipemia were converting the carbohydrates they ate into fat (specifically a type of saturated fat, palmitic acid, that would be blamed for raising cholesterol and causing heart disease), and it was this fat, in the form of triglycerides, and the VLDL particles carrying them, that was accumulating in the blood to excess after meals and causing its milky appearance.

---

* This fat is carried from the gut to the fat tissue by lipoproteins called chylomicrons, which had been discovered in 1924. Just as it does with VLDL particles, lipoprotein lipase on the fat cell membrane breaks down the triglycerides in the chylomicrons into fatty acids, which then flow into the fat cells. Ahrens's patients with fat-induced lipemia were "deficient in lipoprotein lipase activity," he reported; it was the chylomicrons accumulating in the blood that constituted the lipemia they developed after a meal.

Moreover, Ahrens suggested, this carbohydrate-induced lipemia might be "a common phenomenon," perhaps even "an exaggerated form of the normal biochemical process which occurs in all people on high-carbohydrate diets." If this prolonged hypertriglyceridemia—high levels of triglycerides in the blood—persisted, he added, then that, too, might initiate or accelerate atherosclerosis and so be causing heart disease.

In the spring of 1961, Ahrens reported his findings at a meeting of the Association of American Physicians in Atlantic City, New Jersey, just three months after Ancel Keys had been on the cover of *Time* magazine suggesting the entire nation eat less fat. *The New York Times* reported his conclusion "as something of a surprise to many of the scientists and physicians attending the meeting . . . that dietary carbohydrate, not fat, is the thing to watch in guarding against [atherosclerosis and heart disease]."

Ahrens was already bolstering his conclusions by then with the work of two Yale University researchers, Margaret Albrink and Evelyn Man, who were documenting the next link in the mechanistic chain of cause and effect: between the concentration of triglycerides in the circulation and heart disease itself. Since the early 1930s Man had worked with John Peters, a coauthor of the seminal textbook on the chemical analysis of blood serum (*Quantitative Clinical Chemistry*, the first edition of which was published in 1931) and director of the metabolic division in the Department of Internal Medicine at Yale. Peters and Man's Yale laboratory had been one of the very few in the world in the 1930s capable of quantifying and differentiating the types of lipids in the circulation. When Elliott Joslin had begun suggesting that the atherosclerosis he was seeing in his patients was caused by the fat and cholesterol in their diets, and diabetologists had argued for ever more carbohydrate-rich diets, Peters and Man had set out to see if Joslin was right. In 1935, they reported that of seventy-nine diabetic patients they had tested repeatedly at Yale, most did not have high cholesterol levels, and of those who did, most had complications that were likely responsible. "It is quite as impossible," they had written, "to relate either cholesterolemia [high cholesterol] or lipemia [high triglycerides] to fat intake." Moreover, they wrote, "severe arteriosclerosis, with or without hypertension, was evident in patients with serum cholesterols normal, below normal and above normal." This was the reason why both Man and Peters had remained skeptical of the cholesterol hypothesis.

Margaret Albrink had joined Peters's laboratory in 1952 with a medical degree from Harvard and set about comparing the triglyceride and cholesterol levels of healthy individuals (medical students, hospital personnel, and industrial workers) to that of Grace–New Haven Community Hospital patients with heart disease. In 1959, three years after Peters's death, Albrink and Man reported that abnormally high triglycerides seemed to be a far better predictor of heart disease than cholesterol: only 5 percent of the healthy young men they had studied had elevated triglycerides, compared with at most 30 percent of healthy middle-aged men and up to 90 percent of the coronary patients. Their result, they wrote, "suggests that an error in the metabolism of triglycerides is the lipid abnormality operative in coronary artery disease." And the high triglycerides themselves "could not be attributed to dietary fat or excess calories." Two years later, they reported similar numbers from a new group of 212 heart disease patients now compared to over 500 "apparently healthy persons," mostly workers from two nearby factories. The findings were identical. High triglycerides predicted heart disease far better than did high cholesterol, and it was conceivable that when high cholesterol did associate with heart disease it did so as a result of "an increasing concentration of triglycerides" as Gofman's research had also suggested. Albrink, too, had presented her results and conclusions at the 1961 Association of American Physicians meeting in Atlantic City, and while the *Times* did not report on her presentation, she later said it "just about brought the house down . . . people were so angry; they said they didn't believe it."

Albrink and Man's observation that high triglycerides predicted heart attacks better than high cholesterol would also be ignored by the American Heart Association, although it would be repeatedly confirmed by researchers with the requisite technical expertise. These included the future Nobel laureate Joseph Goldstein and his University of Washington colleagues (including Ed Bierman), who reported in 1973 that heart attack survivors in the Seattle area were three times more likely to have abnormally high triglycerides than high cholesterol. When *The New England Journal of Medicine* ran a seminal five-part, fifty-page series on the major disorders of fat and cholesterol metabolism in 1967—authored by Donald Fredrickson, a future director of the National Institutes of Health; Robert Levy, a future director of the National Heart, Lung, and Blood Institute; and Robert Lees of Rockefeller University—four of the five disorders they described were characterized by abnormally elevated levels of triglycerides in the very-low-density lipoproteins.

Fredrickson, Levy, and Lees also warned about the dangers of advocating low-fat diets for everyone for the same reason Ahrens and Gofman had, and now Albrink and Man were: increasing carbohydrate consumption at the expense of fat would make these all-too-prevalent disorders worse. By far the most common of the five, they wrote, which they designated Type IV and described as "sometimes considered synonymous with 'carbohydrate-induced lipemia,'" *had* to be treated with a low-carbohydrate diet, although they did not define how low the carbohydrate content had to be. "Patients with this syndrome," Lees later wrote, "form a sizable fraction of the population suffering from coronary heart disease." In these patients, he added, "it is often useful to redistribute calories to keep carbohydrate intake to a minimum and increase fat and protein intake."

Patients with this particular syndrome, as Albrink and Man would report, also formed a sizable fraction of the population suffering from diabetes. By 1963, Albrink and Man had analyzed the medical records and blood samples from patients who had been treated for diabetes at New Haven Hospital since 1931. The prevalence of atherosclerotic complications in these patients, they reported, had increased dramatically, from one in ten diabetic patients with atherosclerosis in the 1930s to more than one in two in the 1950s. While the cholesterol concentration in the blood of these patients had barely changed over those years, the triglycerides had steadily increased, and those diabetic patients who had atherosclerosis had the highest triglycerides. All of this had happened coincident with the change in diets prescribed to diabetic patients over the intervening years. As these patients had been encouraged to eat ever more liberal carbohydrate diets, Albrink and Man reported in their 1963 paper in *Annals of Internal Medicine*, the triglyceride concentrations in their blood had increased, as had the prevalence of atherosclerosis.

In 1965, Albrink, now on the faculty of the University of West Virginia and working with Paul Davidson, a young physician, reported that they had measured insulin levels in their patients using the radioimmunoassay that Yalow and Berson had developed. The resistance to insulin—high blood sugar despite high insulin as well—that Yalow and Berson had observed in both those with obesity and with adult-onset diabetes, Albrink and Davidson now reported, was also common in individuals with abnormally elevated triglyceride levels. This train of associations from insulin resistance to hypertriglyceridemia to atherosclerosis suggested that all these disorders—heart disease, diabetes, and obesity—were intimately linked with, or perhaps linked *by*, the insulin response to the carbohydrates consumed. In the British medical journals, John Yudkin would cite Albrink's research as support for his sugar hypothesis—the metabolic abnormalities that Albrink, Ahrens, and Gofman were describing were identical to what he was causing in his laboratory animals and human subjects with sugar-rich diets.

Unraveling the physiological mechanisms at work and, perhaps more important, convincing diabetologists to take this science seriously, took another quarter of a century. The Stanford endocrinologist Gerald Reaven deserves much of the credit for both accomplishments.

Reaven had first arrived at Stanford in 1960, a young physician with a medical degree from the University of Chicago. In 1963, he reported that patients who had heart attacks often had poor glycemic control—their blood sugar rose unnaturally high after consuming carbohydrates, the phenomenon known technically as *glucose intolerance*—despite never having been diagnosed with adult-onset diabetes. By 1967, working with John Farquhar, a physician who had come to Stanford after training with Pete Ahrens at Rockefeller, Reaven had determined that the overproduction of triglycerides—Ahrens's carbohydrate-induced lipemia—was triggered by the insulin response to the carbohydrates, and this was the case in heart disease patients, with and without diabetes. The more insulin secreted, the greater the conversion of carbohydrates to triglycerides by liver cells, the greater the concentration of triglycerides in the blood, and the greater the risk of heart disease.

Reaven and his colleagues then set about developing a technique that would allow them to quantify the phenomenon of insulin resistance in their patients. By 1974, three years after the ADA had removed the restriction on carbohydrate consumption in diabetic patients, Reaven reported that the individuals they had tested whose triglycerides rose so dramatically in response to carbohydrate-rich meals were those who were insulin resistant. The two conditions went hand in hand. And this close correlation had been observed, they wrote, "within a group of subjects similar to the kinds of patients seen by the average physician on a diet which attempts to approximate that of the average American."

This condition of being insulin resistant, as Reaven and his colleagues came to understand it, is both central to all the metabolic abnormalities they were observing and a vicious cycle. After eating a meal that contains carbohydrates, the pancreas responds to the rising tide of blood sugar by secreting insulin. As this happens, the cells in the body become resistant to insulin's action. The resistance to insulin leaves blood sugar abnormally elevated, continuing to signal the beta cells to secrete ever more insulin. Now hyperglycemia (high blood sugar) and hyperinsulinemia (elevated insulin levels) occur simultaneously. So long as the pancreas can continue to secrete more insulin to compensate for the insulin resistance, the patient can appear healthy, even as the liver is now overproducing triglycerides and the hyperinsulinemia *and* hyperglycemia are doing damage on their own. ("In this proposal," Reaven and Farquhar had written in their 1967 analysis, echoing Claude Bernard's thoughts on diabetes from a century earlier, "the liver [is] considered the unwitting victim of an excess supply of potential substrate [carbohydrates] for triglyceride synthesis and of a hormone [insulin] which may enhance the synthetic process.")

Type 2 diabetes, in this scenario, is only diagnosed when the pancreas becomes unable to continue secreting sufficient insulin in response to the blood sugar, when the beta cells become exhausted and the supply of insulin fails to keep up with the demand. In both diabetic patients and those individuals for whom the diabetes has not yet manifested itself and been diagnosed, and maybe never will be, the "price paid," as Reaven and Farquhar described it, is "most likely accelerated atherogenesis."

From the mid-1960s onward Reaven and his collaborators published the results of trials testing the short-term effects of diets of varying degrees of carbohydrates and fat on insulin resistance, hyperinsulinemia, and hypertriglyceridemia (carbohydrate-induced lipemia). These consistently confirmed that the more carbohydrates consumed and the less fat, the greater the metabolic disruption. That this condition was made worse by the sugar content of the diet was also a consistent finding in the Stanford trials, as Aharon Cohen and John Yudkin might have predicted. By the late 1970s, Reaven and his collaborators were warning that prescribing carbohydrate-rich diets for diabetes was a mistake and would accelerate the development of heart disease in these patients. At the very least, as Reaven wrote in a 1986 review, diabetologists had little idea of "the appropriate amounts and kinds of carbohydrates [patients with diabetes] should eat. . . . What is required is less advice and more information."

In 1988, the same year that C. Everett Koop, the U.S. surgeon general, had declared that two-thirds of all premature deaths in America could be blamed on the consumption of dietary fat, Reaven won the ADA's prestigious Banting Medal for Scientific Achievement and, in his Banting Memorial Lecture, implicated the carbohydrate content of the diet instead. Reaven reviewed this science now in the context of what he was calling Syndrome X, a condition common to type 2 diabetes, obesity, and heart disease and characterized by a cluster of half a dozen metabolic abnormalities: insulin resistance, poor glycemic control, high levels of insulin, high triglycerides, low HDL cholesterol, and high blood pressure (hypertension). By then, as Reaven noted, multiple studies from around the world had already linked hyperinsulinemia to a higher risk of heart disease and to hypertension, which had been accepted as a major risk factor for heart disease since the Framingham Heart Study had first reported the association in 1961.* Insulin resistance and hyperinsulinemia, though, played "the central role," as Reaven described it, driving the development of all of these chronic disorders: type 2 diabetes, heart disease, and hypertension. "Although this concept may seem outlandish at first blush," Reaven had said, "the notion is consistent with available experimental evidence."

It would take health organizations another decade to accept the science of Reaven's Syndrome X, beginning with the World Health Organization in 1999, which renamed it "metabolic syndrome." None of these organizations, though, would accept the implications unless they could be reconciled with what they had already embraced as true: that heart disease is caused by the saturated fat in our diets raising LDL cholesterol. In the United States, this may have been unavoidable. The only federal agency now publishing guidelines to treat and prevent heart disease was the National Cholesterol Education Program (NCEP), which itself had been founded with the purpose of (and was still dedicated to) educating the public on the dangers of high cholesterol and the benefits of cholesterol-lowering therapies.

In 2002, the NCEP published the third edition of its "adult treatment guidelines" and accepted metabolic syndrome "and its associated risk factors" as a contributor to premature heart disease—as harmful as smoking cigarettes, but not as important as LDL cholesterol, which still trumped all other considerations. It was LDL cholesterol that still remained "the primary target of therapy," which, therefore, determined the ultimate nature of the dietary guidelines as well. While acknowledging that "the greatest potential for management of [metabolic] syndrome lies in reversing its root causes," the treatment guidelines defined a "therapeutic lifestyle" diet as one

---

* By the time of Reaven's Banting Lecture, University of Pennsylvania researchers had demonstrated that insulin signaled the kidney to retain sodium, hence raising blood pressure in response. Lewis Landsberg, an endocrinologist then at Harvard and later dean of the Northwestern University Feinberg School of Medicine, had shown that insulin also stimulated sympathetic nervous system activity and that, too, worked to elevate blood pressure. "Hypertension may thus be viewed as a pathophysiologic consequence of the hyperinsulinemia that complicates insulin resistance," Landsberg would write.

that reduced the intake of saturated fats and cholesterol. The goal of dietary therapy was still "maximum reduction of LDL cholesterol," and only after that was achieved should emphasis shift to metabolic syndrome. Hence, carbohydrates were still expected to constitute 50 to 60 percent of the calories, albeit now with the compromise that the carbohydrates, ideally, should come from whole grains, not refined, and from fruits and vegetables. And, of course, since metabolic syndrome was so closely associated with obesity, those who suffered from it should lose weight and the rest of the public should be advised on how not to gain it.

The cognitive dissonance that might have emerged from accepting two conflicting hypotheses of heart disease as true was resolved in the NCEP report by characterizing metabolic syndrome as an "emerging risk factor" for heart disease and blaming its cause on the rising tide of obesity itself. Thus, heart disease in America, according to the NCEP report, was still officially caused by "mass elevations of serum LDL cholesterol result[ing] from the habitual diet in the United States, particularly diets high in saturated fats and cholesterol," but this risk could be modified by metabolic syndrome in those who had difficulty with either weight or glycemic control.

The primary author of the report—and subsequent similar reports written under the auspices of the AHA, ADA, and NIH—had been a single individual: Scott Grundy, a nutritionist and specialist in the metabolism of blood lipids at the University of Texas Southwestern Medical Center. In interviews, Grundy was more circumspect than his reports had been. He would not say that heart disease was *still* caused primarily by elevated LDL cholesterol, because obesity was now so rampant in this country, but rather that it *had been*, and so all the researchers involved, at one point in time, at least, had been right. "What you're faced with," Grundy told me in 2004, "is a historical change in peoples' habits. Going back to the nineteen-forties, fifties, and sixties, people ate huge amounts of butter and cheese and eggs and they had very high LDL levels and they had severe heart disease early in life, because of such high cholesterol levels. What's happened since then is there has been change in population behavior and they don't consume such high quantities of saturated fat and cholesterol anymore, and so LDL has come down a great deal as our diets have changed. But now . . . we have got obesity and most of the problem is due to higher carbohydrate consumption or higher total calories. And so we're switching more to metabolic syndrome." It was a way to reconcile two conflicting hypotheses of heart disease, such that neither (and, perhaps equally important to these researchers, no one) was necessarily wrong, but, once again, it may have sacrificed the implications of the science by doing so.

The acceptance of metabolic syndrome as a significant risk factor for heart disease raised a series of questions that had confronted diabetologists since their first awareness in the post-insulin era of the epidemic of arteriosclerosis and atherosclerosis. Since both metabolic syndrome and type 2 diabetes are characterized by insulin resistance, hence elevated levels of both insulin *and* blood sugar, and both associate with an increased risk of heart disease and accelerated atherosclerosis, a reasonable supposition was that poor glycemic control might be damaging the arteries directly, not just indirectly through effects on triglycerides and HDL cholesterol. It was an obvious implication of this diabetes-centric hypothesis; diabetologists like Joslin, arguing for good glycemic control, had assumed it to be true implicitly. That hyperinsulinemia itself was damaging the arteries in diabetes, that it, too, was atherogenic, was also a possibility—"consistent with available experimental evidence," as Reaven might have put it—even as diabetologists had been far less comfortable in considering it.

Potential mechanisms to explain how high blood sugar and elevated insulin levels might cause arteriosclerosis and heart disease would be elucidated and worked out in detail, but, once again, only after the fat-cholesterol hypothesis had been widely embraced. Diabetologists would take the work and implications seriously, and the researchers who did the work would be properly acknowledged for their contributions, but always with the caveat that these diabetes-related drivers of heart disease must somehow play only a secondary or subordinate role to that of elevated LDL cholesterol. The research would fuel the pursuit of ever newer drug therapies to reverse these mechanisms but little or no discussion of their implications for diet.

High blood sugar was the obvious suspect. This is the glucose intolerance, the relative inability to control blood sugar after consuming carbohydrates, that is a diagnostic criterion of metabolic syndrome and, of course, diabetes. The ultimate argument against poor glycemic control in diabetes had always been, as R. D. Lawrence had said, that "nature has perfected in the normal such a wonderful mechanism to prevent it." Hence, nature or evolution must have had good reasons.

These began to become clear, serendipitously, in 1968, when a Jewish-Iranian physician named Samuel Rahbar working at the University of Tehran identified an abnormal form of a protein called hemoglobin in blood samples taken from local patients. Hemoglobin is the protein in red blood cells that carries oxygen and ferries it through the circulation. Physician-researchers studying anemias—diseases characterized by a deficiency or dysfunction in red blood cells and hemoglobin, sickle cell anemia most notably—as Rahbar was, had taken to searching out abnormal variants of these hemoglobin molecules in blood samples from their patients. Hemoglobin was also one of the first proteins ever to have its structure decoded,* which made it, as an ADA history would later comment, "the rising star in molecular biology" in the 1960s.

Rahbar had analyzed hundreds of blood samples when he detected a novel form of the hemoglobin molecule in a patient with diabetes. It could have been a coincidence, the result of an unusual genetic abnormality, but when he analyzed the blood from other diabetic patients, he found they had it, too. Rahbar spent the next year at Albert Einstein College of Medicine in New York working with Helen Ranney, one of the world's leading hemoglobin researchers. Together, they confirmed the observation.† While

---

* In 1962, the University of Cambridge molecular biologist Max Perutz would win the Nobel Prize for the work.

† Once again the work was based on a measuring assay that had been developed in the early 1950s, this one by Ranney, who had been studying sickle cell anemia and needed a way to isolate and study the abnormal form of hemoglobin that characterized that disorder.

it turned out that everyone has this abnormal form of hemoglobin, known as hemoglobin A1C, it makes up less than 6 percent of
the hemoglobin in the blood of healthy, nondiabetic individuals.
Patients with diabetes often have twice as much, if not more, which
is why Rahbar's relatively insensitive assays could only detect it in
diabetic patients but not in others.

One possible implication was that hemoglobin A1C might
serve as a measure of the severity of diabetes. By the mid-1970s,
as researchers once again came up with new techniques to more
accurately detect and quantify these hemoglobin variants, Harvard
biochemists led by Frank Bunn established that the conversion of
normal hemoglobin to hemoglobin A1C was the result of glucose
molecules—blood sugar—binding to the hemoglobin, a slow process that proceeded over the four-month lifespan of the red blood
cells. (Bunn and his colleagues made this observation using the
blood from two patients. Bunn later admitted that he was one of
them, having "clandestinely" infused himself with a radioactive
tracer so he could establish how quickly hemoglobin A1C appeared
in his own blood.) The higher the blood sugar, the greater the glucose concentration in the blood, the more likely this would happen.
Hence, the percentage of hemoglobin in the form of hemoglobin
A1C reflects glycemic control over the course of months, not a measure of what might have been eaten in the last meal or the last few
days before the blood sample was taken, as traditional blood sugar
tests are. Put simply: the more poorly controlled the blood sugar
and so the more severe the diabetes, the higher the A1C percentage.

The American Diabetes Association would take another three
decades before it embraced hemoglobin A1C as a diagnostic criterion for diabetes, which it now does. The delay, other than institutional inertia, was the requirement that A1C measurements had
to be sufficiently reliable and standardized such that physicians
anywhere in the country could believe and act on what their lab
reports were telling them. A test that is off by a single percentage
point in its assessment of hemoglobin A1C can be the difference
between an errant diabetes diagnosis (a false positive) or failing to
detect the disease when it is present (a false negative).

When ADA clinical guidelines formally embraced hemoglobin A1C as a diagnostic tool in 2010, epidemiological surveys had clearly established that it reflected not just the severity of diabetes, but the risk of heart disease, stroke, and premature death as well. As implied by the metabolic syndrome science, this was true not just for patients with diabetes, but for patients whose hemoglobin A1C was still considered in the healthy range. Regardless of diabetic status, the higher the hemoglobin A1C, the greater the risk for all these chronic diseases, the diseases that Cleave, Campbell, Yudkin, Burkitt, and Trowell had attributed to western diets and lifestyles.

When Frank Bunn and his Harvard colleagues first reported on what they had learned about the formation of hemoglobin A1C in 1976, they had discussed what made this process so potentially relevant to the diabetic condition. Hemoglobin A1C is created by a glucose molecule bonding to hemoglobin A, the common form of the protein, and doing so *without* the use of an enzyme to make the reaction happen. The role of enzymes is to control the rate of biochemical reactions and assure that the results, the products, as Bunn would explain, "conform to a tightly regulated metabolic program." When they don't, to put it simply, bad things happen. The *glycation* of hemoglobin—the bonding of a glucose molecule to the hemoglobin—is dependent only on how much glucose and hemoglobin are available. This is why the higher the blood sugar, the more uncontrolled the diabetes, the more glycation takes place and the more hemoglobin A is converted to hemoglobin A1C.

This, not surprisingly, raised more questions: If hemoglobin is susceptible to this glycation process, what other proteins in the body might also be susceptible, and is it happening to them? If so, what else might be happening once it does? These questions would begin to be answered by researchers working out of the Laboratory of Medical Biochemistry at Rockefeller University in New York, led by Anthony Cerami, Helen Vlassara, Charles Peterson, and Michael Brownlee, who himself had type 1 diabetes. The glycation of hemoglobin, as Cerami and his collaborators suggested in 1978, provided a "conceptual framework" for understanding "the pathogenesis" of all the diabetic complications. Just as the formation of hemoglobin A1C altered the functional properties of hemoglobin, the binding of glucose to other proteins throughout the body might be doing the same. The Rockefeller researchers would establish that it does.

The process starts with the glucose bonding to a protein, a reaction that's reversible: if blood sugar goes down, the glucose is likely to come off and all is well. But if blood sugar remains elevated, long enough for the now *glycated* protein to proceed further in a series of chemical reactions, the protein rearranges itself in a way that becomes irreversible. The result is what Cerami and his colleagues called "advanced glycation end products," or AGEs. The acronym was intended, because the process that they were describing is also responsible for many of the more visible manifestations of aging.

When these AGEs come in contact with other proteins or other AGEs, the proteins can knit together in a process known as crosslinking. The longer-lived a protein in the body, the more susceptible it becomes to glycation and more susceptible to this cross-linking process, stiffening and darkening as it goes. Among the longest-lived proteins in the body happen to be those in tissues and organs in which diabetic complications manifest themselves, which is not a coincidence, as Cerami, Brownlee, and their colleagues demonstrated. It's the cross-linking of collagen, for instance, one of the longest-lived proteins in the body, that causes the loss of elasticity in the skin with age, as well as the stiffening of joints, the lung, heart, and arteries.

Researchers studying AGE accumulation have compared the process in the skin and in arteries to the toughening of meat or leather as they get older, both of which involve the same chemical processes. This can be seen clearly on autopsy, as Cerami has explained: "If you remove the aorta [the main artery carrying blood from the heart] from someone who died young, you can blow it up like a balloon. It just expands. Let the air out, it goes back down. If you do that to the aorta from an old person, it's like trying to inflate a pipe. It can't be expanded. If you keep adding more pressure, it will just burst. That is part of the problem with diabetes, and aging in general. You end up with stiff tissue: stiffness of hearts, lungs, lenses, joints. . . . That's all caused by sugars reacting with proteins."

By the mid-1980s, it seemed clear that glycation and the cross-linking of proteins and AGE formation played a critical role in diabetic complications: with advanced glycation end products linking together and accumulating in the eyes (retinopathy), in the nerve cells (neuropathy), in the kidneys (nephropathy), and even causing the premature aging of the skin from the cross-linking of collagen. It could also explain atherosclerosis by a variety of mechanisms that would now be worked out.

In 2004, when Michael Brownlee gave the ADA's Banting Lecture, he described what they had discovered as a unified mechanism of diabetic complications. Brownlee's hypothesis had been triggered by his observation that the apparently toxic effects of high blood sugar were not only manifesting themselves outside cells, but inside as well, and not in every cell type in the body but only a subset of specific types: in the retina of the eye, in networks of small blood vessels and capillaries, in the kidneys, in neurons. What these cells have in common, he explained, is that they do not control the inflow of glucose, of blood sugar, from the circulation when blood sugar is elevated. Most cell types can control this process, keeping blood sugar levels constant in the cells, even when blood sugar levels are high in the circulation. The cells that manifest diabetic complications either don't do this or don't do it well.

As a result, high blood sugar in these cells increases the rate at which they oxidize glucose—using it to generate energy—and that in turn results in the production of what are called *reactive oxygen species*. The more glucose oxidized, the more reactive oxygen species are formed. These compounds contain oxygen atoms to which electrons have been attached, converting the oxygen from an atom that is averse to reacting chemically with other atoms and molecules (relatively inert) to one that will do so easily. The result is a quartet of cellular pathways being turned on or *upregulated*, all of which ultimately work to damage the cells from the inside out. One of these pathways causes DNA damage in the nucleus of the cells such that when the DNA is used as the template to create proteins, the proteins themselves are damaged or dysfunctional. (This is a mechanism that also links poor glycemic control to the elevated risk of cancer in individuals with diabetes or metabolic syndrome.) Another of these cellular pathways generates precursors to AGEs, glycated proteins that can then diffuse out of the cell and cross-link with other proteins in the circulation or in, say, the lining of the arteries, to do damage there. The AGEs that circulate through the bloodstream—hemoglobin A1C being one of them—can then adhere to cells, stimulating the cell to release inflammatory molecules and growth factors that would do yet more damage to the veins and arteries.

Researchers would come to learn in the 1980s that one of the molecules most susceptible to oxidation and glycation in the circulation is the LDL particle itself. In this case, both the protein portion and the lipid portion (the cholesterol and the fats) of the lipoprotein are susceptible. The result is oxidized LDL, and it's this oxidized LDL, with its cholesterol, that becomes trapped in the plaques on the artery wall, an early and apparently necessary step in the atherosclerotic process. Oxidized LDL also resists being removed from the circulation by the mechanisms that clear normal LDL, and this is one way in which LDL levels in the blood can become abnormally high. These oxidized LDL particles appear to be "markedly elevated" in both diabetics and in nondiabetics with atherosclerosis and are particularly likely to be found in the atherosclerotic lesions themselves. So while saturated fat might be responsible for increasing the quantity of LDL particles in the circulation or the amount of cholesterol in the LDL particles, it would be the high blood sugar in those with diabetes and metabolic syndrome, a response to the carbohydrate content of the diet, that rendered this LDL atherogenic, that transformed it into an active agent in the pathology of heart disease.

Another way to think of the effect of blood sugar in these processes, using a metaphor embraced by the American Diabetes Association, is as akin to the fuel for a fire that is burning in our circulation. The longer the fire burns and the hotter the flame, the more damage created and the greater the residue left behind. And once the fire is burning, molecules throughout the body participate in the conflagration. Blood sugar is the fuel and the higher the blood sugar, the hotter the flame. "Current evidence points to glucose not only as the body's main short-term energy source," as the American Diabetes Association's Council on Complications explained in 2003, "but also as the long-term fuel of diabetes complications."

Despite the difficulty of demonstrating in clinical trials that the pharmaceutical control of blood sugar would reduce diabetic complications, particularly in patients who were insulin resistant and so had adult-onset or type 2 diabetes, a world of research continued to suggest that the better the glycemic control, the healthier the patient would be. One possible explanation, though, for why the clinical trials continued to show little benefit from better glycemic control is that the method used to achieve it is critical. Perhaps the oral hypoglycemic agents had long-term complications of their own, as the University Group Diabetes Program (UGDP) trial had suggested. More difficult for diabetologists to imagine, perhaps insulin therapy does as well.

Once it was established by Yalow and Berson in the early 1960s that type 2 or adult-onset diabetes was a disorder of insulin resistance and so hyperinsulinemia—high circulating levels of insulin—as well as elevated blood sugar, a handful of researchers set out to investigate the possibility that insulin itself played a role in the diabetic complications. They were interested not just in the microvascular complications but the macrovascular ones as well. Individuals who suffer from type 2 diabetes may spend years or even decades with relatively well-controlled blood sugar, but they do so because their pancreas secretes ever more insulin to compensate for the insulin resistance. Could that hyperinsulinemia be causing problems? Chapters in endocrinology textbooks are as much about the chronic disorders caused by overproduction of a particular hormone—*hyper*thyroidism, for example—as they are about hormonal deficiencies such as *hypo*thyroidism. Why not insulin?

By this thinking, the complications observed in type 1 diabetes, a disorder of insulin deficiency, could still be caused or exacerbated by an excess of insulin, but the excess would come from the insulin therapy itself, not the disease. Asking whether insulin therapy is causing or exacerbating the complications of diabetes is the kind of question diabetologists have understandably been hesitant to address. Prior to the discovery of insulin, as we've seen, patients with this acute form of the disease rarely if ever lived long enough to die from the longer-term complications of the disease. Once insulin arrived and prevented a quick death from diabetic ketoacidosis and coma, and it became clear that the process of arteriosclerosis was vastly accelerated in these patients, diabetologists naturally assumed that this was a consequence of the diabetes itself, not the insulin therapy that kept these patients alive. The insulin their patients were injecting, diabetologists thought, is a naturally occurring hormone that merely replaces what is missing. To some extent, diabetologists still think this way.

This assumption is overly simplistic, however, as Robert Stout, a cardiologist at Queen's University in Belfast, Northern Ireland, would argue beginning in the 1960s. Diabetologists like Joslin initially prescribed minimal doses of insulin to their patients, amounts that were at or even below that typically secreted by the pancreas of a healthy patient (from 20 to 60 units a day, depending on the carbohydrate content of the diet and the technique used to measure the insulin). As diabetic diets became ever more liberal in carbohydrates, insulin doses came to far exceed this amount. It's common now for individuals with diabetes to be taking hundreds of units a day—as much as ten times the amount of insulin secreted by the pancreas of a healthy individual.

The size of the dose isn't the only factor that makes glycemic control with insulin therapy profoundly different in a physiological sense from controlling it by pancreatic insulin secretion, as healthy people do. Insulin is absorbed from the injection site, as Stout explained in a 1982 monograph, "at a rate which is unrelated to meals or blood glucose levels, and at times during the day circulating insulin levels will be higher in the insulin-treated diabetic than in the non-diabetic." In a healthy individual, the pancreas secretes insulin directly into the *portal vein*, a major vessel carrying blood from the gastrointestinal tract to the liver. Roughly half of the insulin secreted never makes it out of the liver. When insulin is given by injection, it circulates throughout the body. One reason why such relatively high doses can seem necessary is to assure that the dose of insulin that arrives in the liver is adequate for metabolizing the carbohydrates that are also being delivered from the portal vein. Hence, not only do patients with type 1 diabetes experience abnormally high levels of insulin in their circulation, at least much of the time—hyperinsulinemia despite their underlying insulin deficiency—but that insulin is not coming from the pancreatic cells, which means it is not stimulating the counterregulatory effects that are a natural and necessary aspect of properly functioning homeostatic systems.

The possibility that abnormally high levels of insulin play a role in atherogenesis and other macrovascular complications had to be taken seriously. When the Canadian cardiologist Lyman Duff had reported in a series of experiments first published in 1949 that rabbits fed high-cholesterol diets did *not* get lesions in their arteries if they were first rendered diabetic, and they did when they were given insulin therapy, the implication was that insulin is a critical ingredient in plaque formation and chronically elevated levels of insulin facilitate the process.

In 1968, Stout reported in *The Lancet* that insulin, unsurprisingly, has similar effects on the walls of the arteries as it does on the fat tissue itself. In the arteries, it stimulates the transport of cholesterol and fats into the cells of the arterial lining—working through very similar enzymes—and it stimulates the synthesis of cholesterol and fat *in* the arterial lining. In 1975, Stout and the University of Washington pathologist Russell Ross (once again, with Ed Bierman as a coauthor) reported that insulin also has what are called *anabolic* effects on the arterial walls: it stimulates growth, specifically the proliferation of the smooth muscle cells that line the interior of arteries, a necessary step in the thickening of artery walls characteristic of both atherosclerosis and hypertension.

If Stout was right, then insulin, whether secreted to excess in type 2 diabetes or injected to excess in type 1, works to accelerate atherosclerosis and perhaps other vascular complications. It also implies, as Stout suggested, that any dietary factor that increases insulin secretion will increase risk of heart disease. This implicated, once again, refined carbohydrates and sugars because they are easily digested and promote greater insulin secretion in response. (In the nutritional terminology, refined carbohydrates have a higher glycemic index than unrefined.) As with Reaven's work on Syndrome X and metabolic syndrome, this research implied that the advice given by the American Diabetes Association to their patients was, indeed, doing more harm than good, that even if restricting saturated fat *did* reduce heart disease risk, the replacement of the fat by carbohydrates and the insulin secreted or injected in response could, at the very least, cancel out any benefit.

The issue, as with many we've been discussing, would never be settled. Diabetologists still shy away from discussing the possibility that insulin can be atherogenic, typically assuming that such a critically important possibility had been experimentally tested in clinical trials and refuted. It hasn't. In the early 2000s, when diabetologists instituted a series of clinical trials to test the hypothesis that keeping blood sugar under control—achieving normoglycemia, the blood sugar levels of a healthy, nondiabetic individual—would reduce the risk of macrovascular complications (heart disease and stroke, in particular) and premature death in patients with type 2 diabetes, they did so by having the patients inject ever higher doses of insulin to do it. When the trials* failed to show any benefit or perhaps even harm, the diabetologists suggested that maybe the intensive control had resulted in more deaths related to hypoglycemia, although the evidence for that was anything but clear and, for at least one notable trial—the ACCORD trial—appeared not to be true. Another possibility was that using such large doses of insulin to control blood sugar comes with its own deleterious complications that may eventually counterbalance any benefits from the improvement in glycemic control.

Some influential diabetologists, however, have continued to argue that high levels of insulin are clearly doing harm. Among the most respected is Ralph DeFronzo of the University of Texas Health Science Center in San Antonio. In the 1970s, DeFronzo helped develop the assay commonly used for quantifying insulin resistance. He's been a prominent authority in the field ever since. When DeFronzo discussed this controversial question in a 2019 review, he pointed out that not only did the amount of insulin necessary to reduce blood sugar to the level in healthy patients cause

---

* Three trials, in particular, as I've discussed and will again: ACCORD, ADVANCE, and the Veterans Affairs Diabetes Trial (VADT), published in 2008 and early 2009.

significant weight gain—almost twenty pounds in six months in patients with type 2 diabetes in the study DeFronzo cited—but the insulin itself would cause all the physiological complications that we've been discussing. Studies both in live patients and in the laboratory, he said, have "demonstrated that insulin, especially at high concentrations, can accelerate the atherosclerotic process by multiple mechanisms."

When DeFronzo gave the Banting Lecture in 2008, his conclusion then was that the American Diabetes Association was wrong to promote insulin therapy for patients with type 2 diabetes, regardless of the circumstances.* For patients with type 2 diabetes, he argued, intensive insulin therapy would always be a mistake, even if the moderate doses that patients were already taking were not doing the job. DeFronzo's solution was to use yet more pharmaceuticals, to target the pathophysiology of the disease specifically—for the insulin resistance and the exhaustion of the pancreatic cells that secrete insulin—rather than considering the possibility that the dietary therapy itself might be misconceived.

Despite all the challenges over the years to diabetologists' thinking about diet, whether as a cause of diabetes or a therapeutic approach, they have found it easier to believe that drug therapy is the better option. They would only pay lip service to the role of diet, as the British diabetologist Edwin Gale and dietitian Lynn Sawyer had suggested the same year DeFronzo gave his Banting Lecture. The idea that a diet that minimized blood sugar and the use of insulin might do a better job of controlling glycemia in the short run, and macro- and microvascular complications in the long run, had still not been tested or seriously considered.

---

* This is the case, DeFronzo said, even if glycemic control remains poor despite dietary therapy and exercise and the use of metformin, a drug that increases insulin sensitivity and so allows for greater blood sugar control with less insulin.

12

What You See Is All There Is

Over the past thirty years, a "modern" philosophy of treatment for [type 1 diabetes] has evolved, that places major emphasis on preservation of a patient's "life style." It assumes that injecting insulin, testing urine, and restricting fancy desserts involve a great degree of personal sacrifice and serve to remind the patient continually that he has a chronic disease and is thus "different" from his peers.

To tell an insulin-dependent diabetic that he is just like everyone else and can lead a normal life confuses him because it doesn't jibe with reality. He wishes that this were so and tends to believe the idealized authority figures (medical personnel) upon whom he is dependent. His bodily perceptions, however, tell him that he is not well and not normal. . . . Many bright and assertive diabetics say, "My doctor says I am fine, but why do I feel so terrible?" It is reassuring when these patients realize that, indeed, there is something wrong, that their diabetes controls many aspects of their lives, and that they must attend to their illness instead of ignoring it in order to feel well. It is better to be told what is wrong and to do all that is possible to set it right, than to be encouraged to ignore the true situation and its needs.

—RICHARD BERNSTEIN,
Diabetes: The GlucograF Method for Normalizing Blood Sugar, 1981

Until there's a cure for diabetes, a diagnosis makes the patients ultimately accountable for their own therapy, a lifelong endeavor. The better the patients do their job, the longer they are likely to live. "The patient with diabetes has to take responsibility for his own metabolic homeostasis," as Robert Tattersall, the British diabetologist and later medical historian, wrote with four colleagues in 1980. From the physicians' perspective, though, this adds a layer of complications to their jobs. Successful diabetes therapy requires that their patients be sufficiently intelligent—or sufficiently mature, if children or adolescents, as those with type 1 diabetes often are—to follow faithfully what can be daunting instructions about diet and drug therapy, but not so intelligent that they challenge the legitimacy of the doctor's orders. This is why the early editions of Joslin's textbook would warn physicians to "beware of the educated diabetic," the patient who thinks he or she knows better. But what if the patients do? Every patient responds to diet and drug therapy differently, implying that patients may know their own bodies' response better than their physicians do. Joslin had an answer for that—"If the patient really does know more than you, extract his knowledge and save yourself"—but what physicians should do with that knowledge if it happened to be in conflict with conventional thinking, he didn't say.

Through the 1960s, this physician-patient relationship—indeed, diabetes therapy itself—was further complicated because neither physicians nor patients could meaningfully quantify the efficacy of the treatment prescribed. Of course, when they were failing, it was all too clear: hypoglycemic episodes or, in the worst case, diabetic ketoacidosis and coma, would provide that information. But they had little feedback to tell them how well they were doing otherwise, and no information on how to improve their therapy in response to doing poorly.

Until 1970, generations of physicians trying to determine how their patients might best control their blood sugar—achieve what they would call *normoglycemia* or *euglycemia*—had no meaningful method to measure it. Diabetologists would argue heatedly about the value of glycemic control without having any idea whether diabetic patients were actually achieving it, as they would often acknowledge. In the pre-insulin era, physicians assumed that their patients could best control their blood sugar by either avoiding carbohydrates entirely, and so eating variations on the animal diet that Rollo had suggested, or semi-starve themselves, in line with Frederick Allen's thinking. Once insulin came along, these approaches, commonsensical as they may have seemed, were left behind.

The presence or absence of sugar in the urine could be used as proxy for the presence or absence of the disease, but it told doctors little about what was happening in the blood, what was causing the complications. It provided patients with no information on low blood sugar (hypoglycemia) and why they were having insulin reactions, and it provided no warnings. The amount of glucose in the urine was at best a reflection, a measure, of blood levels of glucose hours before. Even this was complicated by a concept known as the *renal threshold*, which is the particular level of blood sugar, different for every patient, at which the kidneys will excrete enough sugar into the urine that a test will detect it.

When researchers set out in the 1970s to determine how reliably the urine sugar concentration could be used to monitor the state of the disease—using blood tests taken in the morning to assess *fasting blood sugar* or using hemoglobin A1C to do so—the conclusion was: not very. In one 1978 assessment, the diabetologists who studied 220 children and adolescents with type 1 diabetes summarized what they found in the title of their article: "Good Diabetic Control—A Study in Mass Delusion." Among the many reasons they gave for the failure was that urine tests "do not reflect the true metabolic state." When Robert Tattersall and his colleague Ian Peacock assessed the various techniques for blood sugar monitoring available in 1982, they concluded that urine tests were done (by the many billions each year) only because they were relatively cheap and easy, and because "there was no real alternative. Some information, albeit misleading at times, may be better than none."

Chemists had known how to measure blood sugar itself since the late nineteenth century, but the tests were expensive and required the kind of clinical expertise and proper laboratory equipment that few physicians had. Patients would visit their physicians every month or few months; a blood sample would be taken and sent off to a lab for analysis. Physicians liked the idea of the blood tests because they were a reason for patients to come in and see them regularly, but the results were of limited clinical value.

When determined "at intervals of months or years," as Joslin's textbooks suggested in the 1930s and 1940s, the blood tests gave a "splendid index of the increasing or decreasing severity of the condition." In fact, physicians had no assurance that the results of a test reflected anything other than the blood sugar at the moment that the blood had been taken from the patient's vein, realizing, as they did, that their patients had probably spent the few days before their appointment eating very differently from how they had eaten the rest of the month (or months). So patients were instructed to take enough insulin—typically long-lasting insulin, once it was available—in a morning dose, or with an evening dose as well, to keep their urine mostly free of sugar (and their physicians mollified at their regular checkups), yet not so much that they suffered "frequent or severe reactions." Physician, patient, and family then hoped for the best. Diabetologists and the American Diabetes Association assumed that if their patients faithfully consumed a set amount of carbohydrates every day at each meal and snack, and properly and accurately weighed and counted those carbohydrates so that the proper dose of insulin could be estimated, they would achieve an acceptable compromise between the efficacy of treatment and the quality of their lives.

Whether they did, of course, depended on the severity of the disease and the individual variability of the patients. Hypoglycemia—the insulin reaction—was still the greatest anxiety and the greatest source of day-to-day physical and mental discomfort. Both diabetologists and patients with diabetes would often (they still do) use variations on driving a car, perhaps blind or without breaks, as metaphors for the experience of trying to control their blood sugar with insulin, given all the ways it can go tragically awry. ("The diabetic taking insulin is like a rapidly moving machine," Joslin had said, "which a slight swerve of the wheel will bring to disaster.") Among the many salient questions in diabetic therapy was what might happen on that lucky day when patients (and their doctors) could actually see how their blood sugar responds to each turn of the wheel—each meal consumed or each dose of insulin taken.

When Rockefeller University hosted a symposium in November 1980 to discuss management of insulin-dependent diabetes in the decade to come, the talks focused on recent developments in blood sugar monitoring that had the potential to revolutionize therapy. One of the new developments, pioneered largely by the Rockefeller researchers themselves, was the hemoglobin A1C assay, providing a measure of how well patients were controlling their blood sugar over the course of two to three months. What it didn't do was shed any light on the day-to-day or even hour-to-hour control of blood sugar in response to the foods consumed and the insulin injected. This is why the most promising of the new therapeutic tools discussed and, ultimately, the most revolutionary, was one so seemingly simple that it's hard to imagine it was, until then, a radical idea: let patients measure their own blood sugar and, ideally, do so frequently enough over the course of a day so they can adjust their insulin doses and their diets in response.

Diabetologists would initially consider the idea fanciful, if not dangerous. They feared that patients might have no reason to see them regularly if they could test their own blood sugar. But by the late 1970s, the concept was spreading nonetheless. British diabetologists had been at the forefront of this movement, and Robert Tattersall gave a talk at the Rockefeller symposium on what had been learned from "the English experience." By measuring hemoglobin A1C in their patients, Tattersall said, they had realized that four out of every five of their patients were failing to achieve anything like reasonable glycemic control, that their average blood sugar levels were still two to three times higher throughout the day than that of healthy individuals. That this was happening despite the medical profession's fifty years of experience with insulin therapy, Tattersall said, suggested two profound problems with the therapy. One was clearly the nature of insulin therapy itself. "How . . . can you expect to achieve normoglycemia," he asked simply, "if you give insulin in the wrong place, at the wrong time, in the wrong form, and in the wrong amounts?"

The second was the lack of any information, the absence of feedback, on how blood sugar responded to the insulin injected. Informative as hemoglobin A1C could be, it was still merely an average value, hiding all the variations that averages can always hide. (Fifty is an average of fifty-one and forty-nine and of one hundred and zero, as well as an infinite combination of values in between.) Tattersall described his clinic's experience with one patient whose A1C had been the lowest they'd ever measured in a diabetic patient, equivalent to that of a healthy individual without diabetes. That should have been a sign of excellent blood sugar control. But this patient felt terrible during the day—tired, unable to concentrate, with a poor memory, afraid he would lose his job as a result. So Tattersall and his colleagues took blood samples and measured his blood sugar throughout the night. By doing so they learned that his blood sugar levels dropped dangerously low every night, and that balanced out unacceptably high values during the day. With that knowledge, they readjusted the patient's insulin therapy and he was able to function well again in his job during the day. His subjective experience of his diabetes, which may have been the best measure of glycemic control at the time, improved dramatically.

The other obvious limitation to hemoglobin A1C, as Tattersall noted, is that it only informs about the quality of diabetic control; it provides no information about how to improve it. This is why Tattersall and his colleagues had begun sending their patients home with devices called reflectance meters that allowed the patients, properly instructed, to measure their blood sugar at home from a few drops of blood taken with a prick of the finger. Patients could do it in the morning, on waking up; they could do it before and after meals, and at night before sleep. They could learn how their blood sugar was actually responding to the insulin they were taking, to the food they were eating, and even whatever exercise they were doing. Not only did their patients learn by this feedback how to do a far better job of glycemic control, but it had, Tattersall said, "important effects" on the doctors, as well.

It had confirmed that what they'd been doing, which is what physicians attending to diabetic patients worldwide were still doing, had been failing their patients. "It has shown," Tattersall said, "how badly controlled many of our patients really are and has removed many of the rationalizations we have traditionally used to cover up what is happening in our clinics."

What had become obvious to Tattersall and other diabetologists was not just how high the blood sugars ran in their patients and how often their blood sugar ran high, but how frequent the lows were—the many hypoglycemic episodes their patients were experiencing. Those with insulin-dependent diabetes will often say that no one ever forgets their first hypoglycemic episode, which is why physicians often preferred that their diabetic patients experience their first event in the hospital, under supervised care, as part of learning to live with the disease. Not only would they know what the early stages of hypoglycemia felt like, allowing them ideally to sense it coming and head it off—to eat candy or drink some juice— but that first experience would be powerful motivation to control their blood sugar to avoid such events in the future. And yet, diabetic patients were experiencing these hypoglycemic episodes anyway, averaging one a week, even two a week for those taking insulin in both morning and evening doses. Many were experiencing dangerously low levels of blood sugar while sleeping at night (nocturnal hypoglycemia), influencing profoundly their state of mind and energy levels during the day and, ultimately, their ability to bear the burden of their disease.

This burden had always notably included depression and other emotional disturbances, but diabetologists had never established precisely what was an actual consequence of diabetes—mood swings and "explosive rebellion" clearly seemed like products of hypoglycemia—and what was a natural response to awareness of what the disease was doing to their bodies and would likely do in the future. "Attending a diabetic clinic and seeing one's fellow sufferers with blindness or amputations might make even the most stable personality somewhat hypochondriacal," wrote Tattersall in a 1981 discussion of these issues. Poor glycemic control associated with emotional disturbances of all sorts, but that observation did not disclose what was cause and what effect.

Physicians justifiably feared that hypoglycemia might be responsible for many of the problems their patients were experiencing, perhaps even permanent brain damage or long-term neurological effects, but no clinical trials had been done to address those possibilities. It was easy to imagine that patients who were now given the responsibility of monitoring their own blood sugar daily could find reasons for *increased* anxiety and depression by doing so—including more awareness of how they might be failing in this critical job of controlling their disease. But it was also easy to imagine that such patients would benefit, both physically and emotionally, from having the necessary feedback to successfully control their blood sugar, and so ameliorate the emotional disturbances in the process.

Edwin Gale, who started his career with Tattersall in the 1970s at Nottingham General Hospital, says the medical profession lost fifteen years because physicians simply refused to believe their patients might see the benefit to monitoring their own blood sugar, let alone be capable of following the instructions necessary to do so. This likely is why it was a patient—Richard Bernstein, an engineer in New York—who did it first and then, fortunately, did the hard work of inducing diabetologists, including his own physicians, to buy in. Had he not been in New York City, a major hub of diabetes research and thinking, it might have been another decade still before monitoring caught on.*

Even then, though, the diabetes community tested and then embraced only Bernstein's innovations in monitoring and insulin dosing. Diabetologists found the dietary implications—that avoiding carbohydrate-rich foods was also necessary for achieving healthy levels of blood sugar—too cognitively dissonant, as they would Gerald Reaven's revelations on metabolic syndrome and insulin resistance. They would rarely, if ever, discuss them, leaving Bernstein to go to medical school so that he might be taken seriously as a physician, get his articles published in the medical

---

* In 1961, the British diabetologists Harry Keen and R. K. Knight of Guy's Hospital in London had sent their patients home with the equipment necessary to take blood samples at what would "otherwise be impractical times" for their physicians and then mail the samples off to a lab for analysis. Other than an article in *The Lancet*, nothing ever came of it. The patients, in any case, received no feedback from the technique, only the physicians did, days or even weeks after the blood samples were taken.

journals (although, to his credit, he achieved that even before he became, at age forty-five, a medical student), and write his own books to communicate *all* aspects of what he had learned: not just an effective method to monitor blood sugar and adjust insulin timing and dose to keep it under control, but how to use diet as a tool so that even a patient with type 1 diabetes could have blood sugar little different, if at all, from those of healthy individuals.

While diabetologists in the 1970s were arguing about the benefits and risks of *controlling* blood sugar—keeping its highs and lows within an acceptable range defined by what physicians believed might be possible (unable, as they were, to quantify it)—Bernstein was among the first to advocate for *normalizing* blood sugar, doing what was necessary to have blood sugar no higher or lower than what would be achieved by a healthy pancreas. He did so because he had learned through his personal experience that it was possible. "It's not that I had the ultimate solution," Bernstein would say, "but that no other approach existed that remotely works."

Richard Bernstein's story is a "conversion narrative," as the best-selling author and journalist Malcolm Gladwell would later describe such accounts. Bernstein discovered a therapeutic approach that worked to keep his seemingly intractable disease under control and then assumed it *might* work for others with the same disease. So he wrote books about it. When Gladwell described such accounts in a 1998 *New Yorker* article on obesity and the obesity epidemic, he was implying that such narratives are invented or carefully crafted to maximize book sales. In short, a kind of con. Academic medical researchers, if they pay attention to these accounts at all, will write them off as anecdotal, as they are, testimonials with little meaning, shedding little light on the appropriate therapy for anyone else with the same medical condition. That might be true, although a significant portion of all advances in medicine emerge initially from such anecdotal experiences or observations. In Bernstein's case, though, it is undeniable that he catalyzed a revolution in diabetic therapy and should get credit for doing so. But did he catalyze *enough* of a revolution?

. . .

The idea that patients should monitor their own blood sugar was dependent, of course, on whether a technology existed to do it. The train of innovations that led from at-home urine testing to at-home blood testing began in the late 1930s when chemists at the Miles Laboratories in Elkhart, Indiana, began developing chemical assays that allowed for easier urine analysis: first effervescent Alka-Seltzer–like tablets,* marketed in 1941—they could be dropped in a vial of urine, changing its color in a way that correlated with the glucose content—then, in 1956, a strip of paper coated with the necessary reagents. Patients could dip the paper strip in a test tube of urine and the paper's color change would tell them whether they were in control or not, at least by the standards of the era. A decade later still, chemists at Ames Company, a subsidiary of the Miles Laboratories, realized that the same strips that would change color when dipped in urine could be modified to do so with just a few drops of blood drawn from a finger prick. The Ames Reflectance Meter (ARM), what Bernstein first used in January 1970 and an early version of what Tattersall and his colleagues would give to their British patients half a dozen years later, was simply an electronic device (battery powered, until Bernstein modified his to plug into a wall socket) that read the color change on the strip and translated it into a quantitative assessment of the blood sugar.

Bernstein himself had been diagnosed with type 1 diabetes as a twelve-year-old in 1946. He had been on insulin therapy ever since—one dose of long-acting insulin in the morning, per ADA guidelines, covering a fixed amount of carbohydrates, appropriately weighed and counted, at breakfast, lunch, and dinner and morning and afternoon snacks. He thought of himself, he would say, as an "'ordinary' diabetic, dutifully following doctor's orders and leading the most normal life" possible. He was slightly less normal in that he had graduated from Columbia University with a physics degree, and then become an engineer professionally. By 1969, he was working in a management position, married with three young children.

He was also accumulating already, as those with insulin-dependent diabetes do, the complications of the disease. His health

---

* Alka-Seltzer was the product that made the Miles Laboratories famous.

had declined steadily through his twenties and into his thirties. He later described himself as prematurely aged and chronically ill. He had kidney stones and a stone in his salivary duct; he had a condition known as "frozen shoulders" and neuropathy in his feet. He had a cardiac myopathy, and his vision had deteriorated, a consequence of retinopathy and cataracts. He had elevated levels of the protein albumin in his urine, evidence of kidney disease (nephropathy). He was losing consciousness from hypoglycemic episodes, on average, twice a month. He was hospitalized regularly.

Bernstein believed his diabetes was killing him, which was a reasonable assessment. The life expectancy of a patient with type 1 diabetes in the 1960s was such that Bernstein could not expect to live another decade. When he acquired an Ames Reflectance Meter to monitor his own blood sugar, he did it, he said, because he knew he would die prematurely if he didn't, and because the strain on his family "was clearly becoming untenable."

The original ARM weighed over three pounds and cost $650 (the equivalent of $4,500 in 2021). The Ames Company began selling the device in January 1970, marketing it to physicians for their offices and to hospital emergency rooms, where patients might be admitted unconscious and the ER doctors would have to establish whether they had passed out from alcohol or were in the midst of a diabetic coma (or perhaps both). If this happened on a weekend night—often the case with alcohol—they might not have a laboratory immediately available to run their blood tests, nor perhaps the luxury of waiting to treat the patient for the three to four hours the chemical analysis would take. Ames was averse to selling the reflectance meter to patients, worried about the legal ramifications. Bernstein had seen an ad for the ARM in a medical device trade magazine and had reached out to the company directly. Because Bernstein's wife was a physician—a psychiatrist—he was able to acquire one in her name. By doing so, Bernstein became the first diabetic patient to monitor his blood sugar at home and establish how it responded to meals, to exercise, and the vagaries of everyday life—to learn its natural cycles.

Bernstein approached the challenge of controlling his disease from an engineer's perspective. He measured his blood sugar five times a day. While his insulin shots had kept his urine relatively free of sugar, satisfying his physician that Bernstein's disease was under control, Bernstein learned that his blood sugar was fluctuating wildly over the course of a day. In healthy individuals without diabetes, blood sugar might drop as low as 60 or 70 milligrams/deciliter (mg/dl) and rise to 140 mg/dl after meals. For the great portion of the day and night, blood sugar will remain below 100 mg/dl. Bernstein now learned that his blood sugar was often dropping to 40 mg/dl—hypoglycemia—and reaching as high as 400 mg/dl in the hours after eating. He added a second insulin injection to his one-a-day regime and experimented with eating fewer carbohydrates, which allowed him to use smaller insulin doses, and his blood sugar improved. "The very high and low blood sugar levels became less frequent, but few were normal," he said. He decided this was the best he could do, put the ARM in a closet, joined a gym, and focused on exercise as the much-promoted means to combat diabetic complications.

In 1972, still spending much of his life "either in a hypoglycemic state, entering such a state, or recovering," Bernstein embarked on a year of more methodical experimentation. He measured his blood sugar five to eight times a day and adjusted his diet or his insulin regimen every few days to see the effect. "If a change brought an improvement," he later wrote, "I'd retain it. If it made blood sugars worse, I'd discard it." In the process, he discovered how much a gram of carbohydrate raised his blood sugar, and how much a unit of insulin lowered it; he discovered how exercise affected his blood sugar, how much candy or juice needed to be consumed to forestall hypoglycemia, and how his blood sugar rebounded afterward. According to Charles Suther, who was in charge of marketing the ARM device at Ames, Bernstein spoke with him half a dozen times a month throughout this process; the two became friends. Bernstein, he said, "laid the groundwork for much of what happened then in the United States."

Bernstein's goal was to match his insulin dosing, as closely as possible, to what a functional pancreas would do naturally. He would take *basal* doses of long-acting insulin in the morning on waking and in the evening before bed, but only so much, by Bernstein's thinking, as a healthy pancreas would secrete under the same conditions. He would add *bolus* doses of short-acting insulin before meals based on an assessment of the carbohydrates and protein that would be consumed at the meal, because a healthy pancreas would effectively do the same.

The use of both basal insulin and bolus injections prior to meals was another of Bernstein's innovations, and this too would come to be an accepted practice. During hypoglycemic episodes, Bernstein learned to use only enough candy or glucose tablets to forestall them, never so much that he would overshoot and cause hyperglycemia. Bernstein learned that the fewer carbohydrates he consumed, the easier it was to maintain glycemic control and the lower the doses of insulin necessary to cover his meals.

This revelation was a simple one, but it was and still is at the crux of the diabetes challenge and the controversies that are still ongoing. Blood sugar, as we've discussed, is determined primarily by three factors: the carbohydrates consumed in the diet (including the carbohydrates converted from amino acids in the protein consumed); the hormones secreted in response (of which insulin is the dominant one of many); and the glucose secreted by the liver, which is the one factor that is beyond direct control.

The appropriate dosing of insulin is dependent on the amount of carbohydrates that the insulin has to cover. Hypoglycemia—an insulin reaction or insulin shock—is not technically caused by eating too few carbohydrates, but by the insulin injected to cover the carbohydrates, the result of an insulin overdose, and the failure of the body to compensate as the blood sugar begins to drop. While the ADA was in the process of institutionalizing high-carbohydrate diets for diabetes as a means to reduce the fat content and so, in theory, reduce the risk of heart disease, glycemic control was still considered by diabetologists to be the first principle of nutritional guidance. Bernstein learned that he could do that best by severely restricting the amount and type of carbohydrates he consumed.

Bernstein ultimately encapsulated his philosophy in what he called the "laws of small numbers": "Big inputs make big mistakes; small inputs make small mistakes." This was a reality that physicists and engineers knew intuitively: complex systems, including bio-

logical systems, will respond far more predictably to small pertur-
bations than to large ones. The smaller the inputs—in this case the
doses of insulin and the quantities of carbohydrates consumed—
the smaller the errors in measurements and the smaller the variations in responses. Bernstein, too, used driving a car as a metaphor
to describe his thinking: "You're driving down the road and your
car drifts slightly toward the median. To bring it back into line, you
make a slight adjustment of the steering wheel. No problem. But
yank the steering wheel and it could carry you into another lane,
or could send you careening off the road."

In a healthy individual, the pancreas optimizes its use of insulin
by secreting the hormone only as necessary and only as much as
necessary in response to the anticipated rise in blood sugar levels
and that rising tide itself. Patients with type 1 diabetes, dependent
on insulin therapy, have to do this job by accurately predicting in
advance how their blood sugar will rise in response to a meal and
how the insulin injected to cover it will act. Both entail uncertainty,
from day to day and meal to meal, but that uncertainty can be minimized, Bernstein came to conclude, by minimizing the doses of
insulin injected.

There was simply no way to do it with large doses, a fact that
would become undeniable when researchers led by John Galloway
at Eli Lilly reported in 1981 that blood sugar responds unpredictably to set doses of insulin, not just from individual to individual,
but even in the same individual from dose to dose, depending, for
instance, on the site of the insulin injection and how deeply the
needle is inserted. The fact that insulin injected during the day can
induce dangerous if not fatal hypoglycemia in the middle of the
night, which so worried Tattersall and his British colleagues, is in
part a manifestation of just that uncertainty in insulin action. The
larger the doses of insulin, the sooner the insulin starts working,
the longer it lasts, and the greater the variability in both. By minimizing the amount of insulin necessary, Bernstein concluded, the
patient not only mimicked what a healthy pancreas would do naturally, but minimized the uncertainty in how the injected insulin
would act and the blood sugar respond.

This in turn required minimizing the carbohydrates consumed

to allow the use of only small insulin doses. The easiest way to do that, Bernstein learned, was to avoid entirely what he called "fast-acting carbohydrates"—sugars, grains, fruits, and starchy vegetables that were carbohydrate rich and digested quickly. (In the 1980s, these would come to be known as *high-glycemic-index carbohydrates*.) He assumed that he could get whatever carbohydrates his body required from foods that take a relatively long time to digest—fiber-rich, vitamin-rich green vegetables, in particular—and from the conversion to glucose of amino acids in the protein he ate. Bernstein learned to avoid bread, as he later said, because no matter how much insulin he took to cover it, his blood sugar would initially "go way up."

"Within a year," Bernstein wrote in *Dr. Bernstein's Diabetes Solution,* one of a handful of how-to books he would publish on his program, "I had refined my insulin and diet regimen to the point that I had essentially normal blood sugars around the clock. After years of chronic fatigue and debilitating complications, almost overnight I was no longer continually tired or 'washed out' . . . at last I was able to build muscle as readily as nondiabetics." All of this happened as the amount of insulin he injected daily dropped first by a third and, then, with the introduction of genetically engineered human insulin in 1982, by half again.

Ultimately Bernstein's trial-and-error method led him to recapitulate what Joslin had argued for in the 1920s when insulin therapy was new and its chronic effects unknown (as they still were): the smaller the dose of insulin, the less damage could be done. Joslin had counseled small doses of insulin in the context of Frederick Allen's thinking that patients had to be underfed, if not semi-starved. Bernstein learned that he could keep his blood sugar similar to that of a nondiabetic—normalized, rather than merely controlled—by restricting his carbohydrate consumption alone. Semi-starvation was unnecessary.

Bernstein, as Joslin counseled, would still shy away from a fat-rich diet of the kind Newburgh and Marsh had promoted. He assumed that the consensus forming among nutritionists, heart specialists, and diabetologists that high-fat diets caused heart disease was correct. The diet that he would eat and prescribe in his books, nonetheless, was far higher in fat and far lower in carbohydrates than the American Diabetes Association or the American Heart Association considered healthy. It was the high-fat, carbohydrate-restricted nature of Bernstein's diet that prevented diabetologists from fully testing or implementing his program.

Bernstein gets credit for catalyzing the revolution in self-blood-sugar monitoring not just because he was the first to do it, but because he worked so diligently to get the diabetologists to pay attention. What he didn't do, Ames did, and then the German company Boehringer Mannheim; and they did it, as Charles Suther at Ames would say, because Bernstein had proven its value.* In 1976, Bernstein had written up his experience and his method in an article—"Long-Term Euglycemia Achieved in Labile Juvenile Onset Diabetes—A Modern Technique for Fine Control"—and submitted it to journals, all of which rejected it. The editor of *The New England Journal of Medicine* told him the evidence for the benefits of "fine control" of blood sugar was still ambiguous, which it surely was. Henry Ricketts, a diabetologist, senior editor at *JAMA*, and a former president of the ADA, suggested that few patients would want to "use the electric device" to monitor their blood sugar. "You are a spartan and a paragon, but, I fear, not common clay," Ricketts wrote to Bernstein in his letter of rejection. It's likely that whatever the editors and reviewers thought of Bernstein's evidence and his experience, they were hesitant to publish an article on such a sensitive subject written by a patient, rather than a physician or researcher in academic medicine.

Bernstein responded by networking, and his trial-and-error method of self-blood-glucose monitoring combined with the necessary changes in diet and insulin dosing disseminated slowly through the diabetes community. It did so because Bernstein could prove that it had helped him personally, and it seemed to help other diabetic patients. Bernstein printed two hundred copies of his still-unpublished article, now revised with the help of a medical writer at Ames, and Ames distributed it to diabetologists who used their

---

* Suther left Ames in the late 1970s to join Boehringer Mannheim, which was marketing a strip that would respond to blood sugar and would not require a reflectance meter.

devices. Bernstein became a founder of a local chapter of the Juvenile Diabetes Foundation, only a few years old at the time, and then became a member of its awards committee so he could meet the diabetologists in the region who might be deserving of such recognition. He joined the New York chapter of the American Diabetes Association and got to know its president, Sheldon Bleicher, of Brooklyn Hospital and the State University of New York Downstate Medical Center. As Bernstein tells it, Bleicher was advocating for normalizing blood sugar levels but was skeptical that it was possible with existing technology and devices. Bernstein became his patient, demonstrated his method, and proved otherwise. Bleicher then taught his other patients to monitor their own blood sugar and documented their experience.

After attending a lecture by Anthony Cerami of Rockefeller University on hemoglobin A1C, Bernstein called Cerami to tell him that he was normalizing his blood sugar and would like to have his A1C measured to document what he was doing. Cerami connected him with his Rockefeller colleague Charles Peterson, who, with Bernstein's help, began teaching the patients at Rockefeller the technique and initiated a study to document their progress. Bernstein wrote to Zvi Laron, a diabetologist at Tel Aviv University and founder of the first pediatric endocrinology clinic in Israel, who invited him to a symposium in Israel on juvenile diabetes. Bernstein attended and discussed his experience. Among the diabetologists at that symposium, as Bernstein recalls, were two British diabetologists—Tattersall and Clara Lowy, then of St. Thomas' Hospital in London—who were the earliest proponents of blood sugar self-monitoring in England.

Ames, meanwhile, provided the reflectance meters free to diabetologists, including Thaddeus Danowski, a pediatric endocrinologist at the University of Pittsburgh—who would publish, with his colleague Joseph Sunder, the very first paper discussing the benefits of self-blood-glucose monitoring—and Jay Skyler in Miami, who used the Ames devices at a summer camp for diabetic children and had heard directly from Danowski that self-monitoring and multiple daily injections of insulin could help patients. The end result is that the benefits and potential risks of self-monitoring of blood sugar were tested by influential diabetologists in their clinics, just as oral hypoglycemics had been in the 1950s, and long-acting insulin in the late 1930s, and insulin itself in the years following its discovery.

The Lancet and the British Medical Journal in England and Diabetes and the ADA's new journal, Diabetes Care, in the United States published the first articles on self-blood-glucose monitoring in 1978. The subjects of these early reports were often pregnant women with diabetes, who seemingly had the most to gain from controlling their blood sugar. Diabetes had always been particularly treacherous for pregnancies. Until the era of insulin therapy, it was so likely to end in the death of mother or child that young women with diabetes were advised not to have children. (Even if both survived the birth, the mother could not be expected to live long.) In his Diabetic Manual, Joslin would tell the story of the great French diabetologist Apollinaire Bouchardat saying in 1885 that he had never seen a pregnant woman with diabetes.

Until the invention of long-lasting insulins in the late 1930s, little more than half of the newborns of a mother with diabetes would survive, even with the best possible care. By the 1970s, that number was near 90 percent, but only with careful glycemic control, often doubling or tripling the mother's prepregnancy insulin dose to achieve it. Women with diabetes were still hospitalized three to five weeks before the end of term, if not for the entire last trimester of their pregnancy, to guard against preeclampsia. Babies were frequently delivered by Cesarean because they were long and heavy, likely to be overweight, and "often waterlogged," as Joslin described them. Mothers with diabetes often lost their babies simply because of "the mechanical difficulty of the birth." It was already clear by the late 1970s that the large babies were a consequence of the mother's high blood sugar and the fetal pancreas oversecreting insulin in response, a condition known as fetal hyperinsulinemia. "Provided [the newborns] survive," Jay Skyler and his Miami colleagues wrote, "they are more likely than other children to have congenital malformations: juvenile diabetes mellitus or cerebral dysfunction."

Patients continued to drive innovation. In the United Kingdom, Tattersall and his colleagues developed the confidence to trust their pregnant patients with monitoring their blood sugar at home because a single patient of Clara Lowy's insisted on doing so and confirmed its value. Lowy's patient had tried religiously to control her blood sugar throughout her pregnancy but had still experienced a prolonged and dangerous episode of hypoglycemia. "So at 26 weeks of pregnancy," as Lowy later told it, she advised her patient to admit herself to the hospital to be monitored. "Her response to this request was, 'What are you going to do in the hospital that I cannot do at home?' My reply was, 'Measure your blood glucose.' Her response to this was, 'Why can't I do this at home?'" Lowy taught her patient how to prick her finger and use an Ames Reflectance Meter and sent her home with the device. "The incentive for my patient to get to grips with the technology was overriding. It would mean that she would not be incarcerated in hospital. She learnt very fast. . . . She made about three measurements a day and was not admitted until term."

After that experience, Lowy and her colleagues counseled every pregnant woman in their diabetes unit to do the same. In a 1981 paper, Charles Peterson, working with Lois Jovanovic of New York Hospital–Cornell Medical Center, reported that they had wanted to run a clinical trial comparing patients from the pregnant diabetic clinic randomized to either "standard care," with home monitoring of urine glucose, or intensive care, with self-blood-glucose monitoring. After four deliveries in the standard care group resulted in complications and poor outcomes, Peterson and Jovanovic decided to offer only the intensive program and then to compare those outcomes—in fifty-two women—to those in healthy women without diabetes. Not only did the pregnant women with diabetes keep remarkably stable blood sugar, their outcomes were better than those of the healthy women—the babies were no heavier, they had no "major or minor congenital abnormalities . . . no neonatal incidences of cerebral disfunction."

The benefits seemed undeniable. When Lowy and her colleagues published an article on their experience, they said that they now recommended self-monitoring of blood sugar to all their diabetes patients, pregnant or not. "By so doing we have a group of patients with 'super control' whose prognosis we hope will be substantially better than has currently been possible." Skyler and his University of

Miami diabetologists reported much the same: with their pregnant diabetic patients they were able to "abolish completely" the need for hospitalization prior to their predelivery confinement. Whether pregnant or not, patients taught to monitor their own blood sugar and adjust their dosing and diet appropriately, as Tattersall and his colleagues wrote in *The Lancet* in 1978, had an improved attitude about disease management: "Several patients commented that seeing a graph of their blood-sugar has 'made sense of their diabetes at last.'"

With the clinical experience accumulating, two of Bernstein's academic allies organized symposia to disseminate what was being learned. Bleicher hosted the first in May 1979 at the Brooklyn-Cumberland Medical Center. Ames then paid for the papers from the symposium to be published in *Diabetes Care*. In November 1980, Peterson and his colleagues did the same at Rockefeller. Boehringer Mannheim, for which Charles Suther was now working, paid for that conference and covered the travel expenses for the American Diabetes Association to have two diabetologists from every state in the country attend.

The evidence presented at these meetings was preliminary; physicians reporting solely on what they had learned from the patients, often discussing as few as half a dozen cases. The observations were promising, although inconsistent. These physicians were employing different techniques of self-monitoring—from a few days of multiple blood tests to establish a set routine of insulin timing and dose to the extensive, ongoing self-experimentation of Bernstein's technique. The patients varied widely in their ability and willingness to follow the programs prescribed and their physiological responses to diet and insulin.

One consistent finding was that the patients embraced the opportunity to better understand and control their disease, despite the inconvenience of the finger pricks and the added burden of the multiple daily blood tests. (The physicians worried, justifiably, that not all patients would want to do it or were capable of doing it correctly.) "Children accepted home blood glucose monitoring quite readily and most did not wish to return to urine testing. Most children over the age of 8 performed finger pricks themselves and found this less threatening than insulin administration," the Australian diabetologist Martin Silink reported from his study of thirty children who had been using the technique for several months. Peterson reported that their patients at Rockefeller preferred to do *more* blood glucose measurements each day than required.

The health of patients who followed the guidance clearly improved, even after just a few months of self-monitoring. Bleicher discussed the experience of thirteen patients sent home with the Ames devices, reporting that the relative improvements in blood sugar levels that came with self-monitoring and "aggressive insulin therapy" prevented the earliest stages of retinal damage. An obvious implication was that it might also prevent the later stages, the retinopathies and eventual blindness that were common in insulin-dependent diabetes. Among the patients at Rockefeller, even the "severe depression" experienced by a half dozen of their initial patients resolved almost entirely.

The Rockefeller experience is significant because it was Peterson and his colleagues who worked most closely with Bernstein; they had counseled their patients to follow Bernstein's protocol most closely. They taught their patients—ten in total, ranging in age from fifteen to thirty-six—not just to calibrate insulin doses and timing to cover the foods they were eating, but also to assess how foods influenced their blood sugar, choosing "a diet that optimized blood glucose levels." The Rockefeller approach, like Bernstein's, was a process of self-experimentation, "an empirical approach to meal planning," as it was described:

> Patients are told to document their blood sugars before and after a given meal and also taught the approximate time intervals of conversion of ingested foods, such as simple and complex carbohydrates, fat, and protein into blood glucose. They are not, however, given a specific diet as traditionally prescribed. In addition, they are encouraged to normalize their blood glucose by experimenting with foods and determining which elevate the blood glucose and which do not.

Like Bernstein, the patients self-selected a diet that restricted carbohydrates significantly. They ate, on average, barely half the amount of carbohydrates that the ADA was advising. They ate more fat than the ADA and the American Heart Association considered a safe upper limit, and more protein than nephrologists and diabetologists thought wise for a disease in which nephropathy—kidney damage and kidney failure—were complications.

The results of this experiment in self-selecting a diet to optimize blood sugar were unprecedented: not only did severe depressions resolve over the eight months of the study, but eight of the ten patients achieved the goal of normal or healthy blood sugar—hemoglobin A1C levels below 6 percent—and eight of ten reduced their insulin dose as well. The blood pressure in the patients improved significantly, dropping by 10 millimeters of mercury (systolic) in the first three months and then remaining low; the one patient who had been on antihypertensive medication could discontinue it. Body fat content improved in the patients; those who normalized their blood sugar and lowered their insulin doses also lost weight. Neuropathy in these patients seemed to improve as well, suggesting that, too, might be reversed by Bernstein's protocol. As for hypoglycemic episodes, the patients experienced, in total, seven extended episodes that required hospitalization in the year prior to their experiment with self-blood-glucose monitoring; they experienced three episodes during the eight months of the trial, none of which required hospitalization.

When the ADA journal *Diabetes Care* dedicated an entire issue in 1980 to articles on self-blood-glucose monitoring, Sheldon Bleicher wrote the introduction and said physicians rarely have the opportunity to witness so clearly "the beginning of a new era in medicine. . . . We are about to move into an era where the use of the phrase 'tight control of blood glucose' will be both valid and achievable to a degree not before approximated except for the briefest of time, and that in a laboratory setting."

The 1980s would see an explosion in the use of self-blood-glucose monitoring, as the matters themselves became ever smaller and less expensive for the patient. In 1982, the National Institutes of Health would launch the first phase of a large clinical trial that would test the risks and benefits. In 1987, the ADA estimated that a million patients with diabetes were already using home blood-glucose monitors, and the ADA itself was promoting their use.

But the way self-blood-glucose monitoring would be used and

the way diabetologists came to think of the technology was transitioning as it happened. Bernstein's approach required patients and physicians to think as much about diet as they did about the insulin therapy. That the diabetologists were uninterested in allowing or encouraging their patients to significantly experiment with their diets was clear from the very first reports on self-blood-glucose monitoring. Unlike Peterson and his colleagues at Rockefeller, most of the diabetologists reporting on their early clinical experience portrayed self-monitoring as useful *only* in its ability to inform more correct dosing and timing of insulin injections. Suggesting that patients should adjust their diet in response to what they learned, as Bernstein had, by "choos[ing] a diet that optimized blood sugar control," as the Rockefeller physicians had counseled their patients, was a minor consideration at best. Thaddeus Danowski and Joseph Sunder of the University of Pittsburgh, for instance, who published the first paper on self-monitoring (in 1978, in the first issue of *Diabetes Care*), had their patients monitoring their own blood sugar only to guide them on the proper timing and dosage of insulin injected by a "jet delivery" device. Danowski and Sunder reported the needleless jet delivery device as the breakthrough in diabetes care, not the self-monitoring.

These diabetologists still assumed that their patients would prefer a minimum of dietary restrictions—religiously counting carbohydrates at every meal and snack, rather than abstaining from carbohydrate-rich foods entirely—and that the heart disease researchers had been correct in concluding that dietary fat was the cause of heart disease, and that the AHA and the ADA were correct in limiting fat consumption and allowing for more carbohydrates to be consumed as a result. Any diabetic diet that could be reconciled with the advice of the ADA and the AHA had to be a relatively high-carbohydrate diet. That could be mitigated by insisting the carbohydrates consumed were rich in fiber (or that they had a low glycemic index), as the NIH, AHA, and ADA would counsel when they began to accept the reality of Gerald Reaven's Syndrome X concept in the late 1980s, but the bulk of the calories would still have to come from carbohydrates, regardless of the effect that would have on glycemic control.

Even Peterson and his colleagues at Rockefeller and Cornell avoided being drawn in to what Peterson later called "the carbohydrate wars" by only counseling patients to experiment with their diet, as Bernstein had. They wouldn't counsel them to eat fewer carbohydrates, but they assumed their patients would settle on a lower-carbohydrate diet—as Bernstein had and their patients did—through this process of careful experimentation.

In England, the diabetologists at Nottingham General Hospital and St. Thomas' Hospital who were pioneering self-blood-glucose monitoring—Tattersall, Lowy, Gale, and their colleagues—thought the Americans had embraced too zealously the low-fat message. But they still assumed diabetic patients should eat a balanced or prudent diet, regardless of the added benefits that might be obtained by eating an *unbalanced* diet. "We certainly thought keeping cholesterol under control mattered," Gale later explained, "but basically the advice was eat not too much of anything, eat what a healthy person would eat but controlled amounts at regular times."

As long as diabetologists believed their patients' diets were set, either determined by what healthy people ate or what the diabetes and heart associations believed was necessary to prevent or delay heart disease, it left insulin timing and dose and physical activity as the only variables that could be adjusted in response to what their patients learned about their blood sugar. As a result, the patients whose experiences were reported in those early papers—with the notable exception of the patients at Rockefeller—typically found themselves taking more insulin to control their blood sugar, rather than less. Often they were adding bolus doses of short-acting insulin before meals, to cover the carbohydrates they would eat, to the basal doses of long-acting insulin they had already been taking either once or twice a day. At the Rockefeller symposium, Tattersall said that the final lesson he and his British colleagues had learned from their clinical experience with self-blood-glucose monitoring was that "many of the insulin regimens used in the past were inadequate." Not that the diet had been inappropriate—a distinct possibility—but that the insulin therapy had been inadequate.

· · ·

Two other factors fed into the collective decision by diabetologists to minimize the benefits to blood sugar that might come from dietary restrictions. First, they had inherited the belief from their predecessors that their diabetic patients would not follow their dietary advice regardless of any benefit they might achieve from doing so. That is what studies of patient compliance showed. When Kelly West, then the leading diabetes epidemiologist in the United States, reviewed the situation in 1973—in "Diet Therapy of Diabetes: An Analysis of Failure"—he concluded that the problem was not just with the patient, but with the physicians:

> Some physicians are led to minimize the importance of diet therapy in insulin-dependent diabetes by several different motives. They may believe that regularity of consumption is most important, and therefore restrictive measures are more likely to do harm than good. They may believe that the traditional diabetic diets enhance atherogenesis. They may believe in the theoretical benefits of diet prescriptions but assume that the prescriptions will be followed only negligibly. Others may acknowledge the usefulness of diet therapy in controlling blood glucose levels but doubt the importance of such control. I suspect that a major reason for the de-emphasis of diet therapy is the insecurity of physicians with a method of therapy concerning which they know little.

The second factor was the major advances in insulin and insulin delivery systems in the late 1970s, and the promise of even greater advances to come. Diabetologists considered that the avant-garde of diabetes research. Diet was the purview of dietitians (obesity researchers thought much the same); advances in drug or even surgical therapy was real medical science. It was all too easy for diabetologists to assume then, as they do now, that continued innovation in drug therapy and medical technology would render unnecessary any significant dietary sacrifices for diabetic patients.

In 1980, when *Science* reported on the "insulin wars"—"a period of turbulence caused by new advances in medicine and technology"—self-blood-glucose monitoring was not considered among the advances. Since the discovery of insulin in 1922, the Eli Lilly company had dominated the market for insulin, producing the drug from the pancreases of pigs and cattle collected at slaughterhouses. (Eight thousand pounds of pancreas from more than twenty-three thousand animals were required to make a single pound of insulin.) The molecular biology revolution changed all that. The biotech start-up Genentech produced the first synthetic human insulin, cloning the insulin gene and inserting it into bacteria to convert them, as *The New York Times* reported, "into factories for making human insulin." Genentech and Eli Lilly worked together to bring synthetic human insulin to market, and the FDA would fast-track its approval in 1982. Highly purified pork insulins were also appearing on the market, produced by Lilly in the United States and the Danish firms Novo Industri and Nordisk Insulinlaboratorium (in 1989, they would merge into Novo Nordisk) in Europe. Both synthetic human insulin and these highly purified pork insulins prevented insulin resistance that might have been caused by antibodies to the traditional insulin and allergic reactions that were experienced by one in every twenty diabetic patients.

Considered even more important were the advances in insulin delivery systems. Pancreatic transplants were on the horizon—perhaps by the mid-1980s, according to *Science*—but the promise of effectively curing insulin-dependent diabetes with surgical solutions would not be fulfilled, at least not yet. Insulin pumps were already being used by a few hundred patients. These devices (technically known as *continuous subcutaneous insulin infusion pumps*) did away with insulin needles and syringes and allowed for insulin delivery that was far more aligned with that of a functioning human pancreas. A battery-powered pump infused small, measured amounts of insulin, day and night, as needed through a cannula into the body. Patients could infuse larger doses with the push of a button, as was necessary before meals. The pioneering work testing these devices was being done at the Yale School of Medicine by William Tamborlane and his colleagues and, in London, at Guy's Hospital Medical School by Harry Keen and John Pickup. The technology was developed coincident with the acceptance of self-blood-glucose monitoring. The first publications in the medical literature on the use of these insulin pumps in patients also date to 1978. Based on these early publications—reports, once again, on a few dozen patient experiences—it was clear that insulin pumps, too, could improve glycemic control to something approaching normal.

These reports, though, also reinforced one of the primary anxieties of diabetologists: that the lower the average blood sugar maintained by patients with insulin-dependent diabetes, the closer it was to that of healthy, nondiabetic individuals, the greater the risk of severe hypoglycemic episodes. By 1982, the Centers for Disease Control had reports of thirty-five deaths in the then four thousand to five thousand patients who were using pumps; half a dozen were linked to hypoglycemia. The CDC investigators thought it unlikely that these patients had died because of insulin shock, but it was reason to worry.

When the diabetologist Roger Unger at the University of Texas Health Science Center in Dallas* discussed the potential dangers of what he called "meticulous control of diabetes" in the ADA journal *Diabetes* that same year, he used a driving metaphor to do so: "The risk of a completely normal glycemic profile in a type 1 diabetic," he wrote, "can be likened to that of a normal speed pattern in an automobile with defective brakes." These therapeutic approaches to optimizing blood sugar were promising, he concluded, but they had to be tested rigorously to establish whether the benefits outweighed the risks and, in those tests, precautions had to be taken, particularly to avoid and prevent hypoglycemia.

In 1982, the NIH initiated the Diabetes Control and Complications Trial (DCCT) to establish the risks and benefits of tight blood sugar control—now informed by the use of self-blood-glucose monitoring and hemoglobin A1C measurements. What would be tested, though, was not the optimization of blood sugar control by manipulating insulin doses *and* diet, as Bernstein had done and the Rockefeller diabetologists had demonstrated was seemingly so

---

* In 1987, the name would be changed to the University of Texas Southwestern Medical Center at Dallas.

beneficial, but by "intensive insulin therapy" alone. Specifically, the use of basal and bolus (pre-meal) insulin injections, per Bernstein's protocol, or the use of an insulin pump. Both would be compared to the "typical" or "standard treatment," which was still either one or two insulin shots a day and urine tests alone. Both groups would eat a balanced diet and the intensive insulin therapy group would adjust their doses to their diet as needed. They would not be counseled to reduce the carbohydrates they consumed so as to minimize the insulin required.

The clinical centers involved, twenty-nine in the United States and Canada, could decide whatever "methods of insulin delivery, diet, exercise, and patient self-monitoring could be employed," but the diet prescribed to the patients enrolled as subjects in the trial would be one that aligned with the ADA and AHA guidelines. The dietary goals would restrict fat consumption to minimize, so they hoped, heart disease risk; that took precedence over restricting carbohydrate consumption to optimize or normalize blood sugar.* "For the DCCT protocol," as Peterson recalled, "everybody was so intent on going along with existing guidelines and not offending anybody, that the part about restricting carbohydrates got taken out."

The DCCT investigators announced the results of the trial in June 1993 at the annual meeting of the American Diabetes Association. *The New York Times* hailed the findings on its front page as "the most important discovery for diabetics since insulin." The *Times* quoted Phillip Gorden, director of the National Institute of Diabetes and Digestive and Kidney Diseases, which funded the trial, calling the DCCT "the largest and most important study carried out in the history of diabetes." In short, intensive insulin therapy worked, both delaying the onset and slowing the progression

---

* In 1988, the same year that Gerald Reaven gave the Banting Lecture at the annual meeting of the ADA discussing the role of insulin resistance and hyperinsulinemia in non-insulin-dependent diabetes and metabolic syndrome and, by implication, many of the mechanisms by which carbohydrate-rich diets could do more harm than good, the DCCT investigators amended the protocol of their trial to adopt the carbohydrate-rich diet being promoted by the National Cholesterol Education Program. At least 45 percent of the calories and ideally 50 percent had to come from carbohydrates, and no more than 35 percent from fat.

of the microvascular complications of insulin-dependent diabetes: retinopathy, neuropathy, and nephropathy. (At the Joslin Diabetes Center in Boston, according to one young physician there at the time, "Joslin was right" buttons were widely distributed.)

The DCCT put an end to the seventy years of debate about the value of glycemic control in patients with insulin-dependent diabetes, but it left plenty still open to controversy. In many ways, the study had failed to achieve its goals. Despite all the effort and guidance given to the patients, their blood sugar had remained high— averaging over 150 mg/dl, and often going as high as 200 mg/dl in the hours after breakfast. Only one in every twenty subjects on the intensive insulin therapy *ever* managed to get their blood sugar into a relatively healthy range, what the investigators had defined as a hemoglobin A1C percentage of 6.05 or below. Rather, the average for all the subjects was 7 percent, still in the range considered diabetic.

The intensive insulin therapy was expensive, and the cost of the devices alone was likely to at least double the initial costs of diabetes treatment. The *Times* predicted that the DCCT results would "generate a windfall for manufacturers of medical devices." It also required far more institutional investment, as the DCCT investigators acknowledged: the patients in the intensive therapy arm of the trial had only obtained such promising results by working with "an expert team of diabetologists, nurses, dietitians, and behavioral specialists, and the time, effort, and cost required were considerable." Diabetologists had difficulty imagining that same level of effort and resources being available to patients in the general community.

The intensive insulin therapy also came with serious side effects, just as diabetologists had worried that it might. Weight gain was the immediately obvious one. As early as 1988, the DCCT investigators had reported that patients on intensive insulin therapy gained, on average, more than eleven pounds in just the first year of the trial. Patients receiving the standard therapy gained half as much. Both groups continued to gain weight over the course of the trial. After five years the intensive therapy group averaged ten pounds heavier, and they were significantly more likely to have become overweight or obese, both of which, as diabetologists have always known, are associated with the macrovascular complications of diabetes—heart disease in particular.

What seemed surprising to the researchers is that the subjects in the trial got ever fatter as the years went on, even as they were apparently eating less. As the patients gained weight, the dietitians with whom they worked so closely no doubt had counseled them to eat less, and yet that hadn't stopped the weight gain. In 1988, the DCCT researchers had concluded that "further study" was necessary to determine if it was possible to improve glycemic control in diabetic patients, particularly those on intensive insulin therapy, "without undesirable weight gain." The five years since then had not lessened the need to figure that out.

The intensive therapy had led to another result that diabetologists had feared: an increased likelihood of severe hypoglycemic episodes. The DCCT investigators had defined severe hypoglycemia as those events in which the patient needed help from others to recover. The patients getting intensive therapy experienced three times as many severe insulin reactions as those on standard therapy alone, and a threefold increase in the number of episodes that resulted in a coma or a seizure. Patients getting intensive therapy were more likely to be hospitalized for the hypoglycemia they experienced. The DCCT investigators had been all too aware of the potential risks of what they were observing: "two fatal motor vehicle accidents, one in each group, in which hypoglycemia may have had a causative role. In addition, a person not involved in the trial was killed in a motor vehicle accident involving a car driven by a patient in the intensive-therapy group who was probably hypoglycemic."

The DCCT investigators also reported that the two groups showed no significant difference in how they rated their quality of life. The researchers couched this as a good thing: quality of life hadn't *declined* with intensive insulin therapy "despite the added demands." But it also hadn't improved. If the patients, day to day, felt no better with this intensive intervention than they had without it—and maybe even worse, because of the hypoglycemia—would they bother to do the extra work and spend the extra money to adhere to it? The University of Southern California pediatric endocrinologist and later ADA president Francine Kaufman described the goal of "bringing blood sugar down to the normal range, while avoiding hypoglycemia" as a "tough balancing act." Patients who are two to three times more likely to have severe hypoglycemic episodes and end up in the hospital, and far more likely to have only modest episodes—the kinds of swings in blood sugar that Bernstein had regularly experienced—might not see the benefit of the trade-off. "Many people pull back on treatment when they experience hypoglycemic symptoms," Kaufman wrote. "They sacrifice their future health—which is dependent on avoiding high glucose levels day in and day out over years—to feel safe in the moment."

Another conspicuous issue was whether the DCCT results on type 1 diabetes could be extrapolated to the 90 to 95 percent of diabetic patients with type 2 diabetes. When *The New England Journal of Medicine* published an article by the DCCT researchers reporting in detail on their results, the journal ran it along with a commentary written by Roz Lasker, a physician and endocrinologist who was then deputy assistant secretary of health for the U.S. Department of Health and Human Services. Lasker's cautious perspective served to counterbalance some of the enthusiasm the DCCT study had generated. If intensive insulin therapy had led to greater weight gain and more severe hypoglycemic episodes in patients with insulin-dependent diabetes, Lasker asked, what would it do to patients with non-insulin-dependent diabetes, the ones who are already insulin resistant? "Hyperinsulinemia and insulin resistance, both very common in patients with NIDDM [non-insulin-dependent diabetes mellitus, i.e., type 2], are associated with an increased risk of hypertension, coronary artery disease, and stroke," she wrote, "raising the possibility that insulin itself has atherogenic actions. If this is so, the high doses of insulin that are frequently required to achieve euglycemia [healthy blood sugar] in patients with NIDDM, and the associated weight gain that enhances insulin resistance, could have deleterious effects that offset any potential benefits of intensive therapy in reducing the risk of microvascular complications."

More desirable, Lasker wrote, would be a therapy that improved glycemic control by making the patient more insulin sensitive and that had little risk of hypoglycemia. Restricting carbohydrates rather than fat, as Gerald Reaven had been arguing, might do that, but that's not the advice these diabetics were getting. "Diet and exercise are readily available therapies that satisfy these criteria," Lasker wrote, but then added that such regimens "are rarely successful in the long term." Whether it was because the patients didn't follow them or the disease got worse anyway was a question that hadn't been answered. Once again, the question that Lasker did not ask was whether these programs are rarely successful because they are based on misconceptions about the appropriate diet.

A month after *The New York Times* had first reported on the DCCT results, it ran a follow-up article giving Bernstein the credit for motivating the trial. "'Vindication' for a Diabetes Expert" was the headline. In the years since Bernstein had begun his crusade, he had left his management career behind and gone to medical school. He had been admitted to the Albert Einstein College of Medicine in 1979 to become its oldest incoming student ever. Jay Skyler and Charles Peterson were among the influential diabetologists who had written letters of recommendation for him. Getting a medical degree, Bernstein had told the *Times* in 1988, "was the only way to get credibility" as a diabetes expert.* Since 1985, he had been practicing medicine and counseling diabetic patients how to optimize glycemic control with his program.

Bernstein told the *Times* he was happy to have been vindicated, but he was also concerned by what he considered a significant problem with the DCCT: the study had been motivated by a method to optimize or normalize blood sugar by diet and insulin together, but had become a trial of intensive insulin therapy alone. "Those

---

* Despite Bernstein's concerns about credibility, he had published his first book, *Diabetes: The GlucograF Method for Normalizing Blood Sugar*, in 1981 while still in his first year of medical school, and it sold more than twenty-five thousand copies. "The GlucograF method already has widespread appeal to patients, and many physicians are adopting it without further medical advice. Thus, it behooves all physicians who care for diabetic patients to read this work," read a review of the book in *JAMA*. Bernstein had also published three articles in *Diabetes Care*, one adapted from his book, and been a coauthor on another even before receiving his medical degree.

in the study followed some version of a high-carbohydrate diet and used large doses of insulin," Bernstein told the *Times*, "while our regimen for Type I insulin-dependent diabetes is for a very low-carbohydrate diet and very small doses of insulin." Bernstein believed it was impossible to prevent large swings in blood sugar—and so both hyperglycemia and hypoglycemic episodes—with a high-carbohydrate diet, and the DCCT results seemed to confirm that.

The DCCT study made clear that the patients on intensive insulin therapy, despite all the assistance they had received from physicians, dietitians, and advisors, had not managed to bring their blood sugar under good control. They had better glycemic control with intensive insulin therapy, but the glycemic roller coaster that caused so much of the day-to-day burden of the disease remained—not just the physical burden but the psychological one, the despair, anxiety, depression, and impaired self-image suffered by so many diabetic patients. Intensive insulin therapy had not solved those problems; rather, with the greater likelihood of severe hypoglycemia and weight gain, it may have made them worse.

The DCCT investigators had first published their observations about hypoglycemia and intensive insulin therapy in *The American Journal of Medicine* in 1991. That they had published these results two years before the results on microvascular complications that had garnered such enthusiastic media attention suggested that these diabetologists were worried about what they were seeing in their patients, particularly since a large proportion of these severe hypoglycemic episodes (over 70 percent) were either happening at night, when the patients (and their families) were asleep, or happening during the day without warning. The higher the insulin doses and the greater the patients' weight, the greater the risk of these severe events. The 1991 report had included a discussion not only of the precautions that patients should take using intensive insulin therapy—including "avoid behaviors that may lead to hypoglycemia, such as taking excess insulin, delaying or missing meals"—but also a plea for more research to ascertain the magnitude of the risks and how they might ultimately be weighed against the benefits of intensive therapy.

After that article had been published, *The American Journal of Medicine* published a letter in response from Bernstein. Patients using his method of glycemic control, he wrote, with his guidance alone—no team of physicians, educators, dietitians, and nurses—required significantly smaller insulin doses than those in the DCCT study and saw a reduction in hypoglycemia of all degrees of severity. His patients had been willing to modify their diet as well as their insulin therapy to allow that to happen, abstaining from eating carbohydrate-rich foods. Bernstein suggested that a trial be conducted to "explore the effects of low insulin/low [carbohydrate] on the frequency and severity of hypoglycemia." The DCCT investigators agreed that such research was necessary. Thirty years later (as I write this, and as the eighty-eight-year-old Bernstein still practices medicine from his home in Mamaroneck, New York), it has yet to be done.

13

Low Blood Sugar

The patient should be trained to regard the prevention of hypoglycemia as the highest priority of the self-care program.

—ROGER UNGER,
"Meticulous Control of Diabetes: Benefits, Risks, and Precautions," 1982

When a factor is known that can shift a homeostatic state in one direction, it is reasonable to look for automatic control of that factor or for a factor or factors which act in the opposite direction.

—WALTER CANNON,
The Way of an Investigator, 1945

In 1982, when University of Texas endocrinologist Roger Unger had published his commentary in the journal *Diabetes* on the risks and benefits of what he was calling "meticulous control of diabetes," the risks were clearly what worried him. The potential benefits, by then, seemed obvious, if still unproven—the prevention of microvascular and macrovascular complications. The National Institutes of Health had begun planning the Diabetes Control and Complications Trial, the DCCT, and Unger believed that the trial would put patients with type 1 diabetes in danger. The intensive insulin therapy that the DCCT would test, he argued, would certainly increase the risks of hypoglycemia—as it did—and to believe otherwise was biologically naïve.

Unger had spent the previous quarter of a century studying the hormonal mechanisms by which the human body prevents hypoglycemia. One implication of his research was that intensive insulin therapy would circumvent them, making severe hypoglycemic episodes far more common. Strict precautions had to be taken in the trial, he argued, to minimize the danger: participants in the trial should not try to achieve the glycemic control of a healthy nondiabetic—as Bernstein had, for instance—but rather allow their blood sugar to drift high, in the morning particularly. That way they could use less insulin in their evening dose, and reduce the risk of nocturnal hypoglycemia with all its attendant dangers. Patients assigned to intensive insulin therapy, Unger suggested, should also be instructed to wake every morning at two or three a.m. to measure their blood sugar and assure that it was not dropping precipitously. Finally, even though the hope was that intensive insulin therapy would prove beneficial for all patients with type 1 diabetes, Unger suggested that the trial should exclude any patients who had already demonstrated a propensity for serious hypoglycemic episodes, whether because their diabetes seemed particularly "brittle," as such troublesome cases were then described, or because their physicians deemed them unlikely, for whatever reason, to fully understand and faithfully implement "the principles of diabetic self-care."

While the protocol created for the DCCT embraced many of the precautions that Unger had recommended—the investigators might have done so without Unger's warnings—his concerns had been remarkably prescient. Intensive insulin therapy had tripled the likelihood that a diabetic patient would have severe hypoglycemic episodes, while still failing to stabilize blood sugar in the range that diabetologists considered healthy.

After the DCCT investigators officially published their results, the ADA released a statement enumerating three conclusions about therapeutic implications that the ADA experts said "appear warranted." The first was that intensive insulin therapy would indeed prevent or delay the onset of microvascular complications and it should keep patients free of complications, and so healthier, longer, but it would *not* make the daily life of a patient with type 1 diabetes any more bearable, nor would it make the nights any less perilous. The benefits of avoiding high blood sugar and delaying the damage it did would be traded off against an increase in the psychological burden of hypoglycemia and the risk of severe episodes. "Tight" blood sugar control, the ADA position statement warned, "may have to be sacrificed" because of the danger of serious hypoglycemia. Just as weight gain would turn out to be an unavoidable consequence of intensive insulin therapy, one understandable by the effects of insulin on fat storage and metabolism, so would hypoglycemia.*

In 2016, almost a quarter century after the landmark conclusion of the DCCT, Unger was pointing out what was still an unavoidable fact of diabetes therapy: "Even patients with optimally controlled type 1 diabetes through multiple injections of insulin may exhibit extraordinary glycemic volatility," he wrote in a review with three colleagues, experiencing severe *hyper*glycemia (blood sugar levels as high as 400 mg/dl) and severe *hypo*glycemia (blood sugar "plunging" to 40 mg/dl or below). These were extremes of blood sugar that Richard Bernstein would have recognized from his own experience. This is why diabetologists will still describe hypoglycemia as the "dark side of insulin," as a team of Italian diabetologists wrote in a 2021 review, "the major barrier preventing the full realization of its promise in diabetes management."

Understanding the reasons why is critically important to understanding the risks and benefits of any diabetes therapies, whether dietary or pharmaceutical or some combination of both. As Unger had explained, hypoglycemic episodes, low blood sugar, may technically have been due to insulin remaining active for longer than anticipated, or too few carbohydrates consumed to cover that insulin—a missed or delayed meal, for instance—but also, and of at least equal importance, because of the body's failure to prevent it from happening. If the necessary homeostatic or counterregulatory responses kick in, hypoglycemic episodes never occur. Because the

---

* As Robert Tattersall described these findings in his book *Diabetes: The Biography*, "The 'cost' to the patient was three times more [frequent episodes of] severe hypoglycaemia and an average weight gain of 4.6 kilograms, but importantly quality of life was not reduced."

insulin that is injected or infused in insulin therapy is not secreted by a functioning pancreas, those homeostatic responses are absent. The body has lost its ability to protect itself, in essence, *from* insulin.

This thinking recapitulates Claude Bernard's and his original conception of the constancy of the internal environment. In 1926, the Harvard physiologist Walter Cannon had called this concept "homeostasis." It is subtext and context of all we're discussing. Living organisms are unimaginably elaborate systems—harmonious ensembles, Bernard had said—of regulatory and counterregulatory signals, of feedback loops layered onto feedback loops, all working ultimately to maintain the stability of the environment in which their individual cells live, what Bernard had called the *"milieu intérieur"* and what Unger and modern researchers refer to as the *extracellular fluid*. This fluid includes both the blood in the circulatory system and the *interstitial fluid* that filters through the capillaries from the blood, surrounds the cells, and bathes them in the compounds—oxygen, nutrients, electrolytes like sodium and potassium, and other compounds as well—that are essential for those cells to function and thrive.

In this homeostatic conception, no organ or hormonal secretion works alone, and insulin is certainly no exception. As Unger's friend and colleague Denis McGarry had described in his 1992 *Science* article, "What If Minkowski Had Been Ageusic?," diabetologists had come to think of insulin as the hormone responsible for controlling blood sugar, and they could never free themselves from that very narrow way of thinking. The DCCT had confirmed what a dangerously incomplete conception that was.

While the ultimate argument against chronically high blood pressure, quoting once again the British diabetologist R. D. Lawrence, is "that nature has perfected . . . such a wonderful mechanism to prevent it," the same can be said about *hypo*glycemia as well. Unger did, in essence. Nature had perfected a wonderful mechanism to prevent low blood sugar. High blood sugar had long-term toxic effects—the microvascular and macrovascular complications of diabetes—but low blood sugar could be very quickly fatal. The fact that insulin therapy had effectively created hypoglycemia as a disease state—insulin shock or insulin overdose, as it was once known—was compelling evidence that the problem of the diabetic condition was being grossly oversimplified. "Whereas a single hormone, insulin, sufficed to control the problem of glucose abundance," Unger wrote in 1966, "the more life-threatening problem of glucose lack required an entire battery of glucoregulatory [glucose-regulating] hormones."

When Unger died in 2020, at ninety-six, the Nobel laureates Michael Brown and Joseph Goldstein, also his colleagues at UT Southwestern, described him as a "fiercely independent thinker" who had "to fight to get his revolutionary ideas accepted." His discoveries, they said, "rewrote textbooks and forever changed our conception of glucose metabolism." As with many of the conceptual breakthroughs in our thinking about diabetes, Unger's, too, were the end result of generations of researchers fighting against the insulin-centric, pancreas-centric view of the disease process.

But that did not mean that Unger's insights were generally understood or accepted by physicians and diabetologists. Like the revelations about the role of insulin in fat accumulation, of type 2 diabetes as a disease of insulin resistance, and the work by Reaven, Ralph DeFronzo, and others on metabolic syndrome, they came decades after the discovery of insulin, and the remarkable success of insulin therapy had firmly established the thinking on how these disorders should be treated—insulin and oral hypoglycemic agents were assumed to be the appropriate therapies. What should have been treated as *hypotheses* had been allowed to solidify into dogma. Unger's revolutionary ideas had ceased to be controversial, but their implications for therapy were mostly ignored or considered of little importance.

Researchers from Claude Bernard to Roger Unger, and all those others who took up the study of these vital and yet unimaginably complex homeostatic mechanisms, tended to approach their work with two salient facts in mind about living organisms. Bernard had established both of these experimentally in the mid-nineteenth century and they had played dominant roles in guiding his thinking and later Cannon's as well. The first was that "sugar exists constantly in the blood of animals," as Bernard had phrased it, and the amount of blood sugar remains relatively constant—what is meant by the terms "normoglycemia" or "euglycemia"—regardless of the nutritional state of the animal. Blood sugar will remain stable and in a healthy range, not too high or too low, whether the animal/organism/human has just eaten a meal rich with plant foods and the carbohydrates that were, until Bernard's work, thought to be the source of all the glucose in the blood, or isn't fed at all, or has only consumed animal foods, absent essentially all carbohydrates. In healthy humans, as discussed, blood sugar will very rarely, if ever, dip below 55 to 70 mg/dl even during lengthy fasts. As the body continues to use glucose for energy, to the extent that it does, something is replenishing it, balancing out the utilization, so that a minimal amount of glucose is *always* available in the circulation to be used for energy.

The second salient fact is that the liver, as Bernard had controversially demonstrated and announced in 1848, secretes glucose directly into the circulation. This suggested to Bernard that the liver is the source of that replenishment, the organ that produces and secretes glucose, as necessary, particularly when no plant matter is being consumed. In short, the amount of glucose in the blood at any one point in time is a balance between the sugar being removed from the blood and either stored or used for fuel—utilized by the cells of the body, the process which insulin facilitates, and depleted—and the sugar added back to the blood by the liver. The high blood sugar of diabetes, then, can be caused by the liver secreting too much glucose, just as it can be caused by the relative absence of insulin and the rest of the body using too little.

All of this makes evolutionary sense, Unger would explain (recapitulating how Bernard had also conceived of these complex biological systems), because the brains, muscles, and organs of animals require a constant and reliable supply of fuel. When food is in short supply, they still have the energy and the cognitive faculties required to obtain it. "Only in this way would it be possible," Unger wrote in 1966, "in time of famine, to outrun and capture scarce prey, to outrun or to outfight a rival for scarce food, or to migrate long distances from barren to fertile land areas." Storing calories in fat cells is one efficient way to do it, but if the body and the brain run exclusively on glucose, as researchers believed at the time (and as was in the process of being disproved), it needs to store glucose when and if possible, which it does in liver and muscle cells as glycogen, and it needs a mechanism to synthesize glucose when those limited stores of glucose/glycogen, no more than a day's worth of energy, are themselves depleted.

By the 1920s, researchers had confirmed Bernard's observation that the liver plays a critical role in this process. It is "the means by which the organism 'regulates the quantity of sugar in the blood and renders it independent of the variable conditions of digestion,'" as the Yale medical historian Frederic Holmes wrote. Just as Minkowski had demonstrated the role of the pancreas in diabetes by successfully removing the organ in dogs, Frank Mann of the newly founded Mayo Clinic now did the same with the liver. He perfected a method—technically, a *hepatectomy*—of surgically removing a dog's liver without immediately killing the dog, demonstrating that a direct consequence is hypoglycemia and even hypoglycemic coma. Inject the animal with glucose, Mann reported with his colleague Thomas Magath in 1921, "and the animal immediately and completely recovers." The more glucose injected, the longer the animal appeared normal. "It would seem from this data that it is of vital necessity that a certain minimum percentage of glucose be maintained in the blood," Mann and Magath reported, echoing what Bernard had written half a century before.[*]

In the late 1920s, researchers led by Samuel Soskin at the University of Chicago repeatedly demonstrated the virtual impossibility of producing high blood sugar, and so diabetes, in animals that lacked a functional liver. The implication was that the liver was responsible for secreting into the blood some large portion of all the glucose circulating at any one time. (Approximately 20 percent comes from the kidney, a proportion that increases the longer a person or

---

[*] Banting and Best in Toronto were aware of the work Mann was doing at the Mayo Clinic. Best would later say that Mann's demonstration of hypoglycemia in his hepatectomized dogs was the clue they needed to realize that they were on the right track in isolating insulin, that what they were seeing in their dogs injected with pancreatic extracts was also low blood sugar.

animal goes without food.) The immediate cause of the high blood sugar that is the defining characteristic of the diabetic condition, as Soskin described it in 1941, is not an absolute decrease in the rate at which cells outside the liver take up glucose and use it for fuel, as was and still is a common assumption, but the fact that the liver, which normally responds to rising blood sugar by shutting down its supply, "no longer responds to this inhibitory stimulus." Instead, the liver keeps pumping out glucose, even as blood sugar is rising, and *that's* the cause of the hyperglycemia and the diabetes. If the liver functioned properly in the absence of insulin—which it cannot do—then the blood sugar would remain at a healthy level.

How this process is regulated, and the role of blood sugar, insulin, and other hormones in doing so, was still unknown. What had been established by midcentury is that insulin has no direct action on liver cells other than perhaps signaling them to take up glucose and either use it for fuel or convert it to fat. And if it doesn't, that suggests whatever insulin is doing in the liver is indirect, working through other hormones or mechanisms, but those mechanisms, too, had yet to be identified.

This liver-centric perspective was a controversial one. The Belgian physical chemist and later Nobel laureate Christian de Duve would describe the situation as he found it in the late 1930s at the beginning of his career as "two schools confront[ing] each other, prolonging a controversy that had divided the field of diabetes research [then, already] for more than 50 years." One school, founded on Bernard's work, attributed the high blood sugar characteristic of diabetes "to 'overproduction' of glucose by the liver and believed that the main effect of insulin was to inhibit this phenomenon." This was the "hepatic (liver-centric) theory." The opposite theory endorsed by most diabetologists, de Duve explained, was the "underutilization theory," which "held 'underutilization' of glucose responsible for the elevated blood sugar in diabetes and viewed insulin as primarily stimulating the uptake of this sugar by the muscles." Best and his University of Toronto colleagues had asserted this was true back in 1926—"in accordance with the state of knowledge at that time," as Soskin put it—and that's what diabetologists had continued to believe ever since.

The evidence in favor of the hepatic theory was "virtually

ignored," according to Soskin, because it wasn't the way diabetologists thought about insulin. It still isn't. The great majority of the physicians using insulin to treat diabetic patients thought and still think of the body's cells as being unable to use glucose for fuel without insulin to tell them to do so, and that with an absolute lack of insulin (as in type 1 diabetes) or a relative lack (as in type 2 diabetes), the blood sugar simply remains elevated in the circulation as hyperglycemia. (That's how this is described, for instance, on the website of the National Institute of Diabetes and Digestive and Kidney Diseases.) Physicians, then and still, typically think of the hyperglycemia in diabetes as caused by the failure of insulin to promote blood sugar uptake by the cells of the body,* rather than the failure of insulin to inhibit the otherwise constant production and secretion of glucose by the liver.

When this scientific controversy was resolved by experiment (by de Duve and his mentor, Joseph Bouckaert, in Belgium just as World War II was beginning), "the liver was the winner," as de Duve put it, "at least in quantitative terms!" Whether after meals or during fasts, which include the hours from dinner or late-night snack to breakfast the following morning, the liver plays the dominant role in establishing the level of circulating blood sugar.

This is yet another example of scientific discoveries and advances being driven by the tools available to do the research, the ability to ask the right questions experimentally and so, ideally, get the right answers. The kinds of experiments that these researchers did in the mid-twentieth century to elucidate how the body regulates the availability of energy supplies to maintain homeostasis—how the pancreas and the liver, in essence, are coupled together in this homeostatic system—were very little different in conception from what Bernard had done a century earlier to establish the concept of how the body works to maintain the constancy of the internal environment. They measured the concentration of glucose and other fuels (fatty acids and ketone bodies, in particular) in the blood going into and out of organs and, from that, established the role of each organ in taking up or secreting the fuel. They used the puri-

---

* Insulin facilitates the uptake of glucose by cells but is not necessary for this to happen.

fied hormones, as these hormones became available, to see how those organs responded to addition of that hormonal signal; they found chemical or surgical means to damage or remove the endocrine cells secreting the hormones, to see how the absence of the hormone affected the action of the organs and tissues. When possible they removed the cells from the organs, keeping them alive in the laboratory, so they could observe their response to the presence or absence of different hormones and fuels in that in vitro environment.

From the earliest days of endocrinology, even before that term was born, researchers had established the possibility that insulin does not work alone to control blood sugar and that counterregulatory hormones play equally important but counterbalancing roles. In 1901, a young American pathologist, Eugene Opie, reported in a pair of articles in the *Journal of Experimental Medicine* that damage to specific cell clusters in the pancreas—the islets of Langerhans—seems sufficient to produce diabetes, and that these islets might be the source of the internal secretion that was absent or deficient in diabetes. In 1907, the University of Chicago physiologist Michael Lane reported that Langerhans's islets were composed of two distinctly different cell types, which he named alpha (α) cells, because they could be "fixed" or preserved using alcohol, and beta (β) cells, which couldn't. Lane's terminology is still used today. Lane concluded from this observation that the islets of Langerhans "in all probability have the function of producing a twofold substance which, poured into the blood stream, has an important effect upon metabolism."

With the discovery of insulin and the beginning of the insulin therapy era, diabetologists who thought to measure insulin's immediate effect on blood sugar—in the first ten to fifteen minutes after an insulin injection—would notice that the blood sugar typically went up before it came back down. (Charles Best said they saw this happen with their very first pancreatic extracts, but assumed it was due to the animal secreting adrenaline in response to the injection and "failed to investigate it thoroughly.") In December 1922, just a year after insulin's discovery, two University of Rochester physiologists, Charles Kimball, then still a student, and John Murlin, reported that they had isolated a second substance from pancreatic extracts that clearly raised blood sugar. They called it glucagon—short for "glucose agonist" (the opposite of "antagonist")—and diabetologists mostly assumed it was little more than a contaminant of the insulin preparations. Toward the end of World War II, after the Allied forces had liberated Louvain in Belgium, where de Duve was working, and he could get samples once again of the insulin produced by Eli Lilly in the United States, de Duve and his colleagues proved that those insulin preparations that demonstrated this immediate blood-sugar-raising effect were preparations that were contaminated with glucagon.

Chemists at Eli Lilly then isolated glucagon from pancreatic extracts and began studying its effects in patients with and without diabetes. By the end of the 1950s, with Eli Lilly distributing experimental doses of glucagon (just as the company had done with insulin), physicians were testing it as a means to rescue diabetic patients from insulin overdoses. Psychiatrists treating schizophrenic patients with insulin shock therapy were using it for the same purpose, although with the advantage that they could plan the glucagon injection in advance. "One need only inject the hormone hypodermically about 10 or 20 minutes before the complete awakening of the patient is desired," explained the Northwestern University psychiatrist Benjamin Blackman, who had "terminated over 739 insulin comas with the use of glucagon" by 1961. "The patient awakens smoothly and comfortably."

Diabetologists quickly came to accept and embrace glucagon as a therapeutic agent. It would be prescribed, as it still is, for diabetic patients to keep on hand in case of emergency, a means of reversing hypoglycemia. When eating candy or drinking juice or taking glucose tablets or even a spoonful of sugar or honey fails to suffice, an injection of glucagon can be given by friends or family if the patient is already unresponsive or incapable. Physicians treating patients were grateful for a drug that made the lives of their patients easier (and their lives easier as well). But their interest extended little beyond that, a reason, as we've been discussing, why belief systems in medicine can be so intractable. When Roger Unger entered the field in the late 1950s, trying to understand why insulin therapy did such a relatively poor job of achieving glycemic control, one of his many challenges was not just elucidating the mechanism of insulin's action and those other hormones that might be influencing blood sugar control, but getting the diabetologists to care.

When Unger discussed in his early papers the likely possibilities that might explain why blood sugar control with insulin injections was so erratic, one was that something about the insulin itself was different. Maybe the insulin used for insulin therapy was contaminated during the production process. Once Lilly was producing glucagon-free insulin in the late 1950s, though, that scenario seemed far less likely.

Another possibility is that pancreatic insulin (*endogenous* insulin, because it comes from inside the body) and the insulin of insulin therapy (*exogenous* insulin) do different things in the body because they begin doing their jobs in very different locations, and so have very different fates. The pancreas secretes insulin directly into the portal vein and from there it goes first to the liver, so these two organs—pancreas and liver—are exposed to the highest insulin doses. The insulin has to traverse the liver to make its way into the general circulation, and roughly half of the insulin secreted by the pancreas is extracted by the liver and never makes it out.

The insulin of insulin therapy is injected under the skin, creates a pocket between skin and muscle, and is then absorbed into the circulation over the course of an hour. From the circulatory system, only a small proportion of that insulin makes it to the liver, and only a *very* small proportion to the pancreatic cells. So the pancreas, the liver, and the circulatory system see very different doses of insulin, depending whether the insulin is endogenous or exogenous, secreted by the beta cells in the pancreas or injected elsewhere. The ultimate effect on blood sugar (and fat accumulation and heart disease), as Unger speculated, could be expected to be radically different.

The primary focus of Unger's early research was the liver. As he would come to think of it, the level of glucose in the blood is achieved by a "physiologic balance" between the magnitude of insulin's effect in the liver—whatever that is, which was still to be determined—and the magnitude of insulin's effect in the circulation at large, signaling cells in the organs and tissues to take up glucose and either use it for energy or store it as glycogen or as fat. By injecting insulin into the circulation as in insulin therapy, the liver sees far less of the insulin than it would see from a functioning pancreas; the circulatory system sees far more, and the ideal balance of a functioning homeostatic system is never achieved.

Unger's necessary innovation was to adapt Yalow and Berson's radioimmunoassay to work with glucagon, allowing him to study why glucagon was so effective in raising blood sugar, and how it interacted with insulin.* In a series of systematic investigations through the 1960s, Unger established definitively that glucagon was (1) a hormone, (2) that it was secreted by the pancreatic alpha cells, and (3) was secreted in response to blood sugar dropping, just as insulin, from the neighboring beta cells, was being secreted in response to blood sugar rising.

As Michael Lane had suggested in 1907 when he discovered that the islets of Langerhans were composed of alpha and beta cells, they were secreting what Unger now called "a single bihormonal metabolic regulator." That the alpha and beta cells coexisted, side by side, in the pancreas had profound physiological significance. The very first effect of the insulin secreted by the beta cells is to inhibit the secretion of glucagon by the alpha cells. While insulin then signals cells throughout the body to facilitate and so accelerate their uptake of blood sugar—to increase the utilization—glucagon signals cells specifically in the liver to synthesize more glucose and secrete it into the circulation, to increase the available supply. By reducing glucagon secretion first, the insulin sets the stage for the liver to take up glucose from the blood, rather than adding it in.

Everything insulin does, in essence, glucagon counterbalances and opposes. When the pancreas is functioning properly, we do not get one without the other. Insulin works to build up tissues—it is an *anabolic* hormone—promoting the synthesis and storage of glucose as glycogen in the liver and muscles, fatty acids as fat in

---

* Unger published his first papers on the glucagon radioimmunoassay before Yalow and Berson published their seminal paper on the insulin assay, but Unger had only been able to make his assay by reaching out to Berson for the necessary details. In his first papers on the research, he credited Yalow and Berson for providing the necessary insights.

the fat tissue, and amino acids as protein wherever needed. Glucagon is *catabolic*; it signals the liver to break down the glycogen into glucose, the fat tissue to release fatty acids from the fat stored, and promotes the use of amino acids to provide glucose for fuel, rather than protein for organs and tissues.

"The unique biologic opposition of the two hormones," Unger explained when he gave the Banting Memorial Lecture to the ADA in 1975, "endow [*sic*] the alpha-beta cell unit with the ability to vary glucose flux in a manner physiologically appropriate to the prevailing circumstances while maintaining extracellular glucose concentrations [blood sugar] within remarkably narrow limits, irrespective of those circumstances." In simpler language, insulin and glucagon work together to assure that blood sugar remains stable and that enough fuel is available in the circulation at all times to assure a constant supply of energy to the cells. Glucagon is "the hormone of glucose production," Unger would say, and "insulin is the hormone of glucose utilization." As Unger and other researchers would work out, the cellular machinery in alpha cells that synthesizes glucagon and determines how much will be released is "remarkably similar" to the machinery in beta cells that does the same with insulin. While the beta cell responds to *increasing* blood sugar by secreting its insulin, though, and glucagon inhibits or modulates that secretion, the alpha cell responds to *decreasing* blood sugar by secreting its glucagon, and insulin inhibits or modulates *that* secretion.

During exercise, for instance, the muscles increase their demand for fuel—during intense exercise perhaps tenfold—and so blood sugar starts to drop and glucagon is secreted in response, signaling the liver to produce and secrete more glucose into the circulation, and insulin secretion is inhibited to allow that to happen. During and after consumption of a carbohydrate-rich meal, blood sugar rises, insulin is secreted in response, shunting the glucose in the blood, as necessary, into liver, muscle, and fat, and glucagon secretion is inhibited. The liver shuts off or reduces its secretion of glucose as *that* glucose is now unnecessary. During both the exercise and the meal, the amount of glucose used by the brain—roughly 6 grams an hour—remains constant, which can be thought of as a primary goal of this homeostatic system. By keeping glucose levels in the blood constant, again quoting Unger's Banting Memorial Lecture, "adequate glucose delivery to the central nervous system [is] thus maintained."

By 1970, Unger had extended the implications of the glucagon research to diabetes itself, in what he was calling the "double-trouble hypothesis." Unger and other researchers had demonstrated that patients with diabetes—both type 1 and type 2—manifest not just a relative or absolute lack of insulin, but a relative or absolute *excess* of glucagon. In fact, "in every known form" of high blood sugar, Unger explained, excessive glucagon was present in the circulation: *hyperglucagonemia*, it was called. When other hormones were used to block this excess glucagon from being secreted—somatostatin, for instance, discovered in 1973—the high blood sugar could be "markedly reduced, if not prevented."

Unger's conclusion: "glucagon is the principal mediator" of the high blood sugar that is the defining characteristic of diabetes. Even when insulin is lacking entirely, an animal or a human will not become diabetic unless it is oversecreting glucagon. High blood sugar is caused not so much by the deficiency of insulin and the failure to take up glucose and use it for fuel, but by the "inappropriate" secretion of glucagon, now unopposed by insulin, and the liver continuing to pump glucose out into the circulation when it should have stopped. Low blood sugar—severe hypoglycemic episodes—happens because injected insulin is still active in the circulation. That insulin is not only signaling the cells to take up blood sugar and use it for fuel—"maintain[ing] glucose utilization in the face of impending hypoglycemia," as Unger put it—but continuing, again inappropriately, to inhibit glucagon secretion, the defense against that impending crisis. With glucagon suppressed, hypoglycemia becomes unavoidable.

Once Unger and his colleagues could measure glucagon in the circulation, they could see what was happening to the hormone in their diabetic patients on insulin therapy. As Unger reported in his Banting Memorial Lecture, "massive doses of insulin" were required to control the inappropriate secretion of glucagon in their patients—hence, the inappropriate fluctuations in their blood sugar. That was true for either type 1 or type 2 diabetes. Most of their young diabetic patients, Unger reported, required more than 200 units of insulin a day to reduce plasma glucagon levels to healthy levels—"certainly far more" insulin, he said, "than is secreted by nondiabetics" (20 to 60 units, as noted, depending primarily on diet). The glucagon levels in these young patients fluctuated significantly from hour to hour. While the pathology of type 2 diabetes is very different from that of type 1, Unger said, they are both characterized by having far too much glucagon circulating in the blood and all the problems of glycemic control that go with it.

If Unger was right, and the rewriting of the textbooks to explain the true cause of high blood sugar suggests he was, then the control of inappropriate glucagon secretion—both too little and too much—would always be necessary to stabilize blood sugar in individuals with diabetes, whether type 1 or type 2. Insulin therapy alone would not do it, no matter how fast acting or intelligently delivered the insulin might someday be. If researchers found a way to safely restore pancreatic insulin secretion so that it could be maintained for a long, healthy life, to replace or regrow the defective beta cells, that might solve the problem. But so long as the insulin was injected or infused elsewhere in the body, it would act mostly unopposed by glucagon; blood sugar would swing inappropriately high and inappropriately low, regardless of the intensity of the insulin therapy. That's why Unger had predicted, correctly, that intensive insulin therapy in the DCCT would increase severe hypoglycemic episodes.

The DCCT had initially been motivated by Richard Bernstein's experience and the patients with type 1 diabetes who were put on Bernstein's protocol by the diabetologists at Rockefeller University in the late 1970s and early 1980s. These patients had managed to achieve glycemic control comparable to that of healthy individuals. In one trial of pregnant women with diabetes led by Lois Jovanovic and Charles Peterson and published in 1981, the expectant mothers lowered their hemoglobin A1C to that of healthy women of the same age. They had fewer complications than did the healthy pregnant women to whom their experience was compared. Although these patients weren't told explicitly how to eat—as Peterson would later say, they didn't want to get involved in the "carbohydrate wars"—they were told to avoid foods that raised their blood sugar, which meant they avoided carbohydrate-rich foods.

The experiences of these patients clearly suggest a way around the glucagon-hypoglycemia problem. It's always possible that Bernstein and the Rockefeller patients were unique or exceedingly lucky or so devoted—mothers-to-be, after all—that their experiences cannot be considered relevant to most or even many diabetic patients. That uncertainty is unavoidable. As Henry Ricketts at *JAMA* had suggested of Bernstein (while rejecting his first article on his protocol), perhaps *all* these individuals were spartans and paragons, not common clay.

But what if they were representative, as Bernstein and the Rockefeller researchers believed they were, and some large proportion of diabetic patients might be willing to follow a spartan or restrictive program if it made them feel better, if it reduced the physical and psychological burden of the disease? Diabetologists, though, had what they thought were very good reasons to dismiss the very idea of such low-carbohydrate, high-fat diets for their patients. Many of these dated, as we discussed, to the era prior to insulin therapy, and to Joslin's admonitions. Most would prove to be misguided as physiologists and diabetologists in the mid-twentieth century continued to make progress elucidating the biological basis of human metabolism.

14

High-Fat Diets

It is noteworthy that although diabetics have been found to thrive on high fat diets no instances have been reported in which they manifested a marked craving for fat or substances with a high fat content.* On the contrary it has often been observed that they manifest a great appetite for sugar and candies. It is possible that human beings are not capable of making such beneficial dietary selections as are made by rats, on account of defective taste mechanisms. It is also likely, however, that cultural ideas in regard to the eating of pure fat may deter them from eating large amounts of lard or suet, or drinking great quantities of olive oil.

—CURT RICHTER AND COAUTHORS,
"Further Observations on the Self-Regulatory Dietary Selections
of Rats Made Diabetic by Pancreatectomy," 1945

Once diabetologists were persuaded to fear high-fat diets, they paid little attention to evidence that suggested their beliefs were unfounded. Researchers would continue to learn about the potential therapeutic benefit of very-low-carbohydrate, high-fat diets for diabetes, but the articles these researchers published would have little to no influence. The conclusions would not be

---

* Richter and his colleagues were apparently unaware of Harold Himsworth's 1935 survey, discussed in chapter 9, in which he reported just this kind of fat craving in the years leading up to his patients' diagnosis of diabetes.

refuted; rather they would pass unnoticed. This would be the case even when the researchers involved were among the most influential in the field. Diabetologists had come to consider these issues settled. Acknowledging that they were not could be inconvenient, if not embarrassing. Yet the research spoke directly to the question of how individuals with diabetes should eat to maximize their health and well-being.

Consider the story of Arnold Durig, for instance, who was one of the great physiologists of the early twentieth century. Durig was born in 1872 in a village in the Austrian Alps. He studied medicine at the University of Innsbruck, served as a chief staff surgeon in a military hospital during the First World War, and then was appointed director of the Institute of Physiology at the University of Vienna, where he spent the remainder of his career. In 1938, he was forced to retire when the Germans invaded Austria. Throughout his life, Durig's passion had been mountain climbing, and much of his seminal research was on the adaptation of human metabolism and respiration to the thin air and cold temperatures of high altitudes. He continued to climb into his old age.

Durig was a "mild diabetic," as his condition was described by his former student Walter Fleischmann (also a distinguished Viennese physiologist), "consistently spilling sugar and occasionally showing acetonuria," another word (along with ketosis, ketonuria, and ketoacidosis) for a measurable rise in the concentration of ketone bodies in the urine. "He never used insulin," Fleischmann said. "As a man in his fifties he used to do a lot of mountaineering during his vacations, often climbing or skiing in altitudes from 6,000 to 9,000 feet as long as 10 hours a day. He told me and other members of his staff that during these tours his only food consisted of lard with only an occasional dried prune to quench his thirst. He claimed that this diet was essential for his ability to climb the highest peak in the Austrian Alps."

The Durig story is told in a footnote in a 1945 article written by Curt Richter of Johns Hopkins Medical School, a pioneer and, at the time, the single most influential researcher in a scientific discipline known as physiological psychology (now more commonly known as behavioral neuroscience). The discipline of physiological psychology had emerged from Claude Bernard's thinking on homeostasis—Walter Cannon, who coined that term, was another pioneer in this research—and the assumption that not only physiology but fundamental behaviors work ultimately to maintain the stability of the internal environment. Physiological psychology experienced its heyday in the mid-twentieth century, before the fashion in biological research moved on to molecular biology and genetics and all that has followed.

Among Richter's many significant contributions to his discipline were a series of reports documenting how animals change their behavior—specifically, what they preferentially eat and drink—in response to physiological dysfunction. As Richter and his colleagues reported, they will develop a preference for the foods or beverages that will restore the necessary conditions of the internal milieu. In 1936, Richter and his Johns Hopkins collaborators reported that rodents that have had their adrenal glands removed—the "chief physiological means of regulating sodium metabolism"—develop a pronounced taste for salt water. They will preferentially drink salt water rather than fresh, replenishing the sodium being lost in their urine and keeping themselves alive by doing so. In 1939, Richter and his colleagues reported that animals that had their thyroid glands removed—"the physiological mechanisms for the maintenance of a constant calcium balance"—respond by preferentially consuming a calcium solution that they otherwise would shun. Again, they keep themselves alive by doing so.

Richter then moved on to the question of how animals respond to a pancreatectomy, the removal of the pancreas, and so the chief physiological means of regulating carbohydrate metabolism. Richter published two articles on this research, on "the self-regulatory dietary selections of rats made diabetic by pancreatectomy."

Since the pancreas plays such a critical role in carbohydrate metabolism, Richter wrote, "we were interested to learn, whether pancreatectomized rats seek fat or carbohydrate and whether their selections actually help to maintain the constancy of their internal environment." When Richter and his collaborators fed their pancreatectomized rats the laboratory diets of the era—60 percent carbohydrates and 14 percent fat—the animals manifested the expected symptoms of diabetes, the excessive thirst and hunger, loss of weight and hyperglycemia. When the animals were allowed to choose their own diet from separate containers of purified ingredients, they selected a diet almost exclusively of fat, and their symptoms vanished. "It was not possible to determine," Richter wrote, "so long as the rats remained on the self-selection diet, whether they actually suffered from diabetes."

These experiments recapitulated what diabetologists had believed until Frederick Allen's research in the 1910s, and confirmed the conclusions of Newburgh, Marsh, and Petrén in the early days of the insulin era. That's why Richter had told the Durig story—an anecdote about an individual with diabetes thriving on a diet of mostly or all fat. As Richter interpreted the experience, Durig's mountain climbing had depended on his fat diet because his body could utilize the energy from the fat, even in the relative absence of insulin. It could not do so with carbohydrates.

Richter was not alone in reaching this conclusion. It had become a common observation made by researchers studying the new animal models of diabetes that had begun to appear in the late 1930s: dogs, for instance, that over the course of months were made diabetic by injection with extracts of pituitary hormone (known to have hyperglycemic properties), as Frank Young of the National Institute for Medical Research in London reported in 1937; rabbits and laboratory rodents made diabetic by injections of the chemical alloxan, as reported in 1943 by the Glasgow pathologist J. Shaw Dunn and his colleagues. With these new models, the researchers had set out to understand how diabetic animals might respond to different diets, and specifically how their responses differed from those of the pancreatectomized dogs that had been used since Minkowski's era, or the partially pancreatectomized dogs that had led Frederick Allen to promote his semi-starvation diet therapy and the thinking that Joslin had found so compelling.

All of these diabetic animals seemed to thrive on diets of mostly to all fat; the symptoms of the diabetes—the thirst and voracious hunger, the glucose and ketones in the urine—vanished so long as fat was mostly if not all the animals ate. "We have been able to demonstrate that alloxan diabetes in the rat can be brought under complete control by diet," wrote Joshua Burn and his colleagues from Oxford University in the *British Medical Journal* in 1944, describing experiments in which they had slowly increased the proportion of fat consumed by their diabetic rodents until it constituted 70 percent of the calories the animals were eating. At Frank Young's lab in London, feeding their diabetic dogs raw meat or even raw meat with pancreas and some glucose (the "Toronto diet," they called it) brought blood sugar and the diabetic symptoms under better control than did the standard carbohydrate-rich laboratory food. On a diet exclusively of beef suet, which is over 99 percent fat, the dogs seemed to thrive. "It is unfortunate," Young had written with Henry Marks in 1939, "that in the past so few experiments have been made concerning the influence of administering pure fat to diabetic patients or animals. Because of the fear of precipitating coma, a good proportion of protein food such as meat was usually added to the fat, the addition of which may have completely obscured the influence of the fat. Those courageous clinicians who, like Petrén, gave food containing a high proportion of fat, but no flesh, observed a dramatic fall in ketonuria [ketones in the urine]."

Even Charles Best at the University of Toronto, codiscoverer of insulin, had come to believe that high-fat diets might be uniquely beneficial for diabetes. Best and his Toronto colleagues reported in 1939 that feeding healthy rats a high-fat diet for as little as a week induced "a very definite change" in the insulin content of the rats' pancreas; it would drop to less than half that of animals fed on their usual laboratory diet. The same decrease in insulin was observed when the animals were fasted or given insulin injections, suggesting to Best that all these interventions were "resting" the pancreas; less insulin was required during fasting, high-fat feeding, or insulin therapy, so the pancreatic beta cells produced less. "Rest restores the cells and makes possible a better islet function in the future," Best suggested in *The New England Journal of Medicine* in 1940. "The resting procedures, then, serve two purposes: they prevent the degenerative changes from occurring in cells not already affected; and they permit the restoration of those exhausted cells which still retain their ability to recover."

Best considered the observation sufficiently compelling that he suggested in *The New England Journal* that diabetologists should test high-fat diets as a means of preventing the disease, specifically in children with a family history of type 1 diabetes. These children would be relatively likely to acquire diabetes as they passed through adolescence, and any delay in its onset would be to the good. Such children, Best suggested, would be divided up into two "large and comparable groups," one of which would eat "a normal diet and the second a diet as low in carbohydrate and protein and as rich in fat as is feasible," perhaps with prophylactic insulin injections as well. They would then be compared, apparently for as long as necessary, to determine if diabetes was less likely to appear in the children who had their pancreas perpetually eased of the strain of carbohydrate consumption. "We appreciate the difficulties inherent in this type of clinical investigation," Best wrote, "but believe that the goal justifies the endeavor."

Best and his Toronto coauthors appear to have been virtually alone in thinking so. There never was such a trial. The very idea of it ran against all the major trends in diabetes therapy at the time, perhaps none more so than the belief that children with diabetes should be encouraged and allowed to lead as normal a life and to have as normal a childhood as their disease permitted. "Such children would be brought up with the sugar menace always over their heads," a *BMJ* editorial opined in response, "with a diet which would make eating an unpleasant task, with the hospital or doctor constantly on their track, and, as they grew up, with rebellion and resentment in their hearts."

Research, nonetheless, continued to chip away at the belief system that had fueled the resistance to high-fat diets since Frederick Allen's work. The same revolution in laboratory assays and tools that had begun in the 1930s and transformed the understanding of the liver and glucagon's role in the regulation of blood sugar now elucidated the role of ketone bodies and ketosis in both health and diabetes. Once again, the researchers were asking, as Claude Bernard had, how humans and animals survive periods of fasting; how they provide their cells with the necessary energy even when little or none is consumed.

Among the many reasons diabetologists had shied away from

high-fat diets in the insulin therapy era was the belief, disseminated widely by Joslin, that the fat would precipitate first ketosis or ketonuria—the appearance of significant amounts of ketones in the urine—and then ketoacidosis, coma, and death in their diabetic patients. With the great proportion of diabetic hospitalizations due to diabetic ketoacidosis, the apprehension was justifiable. Indeed, physicians had first become aware of the existence of ketones when they were discovered in the urine of patients in diabetic coma. It was a reasonable assumption, although incorrect, that the coma itself was the result of the overproduction of ketones, and that any diet or process that stimulated ketone synthesis was likely to end badly.

Diabetologists therefore considered ketone bodies "noxious substances" and the state of ketosis to be avoided at all costs. "The clinician thinks [of ketone bodies] as being evil omens and therefore representative of pathology," is how this was described in 1973 by the Harvard physiologist and diabetologist George Cahill in an article that focused on the substantial evidence to demonstrate otherwise. Since ketone bodies are a product of fat metabolism and ketoacidosis, as Joslin wrote in the early editions of his textbook, and could "disappear as if by magic when the subject begins to take and burn carbohydrate," this had been another rationale for believing that some minimal amount of carbohydrate was an absolute necessity in diabetic diets.

What made this thinking so paradoxical, at least in retrospect, is that researchers studying human metabolism had also known—since the 1890s, at least—that ketone bodies were produced naturally in healthy subjects. "If a healthy individual lives for three successive days upon a carbohydrate-free diet," wrote Joslin in the 1923 edition of his textbook, "the urine voided upon the subsequent morning will show the presence of diacetic acid and acetone [i.e., ketone bodies]." The same ketosis was observed when healthy individuals went more than a day or two without eating, and after exercise, particularly strenuous exercise. Still, diabetologists had come to think of ketone bodies as the product of the incomplete oxidation of dietary fat, a metabolic process, they believed, that required carbohydrates to function properly. They would cite Rollin Wood-

yatt's early work, as Joslin did, saying "fat burns in the flame of carbohydrate, but without it, it smokes." And they would equate this
smoke of ketosis, as Woodyatt also had, to an automobile engine
smoking "with too much oil in the cylinders."

Physiologists eventually came to understand how incorrect
this thinking was, but it took them half a century to do so. Led by
researchers at the Scripps Metabolic Clinic in La Jolla, California,
and taking advantage, once again, of new laboratory assays—in this
case, more sensitive methods for measuring ketone bodies in the
blood and urine—they had established that ketone bodies are a
dominant source of energy for animals (or humans) that have to rely
on either dietary fat or their own body fat for fuel, and that muscles
and other organs freely use both ketone bodies and fatty acids to
generate energy. In 1958, researchers from the United States and
Scotland reported that healthy young men *always* have measurable
amounts of ketone bodies in their blood and their urine, just less
than had previously been measured in other physiological states.
Some organs, the heart particularly, seemed to run with greater
efficiency when relying on ketone bodies for energy, and researchers had established that under certain conditions—multiday fasts,
for instance—ketone bodies might supply more than a third of all
the energy used by either animals or humans.

This raised another question: Why would the body expend
energy converting fat to ketones if these same tissues and organs
could also use, as they did, the fat itself for energy?

When the Columbia University professor emeritus Theodore
Van Itallie reviewed this science in 2003, he said that the answer
should have been obvious all along. Studies of the fuel use of the
human brain had established that under normal conditions and a
typical carbohydrate-rich diet, the brain will use 100 to 150 grams
of glucose every day for energy. This observation led generations
of nutritionists and dietitians to assume, incorrectly, that humans
require 100 to 150 grams of carbohydrates daily and that any diet
containing fewer carbohydrates would be deficient.

Another source of glucose during periods of fasting would be
the amino acids in the protein in the body's muscles and organs.
But researchers had established as early as 1915—in studies on the
metabolism of fasting carried out by Francis Benedict of the Car-

negie Institute of Washington—that the body did not metabolize enough protein to satisfy the demands of the brain for glucose.* At the maximum rate that muscle tissue will break down during fasts, it was hard to imagine how humans could think, let alone live, after just two weeks without food. Yet young healthy men had been known to survive without food for two months. Researchers had also established that fatty acids in the blood will not cross into the brain over the network of tissues and blood vessels known as the blood-brain barrier. Hence, some other fuel had to be available. "If the nutrition scientists of the time had given more thought to the meaning of this unexplained 'prolongation' of life during total starvation," Van Itallie wrote, "they might have reasoned that during extended food deprivation the brain *is* able to use ketones generated in such abundance as an alternative energy source."

Suspecting that ketones are the fuel, George Cahill and his colleagues at Harvard measured the ketones in the blood supply going into and then out of the brain of subjects experiencing extended fasts. The measurements confirmed that ketone bodies were supplying much of the fuel the brain was using. During the first days of a fast, as Cahill and his colleagues came to learn, the brain uses glucose from any glycogen stored in the liver and then from the amino acids of the protein converted by the liver into glucose as the protein begins to break down. The muscles meanwhile shift to using fat for fuel. As more and more protein is used as a source of glucose, the liver ratchets up its synthesis of ketone bodies. If the fast continues from days to weeks, the brain shifts its dependence on glucose for fuel to the ketones. By doing so, the protein in the muscles and tissues, required by the body to function, is conserved. And as the brain shifts to using mostly ketone bodies, as Cahill reported, "intellectual capacity and performance are unimpaired."† When the

---

* When fat is mobilized and oxidized during fasts, the glycerol molecule (from the triglycerides) is converted to glucose and can be used by the brain. Lactate, too, produced by muscles during exercise, is taken up by the liver and converted to glucose.

† It was the brain's shift from using glucose to ketone bodies for fuel, as Cahill noted, that apparently explained the success of ketogenic diets as a therapy for childhood epilepsy, first reported in 1921 by Russell Wilder at the Mayo Clinic. This is a likely explanation for why Wilder was particularly open-minded on the subject of the risks and benefits of high-fat diets for diabetes.

NIH neuroscientist Louis Sokoloff reviewed this research in 1973, he pointed out that the brain is "not the inflexible, selective organ it was once thought to be" but rather uses what's available for fuel. The more glucose in the circulation, the more the brain will use; the more ketones, the more the brain will burn those. "Ketone utilization is regulated in [the] adult brain," Sokoloff wrote, "entirely by the supply of [ketones] brought to it in the blood."

This process of fuel partitioning—the shift in fuel supplied and metabolized by tissues and organs as the body transitions from metabolizing the energy it consumes to the energy it has stored—is controlled primarily by insulin and glucagon. (The role of glucagon in regulating ketone production was elucidated by Roger Unger's colleague and friend Denis McGarry, with his and McGarry's mentor Daniel Foster, beginning in the late 1960s.) In a diabetic state, with insufficient insulin and excess glucagon, the fat tissue can no longer maintain the fat it has stored—requiring insulin to do so—and releases fatty acids into the circulation; the liver converts them into ketone bodies. Because the glucagon is also signaling the liver to secrete glucose into the circulation and this is happening unrestrained, the result is the intense metabolic disruption of diabetic ketoacidosis, with high levels of both ketone bodies and glucose in the urine—hence, ketonuria and glucosuria. In contrast to the fasting individual, as Cahill described it, the diabetic patient "gets into trouble because of the rapidity of these changes," not because the production of ketone bodies and the state of ketosis is inherently pathological.

As for the clinical implications, if the brain will use ketones for energy when they're available, then diabetic patients, properly managed, might be able to thrive without carbohydrates in their diet and do so without the otherwise ever-present danger of hypoglycemia, of insulin shock. When the brain oxidizes ketones for fuel rather than the great bulk of the glucose required, it might well be protected from hypoglycemia. Low levels of blood sugar might not stimulate the compensatory responses—particularly secretion of adrenaline and other stress hormones—that they would when the brain is dependent entirely on that blood sugar for its fuel.

This prospect was ultimately tested in the early 1970s by Ernst

Drenick of the University of California at Los Angeles. Drenick had spent the previous decade studying the use of fasting—"prolonged starvation," he called it—as a means of reducing weight in severely obese patients. In 1972, Drenick published the results of a study in which he and his colleagues had set out to determine whether individuals whose brains were fully adapted to using ketone bodies for energy would be protected from insulin shock or overdose. Drenick and his colleagues fasted severely obese subjects for two months before injecting them with insulin to lower their blood sugar, without the requisite carbohydrates to cover the insulin. And yet their fasting patients experienced no symptoms of hypoglycemia. While patients with diabetes had been known not to experience severe hypoglycemia until their blood sugar fell below 50 mg/dl (and "normal young individuals," according to Cahill, as low as 30 mg/dl), Drenick's fasting obese subjects, adapted to using ketones as fuel, had their blood sugar drop to below 10 mg/dl without experiencing any symptoms.

This was the kind of experiment that few researchers in the world either could or would do. Institutional review boards today would be unlikely to allow such an experiment to be conducted on any human subjects, let alone those with diabetes. And yet, if reproducible, the implications for diabetes therapy would be profound: "Brittle diabetics, subject to recurrent symptomatic insulin reactions, may possibly benefit from eating ketogenic diets," Drenick suggested.

When Cahill reviewed this science in 1980, he noted that the same protection against severe hypoglycemia, against insulin shock, had been demonstrated in animal models of diabetes. "Both the potential danger as well as the ethics involved probably prevent a direct assessment in man as to whether ketoacids can provide sole substrate or almost sole substrate" for the brain, Cahill wrote, but it was clear that the experience of severe hypoglycemia in diabetic patients "is not simply a function of circulating glucose levels." Rather, low blood sugar, even very low blood sugar, might well be benign if the body is living off mostly fat and protein—as it does in starvation and with the kind of diet that Bernstein had by then proved to be an effective therapy.

. . .

When Claude Bernard in the late 1870s and Walter Cannon half a century later worked out their thinking on what Cannon came to call homeostasis, the relative stability of blood sugar levels in a healthy organism had presented a critical stimulus to their ideas, one of the observations that they sought to understand and explain. By 1929, when Cannon published his seminal paper on "Organization for Physiological Homeostasis," he was able to use the body's response to insulin-induced hypoglycemia as an example of such homeostatic regulation.

"If known [blood sugar] elevating agencies," Cannon had explained, referring to the sympathetic nervous system or the adrenal glands releasing adrenaline,

> are unable to bring forth sugar from storage in the liver, the glycemic level falls from about 70 to about 45 [mg/dl], whereupon serious symptoms (convulsions and coma) may supervene. The range between 70 and 45 [mg/dl] may be regarded as the *margin of safety*. On the other hand, if the depressing agency [insulin or the nervous system stimulation of insulin secretion] is ineffective, the glycemic level rises to about 180 [mg/dl] and then sugar begins to be lost through the kidneys. The range from 100 or 120 to 180 [mg/dl] may be regarded as the *margin of economy*—beyond that, homeostasis is dependent on wasting the energy contained in the sugar and the energy possibly employed by the body to bring it as glucose into the blood.

This conception of homeostasis is the context of all we've been discussing. Endocrine disorders, of which diabetes is one, are disorders of homeostasis: a critical element in an unimaginably complex system is failing to function properly. Too much or too little is being secreted, or the hormone secreted is not working as it should. Maybe the hormone itself is defective. Perhaps the problem is with the target tissues and cells, which are not responding as they should. The immediate symptoms and eventual complications are all the downstream manifestations of this dysfunction, the body trying and ultimately failing to restore the conditions of the internal milieu that are necessary for health.

We saw how the treatment of diabetes might have differed had physicians been able to diagnose the disease by the relative presence or absence of insulin in the circulation—a fundamental defect—rather than sugar in the urine or high levels of sugar in the blood, which are among the many downstream effects of the insulin dysregulation. The limited laboratory technologies available to diabetologists at the time prevented a diagnosis by insulin levels. That had profound consequences that are still with us.

Robert Tattersall had made this point in 2009 in *Diabetes: The Biography*: diabetes is the only major endocrine disease that is still *not* diagnosed by the relative presence or absence of the critical hormone. Until a method had been developed to measure hormone levels in the blood with accuracy—Yalow and Berson's radioimmunoassay in 1960—this had been unavoidable. Endocrinologists working with other diseases would embrace the new assay to use for diagnosis; the diabetologists never did. Instead, they continued to diagnose the disease based on glycemic control alone—as they still do*—and all their therapeutic modalities are targeted at trying to normalize *that*, rather than normalizing insulin physiology.

With the availability of insulin therapy in the early 1920s, diabetologists thought of themselves as replacing what was missing (restoring homeostasis, if they thought about it in that sense): adding back the insulin the body needed for healthy functioning and, by doing so, restoring blood sugar to healthy levels. But the physiological revelations of the mid-twentieth century had exposed the shortcomings and biological naïveté of that thinking. Even for those patients who are insulin dependent, who have type 1 diabetes, the complications they experience—the day-to-day roller coaster of hypoglycemia and hyperglycemia and the long-term microvascular and macrovascular issues—can be explained by all the ways insulin, injected or infused by a pump, fails to mimic all the many

---

* "Diabetes is defined by hyperglycemia," as the American Diabetes Association wrote in a 2021 report.

actions of insulin secreted from the pancreatic beta cells themselves. The hypoglycemia and the weight gain of insulin therapy may very likely be unavoidable consequences of insulin therapy. The microvascular and macrovascular complications may be as well.

As for the great majority of those with diabetes who are not insulin dependent, who have type 2 diabetes, the revelation of Yalow and Berson's assay had been that they have too much insulin. They are both insulin resistant and hyperinsulinemic, and yet they would still be prescribed insulin therapy—even more insulin—if necessary, for glycemic control. Their complications, too, including the excess weight that associates with their obesity, could be explained by their hyperinsulinemia, as Yalow and Berson had suggested in 1965, and yet restoring insulin to nondiabetic levels was not the target of therapy. When researchers explained all the metabolic abnormalities that associate with type 2 diabetes as a likely response to this same dysregulation of insulin signaling, and all of those as caused or exacerbated by the carbohydrate content of the diet—perhaps sugar and white flour most of all—the therapeutic approach remained essentially unchanged. Rather, as new insulins and new methods of delivery, new oral hypoglycemic agents were developed, physicians and diabetologists assumed that these would do a better job of mimicking what the bodies of healthy individuals do normally.

Ultimately, the success or failure of this therapeutic thinking would be tested by a series of large clinical trials that were launched from the 1970s onward. A critical question was whether treating type 2 diabetes by drug or diet would prevent the slow physical degeneration that characterized the disease. This would be the benchmark for any proposed therapies, particularly dietary therapies.

In 1977, British researchers had set out to answer this question with the United Kingdom Prospective Diabetes Study (UKPDS), considered another of the seminal trials in diabetes therapy. Over the next fourteen years, some four thousand patients with newly diagnosed type 2 diabetes were enrolled into the trial. All of the patients were taught to consume diets consistent with the ADA guidelines—"low in saturated fat, moderately high fibre ... about 50% of calories from carbohydrates; overweight patients were advised to reduce energy content." After three months of this dietary therapy, the patients were then randomly assigned to continue on the diet alone or to add drug therapy—either an oral hypoglycemic agent, an insulin-sensitizing drug known as metformin, or insulin itself. They would remain on the diet or the single drug— "monotherapy," the investigators called it—as long as they maintained relatively healthy blood sugar levels, a hemoglobin A1C, for instance, below 7 percent, consistent with ADA guidelines. They would still technically suffer from diabetes, but it would not be as severe as previously. When they could no longer accomplish even this level of glycemic control, they would be prescribed multiple drugs or more intensive insulin therapy. They would then be followed for nine years to see how they fared.

The UKPDS investigators published the first results from the trial in *The Lancet* in 1998: drug therapy in these patients with type 2 diabetes, including intensive insulin therapy, had reduced the risk of microvascular complications compared to the diet alone. This was considered the great success of the trial, demonstrating in patients with type 2 diabetes what the Diabetes Control and Complications Trial had achieved in type 1 diabetes in the early years of that decade. This delay or prevention of microvascular complications was achieved, once again, while increasing the prevalence of hypoglycemic episodes and weight gain. The drugs had no measurable effect on macrovascular complications. That the intensive insulin therapy did not seem to make heart disease more prevalent was considered a good thing and evidence that insulin therapy itself was benign. "The UKPDS is another landmark diabetes study proving the value of metabolic control," as the American Diabetes Association phrased it in a 2002 review.

As for the ability of either the ADA diet alone or drugs to halt the progression of the disease, the UKPDS investigators published those results in 1999: neither drugs nor the ADA diet nor both together had been able to prevent diabetes from getting worse as the trial progressed. The proportion of patients who could maintain the targeted level of hemoglobin A1C—below 7 percent—

steadily decreased over the years. After nine years, fewer than one in ten of the patients who had been randomized to the diet alone was still maintaining this level of blood sugar control; only one in four of those on the single drug therapies. "The increasing failure of monotherapy," the investigators reported, "is consistent with the progressive deterioration of [beta]-cell function."

This was the bad news, and the basis of the assumption that diabetes is a chronic progressive disease, one that requires ever more drugs or increased doses to keep blood sugar relatively stable and prevent or delay diabetic complications.

The UKPDS was another in a burgeoning series of large clinical trials that would only test conventional dietary therapy. What it had demonstrated was that diets consistent with the ADA prescriptions—low in fat and saturated fat, relatively rich in carbohydrates—could not prevent progression of the disease. Indeed, one possible interpretation of the trial was that the ADA dietary prescription, as Reaven might have predicted, was causing or exacerbating the disease progression. Because only the ADA's dietary therapy had been used in the trial, this possibility could not be tested. The question the trial did not ask, *which must be asked*, is whether the complications observed in the trial and the progression of the disease itself is as dependent on the composition of the foods consumed as they on are the nature of the disease itself.

In the United States, the federal government had only acknowledged that diabetes was a problem deserving significant research attention in 1974, when Richard Nixon, in the midst of the Watergate scandal, signed into law the National Diabetes Research and Education Act. At the time, the NIH budget for diabetes research was less than 2 percent what it was for cancer. The new law called for the creation and funding of diabetes research and training centers throughout the country, and the establishment of a National Commission on Diabetes to formulate a long-term plan of attack. A year later, when the eighteen members of the commission—including George Cahill and directors of seven of the National Institutes of Health—submitted their report to Congress, they began it, not surprisingly, with a discussion of the dire nature of the problem.

Since the mid-1960s, the commission wrote, diabetes prevalence had increased by 50 percent. Ten million Americans were afflicted. Physicians were diagnosing six hundred thousand new cases each year, and diabetes was responsible for three hundred thousand yearly deaths. It had become the fifth leading cause of death in the country. "Few substantial improvements in therapy have occurred since the discovery of insulin more than 50 years ago," the commission wrote. "Despite considerable research in diabetes and metabolism, the biochemical basis of the disease remains unknown." Among the commission's recommendations was that funding for diabetes research and treatment be increased significantly. By the end of the decade it was up over $100 million a year.

In the early 1980s, diabetologists had begun laying out the strategy for the first of the landmark clinical trials that the NIH would fund, the Diabetes Control and Complications Trial, discussed earlier. These large and ambitious clinical trials would test the fundamental tenets of diabetes therapy, embraced by diabetologists for more than half a century; specifically, that their patients could prevent or delay not just the microvascular complications of the disease but the macrovascular complications that would cause premature death if they could achieve and sustain two goals of equal importance. One was to control their blood sugar; the closer they could get to the blood sugar of a healthy, nondiabetic individual, the healthier they would be. The other goal, more relevant, specifically, to patients with type 2 diabetes, was to maintain a healthy weight. If they could remain relatively lean, they could avoid or at least prevent the complications of the disease; if they were overweight or obese when they were diagnosed, then weight loss and sustaining that loss was a primary goal.

The clinical trials that would test these assumptions in type 2 diabetes began in the early 2000s. Thousands of subjects were recruited and then followed for the many years necessary to establish whether those patients who successfully made the effort to control their weight and their blood sugar experienced fewer heart attacks and strokes than those who didn't. Would they live longer, healthier lives?

An implicit and critical assumption of these trials was that it does not matter how patients achieve these goals—whether by drugs or diets or what kind of diets—only that they do. When the results were later interpreted and disseminated, they were interpreted primarily in this context—not whether or not glycemic control by intensive drug therapy prevents or delays macrovascular complications, but whether glycemic control does *at all*; not whether a weight control program of exercise and a calorie-restricted version of the ADA's low-fat, high-carbohydrate diet prevents and delays macrovascular complications, but whether weight control does *at all*. As a result, diabetologists, having failed to confirm the hypotheses they had set out to test, somehow retained their preconceptions, while shedding no light on other potential means of achieving the desired outcomes. They kept doing what they'd done all along, even now that they had good reason to think it would fail.

It is important to review these trials, once again—what they had tested and what they had not—now in the context of these fundamental tenets of diabetes therapy.

The first three trials to be published—in 2008 and 2009—were all tests of Joslin's glycemic control hypothesis that intensive blood sugar control will delay or prevent heart disease, stroke, and premature death. These were the ACCORD (Action to Control Cardiovascular Risk in Diabetes) trial, conducted at over eighty medical clinics in the United States and Canada; the ADVANCE trial, a twenty-country study; and the Veterans Affairs Diabetes Trial (VADT), conducted at Veterans Affairs medical centers throughout the United States. Together, these trials comprised some twenty-three thousand diabetic patients. Half of these patients continued with the standard diabetic therapy of the era, with the goal of bringing their hemoglobin A1C down below 8 percent, which was high enough to be considered diabetic but consistent with what diabetologists had come to accept in their patients with the standard therapy. The other half of the patients would be randomized to "intensive glycemic control" with the goal of getting their hemoglobin A1C below 6 percent (ACCORD and ADVANCE) or 6.5 percent (VADT). These were at the high end of what would be considered healthy blood sugar or the very low end of the diabetes scale. In both groups, multiple drugs would be used as necessary, including insulin.

That intensive glycemic control will prevent or delay cardiovascular complications had been among the core tenets of diabetes therapy since the discovery of insulin, and yet all three trials failed to confirm it. In the ACCORD trial, intensive glycemic control was associated with an *increase* in deaths: the more intensive the glycemic control, the more likely the patients in the trial were to die prematurely. Hypoglycemic episodes, as we discussed, were also more common with the intensive glycemic control, and the patients gained more weight.

Diabetologists were back to where they were in the early 1970s when the UGDP trial had failed to show any benefit from the use of the oral hypoglycemics or insulin of that era. Now three more trials had offered up a referendum, in effect, on "one of the most contentious issues in medicine," as the Harvard diabetologist David Nathan described it. On the subject of whether lowering blood sugar prevents or delays heart disease or strokes, as William Duckworth, co-chair of the Veterans Affairs trial, said, "these three studies are unfortunately unanimous in saying, 'No, it doesn't.'" None of the three trials had tested the benefits of achieving normoglycemia by diet.

Three years later, the $200 million Look AHEAD (Action for Health in Diabetes) trial was ended prematurely when it was clear that the hypothesis it was testing would not be confirmed. We discussed Look AHEAD earlier in the context of the fat-cholesterol hypothesis of heart disease, but the trial had been designed, specifically, as a test of the second of the fundamental tenets of diabetes therapy: that overweight or obese patients with type 2 diabetes who lost weight and sustained that weight loss would also delay or prevent macrovascular complications. Many diabetologists would have been hard-pressed to say whether it is more important for their overweight patients to achieve and maintain glycemic control or to achieve and maintain a healthy weight. But surely the latter is of critical importance. Joslin had argued that it was the most important, that maintaining a healthy weight in older, overweight patients would prevent the diabetic complications and maybe even return them to health.

Beginning in 2001, the Look AHEAD investigators had recruited over five thousand participants—overweight or obese patients with type 2 diabetes—and randomized half to what they called the "intensive lifestyle intervention," or ILI. These patients were given intensive counseling to eat calorie-restricted versions of the ADA's recommended low-fat, carbohydrate-rich diet, to engage in three hours of moderately intensive exercise every week, and virtually any other help that was necessary, short of bariatric surgery, to get them to achieve the requisite weight loss. In the first year of the trial, the participants lost on average 8 percent of their body weight (sixteen pounds for a two-hundred-pounder). The Look AHEAD investigators considered it a major accomplishment that their patient participants had managed to sustain a weight loss of 6 percent (twelve pounds for a two-hundred-pounder) for the duration of the trial—more than nine years on average. By then medical authorities were arguing that if those suffering with obesity could lose even 5 percent of their body weight they would experience meaningful health benefits. (Although when the Look AHEAD trial results were first announced in October 2012, David Nathan had suggested to *The New York Times* that they would be well advised to "put 'meaningful' in quotes.") The weight loss wasn't without apparent benefits—it associated with better physical functioning and quality of life, and modest reductions in sleep apnea, urinary incontinence, and symptoms of depression—but those benefits did not include living longer or experiencing fewer heart attacks or strokes.

The Look AHEAD investigators did acknowledge in the article they published on their results in *The New England Journal of Medicine* that they had only tested one "specific lifestyle intervention that focused on achieving weight loss through caloric restriction and increased physical activity," and considered that a limitation of the study. "It is unclear," they added, "whether an intervention focused on changes in dietary composition, for example, the Mediterranean diet, might yield different outcomes."

That was the extent of the discussion, though. The Look AHEAD investigators had suggested the Mediterranean diet because that would replace some of the carbohydrates in the diet with olive oil—monounsaturated fats—and so provide a compromise between what Reaven had been suggesting was true of the ADA's high-carbohydrate diets and what these researchers had come to believe was true of high-fat diets. But in the context of diabetes this remained only an untested assumption.

The potential alternative had always been a dietary therapy that served the purpose of restoring homeostasis and the conditions of life that are conducive to health and vigor. The high-fat diet, absent most or all carbohydrates, was always an obvious candidate. It would not remedy insulin deficiency for those who have type 1 diabetes—they would always be dependent on insulin therapy—but it would minimize the need for that insulin and perhaps allow the body to maintain homeostasis at conditions as close as possible to those that would occur normally in healthy, nondiabetic individuals. For those with type 2 diabetes, it offers the possibility of fully restoring homeostasis, of returning insulin levels to those of healthy individuals, and, with it, perhaps reversing the condition.

Not only had human populations demonstrated that they could thrive on such diets—as Frederick Pavy and Joslin and others had commented about the Inuit and the gauchos of Argentina—but researchers and diabetologists had pointed out that any individual who went more than a few days without food was doing essentially the same: obtaining the energy their bodies used from a fuel mixture of mostly fat and some protein, albeit from their own body stores (endogenous) rather than from food itself (exogenous). The implication, as Curt Richter and others had suggested in the 1940s, was that the homeostatic disruption that appeared in response to the carbohydrates in the diet and the relative lack of insulin needed to properly metabolize those carbohydrates would not manifest itself in a diet that provided sufficient energy—plenty of fat—but few or no carbohydrates. That, too, was only a hypothesis, but it could be tested.

15

Very-Low-Carbohydrate Diets

Overriding all rules should be the maxim that the right line of treatment for each individual diabetic is that line of treatment on which he is best able to maintain health and vigor over a prolonged period. A lifetime of study may be necessary to make sure that years of apparent good health do not conceal the insidious development of arteriosclerotic or degenerate processes.

—ROBERT MICKS,
"The Diet in Diabetes," 1943

There are two cosmic truths that we should take home that relate to all of our science. That it took so many years to get a low-carbohydrate diet accepted as a possible thing to experiment on and talk about was a shock to us. The scientific community absolutely stonewalled that for so long, and we allowed them to stonewall it. It is a shame, and in a way, we're both guilty, the stonewallers and the people who knew better but weren't able to put up enough opposition to force the issue.... The second thing is another example of the way we accept that the naysayers have fixed their minds. When I was growing up, everybody knew that eating too much sugar was what caused diabetes. And then we went through the stage where eating glucose and sugar had absolutely nothing to do with diabetes and now suddenly we're coming back to that again. Maybe there's a middle position in here. But, again, the strength of the community, the [fixité] of the milieu of the scientist's brain, is the biggest enemy of progress.

—JESSE ROTH,
former scientific director of the National Institute
of Diabetes and Digestive and Kidney Diseases, 2004

The first cycle of clinical trials that would test diabetic diets of different proportions of fats and carbohydrates, and so test the ADA dietary recommendations themselves, would begin in the mid-1980s. The implications of Gerald Reaven's work at Stanford on what he was calling Syndrome X was the initial motivation, and Reaven would essentially force the issue. The high-carbohydrate diet promoted by the American Diabetes Association, as Reaven wrote in the journal *Nutrition Reviews* in 1986, is "at best, unlikely to have any useful metabolic effect, and, at worst, may aggravate the defects in carbohydrate and lipoprotein metabolism present in diabetes. The only conclusion consistent with this point of view is that widespread changes in dietary recommendations be avoided until they can be strongly supported by rigorous data. What is required is less advice and more information."

Diabetologists and public health authorities saw such trials as beside the point. They had come to believe that individuals, whether with diabetes or overweight, had little interest in complying with a dietary intervention for long enough to make a meaningful difference in their health. That might be true, but these trials would be tests of the core tenets of the diet-disease relationship, tests of hypotheses. They could not test them rigorously, because the constraints of the trials left too many factors uncontrolled—did the subjects in the trial, for instance, really eat the diets they were told to eat—but they nonetheless challenged what diabetologists, nutritionists, and public health authorities assumed to be true.

Reaven's group at Stanford would publish the first of these trials, and then they were joined most prominently by researchers at UT Southwestern—Roger Unger, Abhimanyu Garg, and Scott Grundy. In designing small clinical trials to test the risks and benefits of the ADA guidelines, these researchers were still not testing high-fat diets—not 70 or 80 percent fat as Newburgh and Marsh had used, or even more as Petrén had prescribed to his diabetic patients in Sweden. That was still considered unthinkable given the nutritional consensus of the era. The diets that would be tested would be 45 to 50 percent fat, roughly what diabetic patients might have been counseled to eat in the 1960s, prior to the wide dissemination of the idea that a healthy diet is de facto one that is low in fat and high in carbohydrates.

The UT Southwestern researchers were motivated not only by Reaven's work and the potential dangers of carbohydrate-rich diets for diabetic patients, but by Grundy's research as well. Grundy was a physician and nutritionist whose specialty was lipoprotein and cholesterol metabolism. In 1985, he reported that monoun-saturated fats—most notably, the oleic acid in olive oil—might be uniquely beneficial for preventing heart disease. When substituted for carbohydrates in the diet, Grundy reported, these monoun-saturated fats not only lower LDL cholesterol, but do so without raising triglyceride levels or lowering HDL cholesterol, as low-fat, high-carbohydrate diets were known to do. (Since olives and olive oil were staples of diets in the Mediterranean nations, this thinking was also embraced by those who had come to believe that Mediter-ranean diets, however they might be defined, are uniquely healthy.) For this reason, the higher-fat diets tested in the UT Southwest-ern trials were high in monounsaturated fats, so that they might also improve cholesterol profiles, not just the metabolic abnor-malities that associated with the diabetes itself, as Reaven was speculating.

These trials were small—initially fewer than a dozen mildly diabetic patients; eventually forty-two participants in a trial that the Stanford and UT Southwestern researchers would do collab-oratively and publish in *JAMA* in 1994.* That was the nature of most nutrition trials at the time: they were of limited duration, a few weeks to months. The underlying assumption of such trials is that eight or ten diabetic patients can be considered typical of most to all diabetic patients, and what happens in a few weeks to a few months will be a reliable indicator of what is likely to happen over years to decades. These limitations, though, meant the trials were testing the effects of these diets only on the metabolic abnor-malities thought to cause the microvascular and macrovascular complications of diabetes. These trials would not include nearly

---

* Along with researchers from the University of Minnesota and the Veterans Affairs Medical Center in San Diego.

enough subjects nor run long enough to test the effect of diet composition on the complications themselves, and whether patients could live longer and healthier lives by eating, say, the diets with more fat. The trials were what researchers would call preliminary studies—their results suggestive, but requiring further tests and confirmation.

Still, the results were consistent. As Reaven had predicted, the high-carb ADA diet exacerbated the defects in fat and carbohydrate metabolism that were characteristic of type 2 diabetes and, these researchers assumed, increased the risk of heart disease and other complications in diabetic patients. Compared to the higher-fat diets, the ADA's high-carb diets, wrote Reaven and his collaborators in *JAMA*, "caused persistent deterioration of glycemic control and accentuation of hyperinsulinemia, as well as increased plasma triglyceride." In short, it made the diabetes and its accompanying metabolic abnormalities worse.

The ADA did not ignore this research, but the nutritionists recruited to assess the evidence and write the guidelines would not see it as sufficient to refute what the association had been preaching since 1971. Rather, new editions of the guidelines would accentuate the need to individualize diets for people with diabetes, primarily trading off moderate increases in monounsaturated fats against a "more moderate carbohydrate intake" as necessary. This would be the case for those patients for whom "triglycerides and very-low-density lipoprotein cholesterol are the primary concern." For those at a healthy weight, or for those with elevated LDL cholesterol, the ADA still recommended following the National Cholesterol Education Program guidelines—less than 30 percent of calories from fat, and 50 to 60 percent from carbohydrates. And for those for whom "obesity and weight loss are the primary concerns, a reduction in dietary fat should be considered," even if those patients with diabetes who are obese and might need to lose weight are those for whom triglycerides are mostly likely to be a concern. In that context, nothing had changed. The diabetic diet would still be relatively carbohydrate rich, with all the drug therapy that would require, until remarkable evidence might persuade these diabetologists to change course.

. . .

In the next cycle of clinical trials that began in the late 1990s, physicians had finally found a reason to test diets that were fat rich and very low in carbohydrates, similar to Bernstein's protocol or even the very-high-fat diets of Newburgh and Marsh three-quarters of a century earlier. These physicians, though, would not be diabetologists or endocrinologists, and their initial motivation would not be diabetes. Rather they were family physicians, internal medicine specialists, cardiologists, and pediatricians confronted with the increasing prevalence of patients struggling with obesity and all the chronic disorders and complications that accompany that disorder, type 2 diabetes among the most obvious and most harmful.

Researchers at the Centers for Disease Control (CDC) had first reported in 1994 that rates of obesity and overweight in the United States had been increasing dramatically in the previous decades. This had been documented in a series of government health surveys—first the National Health Examination Survey in the early 1960s, and then three National Health and Nutrition Examination Surveys that followed. The percentage of Americans who met the technical requirements for obesity (a body mass index, or BMI, greater than 30) had increased only slightly from the early 1960s through the late 1970s, from 12.8 percent to 14.5 percent, but then it had risen precipitously over the 1980s to 22.5 percent, and higher still by 1994. Similar trends were observed in children, adolescents, and adults. By the end of the decade, the same trends, although typically not so dramatic, were being reported in populations worldwide.

Obesity researchers had seemingly been caught by surprise. When *The New York Times* reported in 1994 on the first CDC report, the paper quoted the Columbia University diabetologist and nutritionist Xavier Pi-Sunyer describing the obesity numbers as "incredible." "If this was about tuberculosis," he said, "it would be called an epidemic" (as it would be and has been ever since). That this happened coincident with a diet industry generating between $40 billion and $50 billion a year in revenue, as the *Times* estimated, not to mention decades of advice from physicians, health organizations, and federal health agencies counseling overweight Americans to avoid eating "too much" and maintain a healthy weight, made it all that much more perplexing. The immediate challenge to obesity researchers was to understand what was happening.

They knew that their standard advice to combat obesity—eat less and exercise more—was of little use. One of the seminal papers in obesity research, as discussed earlier, had been published in 1959 by the psychiatrist Albert Stunkard, making exactly this point. Stunkard, who was then at New York Hospital and would later move to the University of Pennsylvania, had reviewed his own clinical experience with obese patients and the existing medical literature and concluded that the conventional dietary therapy was "remarkably ineffective." In his own clinic, Stunkard reported, only twelve of one hundred obese patients for whom he had prescribed balanced diets of 800 to 1,500 calories per day—semi-starvation diets, as they were then known—had managed to lose even twenty pounds. Only one had lost as much as forty, and only two had managed to sustain their weight loss for as long as two years. That amounted to a 98 percent failure rate.

These numbers implied that obesity, like diabetes, is a chronic, intractable disorder. Stunkard said as much himself after he spoke in 1969 at President Richard Nixon's White House Conference on Food, Nutrition, and Health. Simply put, the harms caused by counseling patients to lose weight could outweigh the very small potential benefits. "Obesity is a chronic condition, resistant to treatment, prone to relapse, for which we have no cure," he wrote. "Many obese persons today might well be better off if they learned to live with their condition and stopped subjecting themselves over and over to painful and frustrating attempts to lose weight." When the CDC had published its 1994 report on the nation's emerging obesity epidemic, Pi-Sunyer wrote an accompanying editorial for *JAMA*—"The Fattening of America"—and was still echoing the same pessimistic assessment. "The imperative is to prevent weight gain, not to try to lose it after it has accumulated," he wrote. "Decreasing food intake and increasing activity seems an easy formula, yet it is proving extremely difficult to implement."

Pi-Sunyer might have said that such a formula had *always* proved extremely difficult to implement. For as long as physicians had written about obesity, they had portrayed it as irreversible, clearly resistant to "simple sobriety in eating and drinking," as the British physician Thomas Hawkes Tanner had written in the 1860s in his textbook *The Practice of Medicine*. The working assumption was that if an individual burdened with obesity could achieve and maintain a healthy weight merely by eating less, they'd have very likely done it before they found themselves a patient, being so instructed by a physician.

With obesity, though, as with diabetes, there had always been an alternative means of dietary therapy that had seemed to hold promise. That obesity was caused not by the amount of food eaten, but by the type, had always been a viable proposition. Just as those with diabetes had been counseled in the pre-insulin era to abstain from carbohydrates and eat a diet composed mostly if not exclusively of animal products, similar dietary protocols had been proposed for patients with obesity.*

As early as 1825, the French lawyer and bon vivant Jean Anthelme Brillat-Savarin, writing in *The Physiology of Taste*, considered among the most famous books ever written about food, had claimed that obesity was caused by the starches, grains, and sugars in the diet and that "more or less rigid abstinence" from these foods was the only viable solution. Through the nineteenth century and the first decades of the twentieth, carbohydrate-restricted, high-fat diets of varying degrees of severity were promoted widely by influential physicians in both Europe and the United States. When Sir William Osler, for instance, discussed obesity in the 1890s in the early editions of his seminal textbook, *The Principles and Practice of Medicine*, he listed three versions of carbohydrate-restricted, high-fat diets as the primary means of treatment. These diets would be reinvented in successive generations as physicians stumbled on the approach and came to believe that more or less rigid abstinence from carbohydrates was effective. Through the 1960s, the idea that

---

* Even Tanner had believed this approach was promising, that it was a "more sure basis" for obesity treatment than any other.

obesity was caused by the quantity and quality of carbohydrates we eat, that carbohydrates were somehow uniquely fattening, as Brillat-Savarin had suggested, competed with the idea that it was caused by the overconsumption of all calories—gluttony—and that "a calorie is a calorie," as physicians would be taught in medical school.

The treatment of obesity by dietary therapy was another line of research that emerged as a more serious endeavor in the post–World War II era. Physicians and nutritionists moved from studying vitamin- and mineral-deficiency diseases that had captured their attention prior to the war to what they thought of as diseases of excess—obesity being the most obvious—that were becoming increasingly common after. While researchers studying fat metabolism, as we discussed, focused on the role of insulin in fat storage, physicians, nutritionists, and dietitians took to experimenting with different dietary approaches to weight loss, from fasting, to meals of different size and frequency, to varying proportions of proteins, carbohydrates, and fats. As it turned out, the diets that seemingly did the best job of reducing weight without hunger were also the diets that minimized the secretion of insulin: those that prescribed more or less rigid abstinence from carbohydrates.

As obesity began to be treated as a legitimate subject of medical research, academic institutions and medical associations also took to hosting ever more frequent meetings focused exclusively on obesity. By 1973, seven such conferences or symposia had been held and their proceedings published: at Harvard University (1952); in Falsterbo, Sweden (1963), hosted by the Swedish Nutrition Foundation; the University of California, San Francisco (1967); in London, hosted by the fledgling Obesity Association of Great Britain (1968); in Paris, hosted by European nutrition and dietetic associations (1972); at the National Institutes of Health in Bethesda, Maryland (1973); and in London again (1973), hosted by Queens College. In each, the only presentations describing a successful dietary therapy for weight loss were on carbohydrate-restricted, relatively high-fat diets, what Brillat-Savarin had recommended in 1825 and very similar to what physicians in the pre-insulin era, up until Frederick Allen's work, had considered the best dietary therapy for glycemic control.

At the 1973 NIH conference, the only presentation on dietary therapy had been given by Charlotte Young, a professor of nutrition at Cornell University. Young had reviewed the thinking on obesity and diet going back to the nineteenth century. She also reported on the diet trials that she herself had been doing since the 1950s, first at Michigan State University, where she had collaborated with Margaret Ohlson, who had been her mentor and chair of the University's Foods and Nutrition Department. "Weight loss, fat loss, and percent weight loss as fat," Young reported, "appeared to be inversely related to the level of carbohydrate in the diets." Put simply, the fewer carbohydrates and the more fat in these diets, the more weight and the more fat the subjects lost. "No adequate explanation could be given for the differences in weight losses," Young reported, and yet all the carbohydrate-restricted diets "gave excellent clinical results as measured by freedom from hunger, allaying of excessive fatigue, satisfactory weight loss, suitability for long term weight reduction and subsequent weight control."

The NIH conference had been organized, in part, to establish a research agenda on obesity for the coming years. As such, Young's presentation should have had at least some influence on how the community of obesity researchers conceived of dietary therapy and what research was necessary to further test these approaches. It had little influence. With both the AHA and the ADA embracing the idea that dietary fat caused heart disease and prescribing low-fat, carbohydrate-rich diets as preventive measures, these organizations assumed the same dietary prescription would work for overweight people as well.

One remarkable consequence of this thinking was the complete disassociation between what nutritionists, dietitians, and physicians were presenting in these obesity conferences as an apparently effective means of weight control and the advice being disseminated by the health organizations themselves. The presentations reported on the clinical efficacy of high-fat, carbohydrate-restricted diets and the failure of calorie-restricted diets for weight loss. The health organizations, in the belief that both heart disease and diabetes associated so closely with obesity that the excess body fat could be a cause, disseminated the opposite message: weight loss could only be achieved and maintained through calorie-restricted diets, and these should be carbohydrate rich and low in fat. Just as the ADA had been willing to sacrifice some level of glycemic control in diabetic patients with its prescription of high-carbohydrate, low-fat diets, assuming these diets would prevent or delay the atherosclerosis that seemed part and parcel of the diabetic condition, now they would potentially sacrifice weight control as well.

The situation was complicated further by the acceptance and promotion of carbohydrate-restricted, high-fat diets by physicians working outside academia. When these physicians, based on their clinical experience, concluded that the thinking from Brillat-Savarin onward had been right about the benefits of carbohydrate-restricted diets and disseminated their thinking through popular diet books—claiming little more than what was presented in the obesity conferences, albeit with a notable lack of both caveats and humility—the physicians and researchers who had been thought leaders in academia and considered themselves the gatekeepers of responsible medicine responded harshly.*

These academics took the position that they were protecting the public from quackery, that physicians have an obligation to do no harm, and that the physicians who wrote diet books for the general public had transgressed their Hippocratic obligation. Physicians wrote those books, though, advocating carbohydrate-restricted, high-fat diets because that's what worked for their patients (and often themselves). Nonetheless, the authoritative responses were

---

* In the late 1960s and through the 1970s, as this schism manifested itself, the influential figures in the study of human obesity in the 1970s, those who wrote textbooks and gave presentations at the conferences, could be divided into two groups. There were those who believed carbohydrate-restricted diets were the only effective means of weight control—Denis Craddock, Robert Kemp, John Yudkin, Alan Howard, and Ian McLean Baird in England and Bruce Bistrian and George Blackburn in the United States—and developed variations on these diets with which they could treat patients. These men struggled to maintain credibility. Then there were those who refused to accept that carbohydrate restriction offered anything more than calorie restriction in disguise—John Garrow in England, and the Americans George Bray, Albert Stunkard, Theodore Van Itallie, George Cahill, Fred Stare, and Jules Hirsch. These men rarely treated obese patients themselves, but they were far and away the most influential authorities in the field, and they repeatedly suggested that no diets worked and so little was to be learned by studying them.

that they were "nutrition nonsense and food quackery," as a book review in *The Journal of the American Medical Association* said about one such best-selling diet book, *Calories Don't Count*, in 1961, and "bizarre concepts of nutrition and dieting [that] should not be promoted to the public as if they were established scientific principles," as an American Medical Association review in *JAMA* said about a similar book, *Dr. Atkins' Diet Revolution*, in 1973.

The latter had been written by the New York cardiologist Robert Atkins, whose dietary approach was little different from what Charlotte Young was describing that same year at the NIH conference as uniquely effective. Despite the anonymous authors of the 1973 AMA editorial accepting that such carbohydrate-restricted, high-fat diets, "unrestricted in calories" seemed to help patients lose weight, the editorial still denounced such an approach as "grossly unbalanced" because it "interdicts the 45% of calories that is usually consumed as carbohydrates" and so cannot "provide a practicable basis for long-term weight reduction or maintenance, ie, a lifetime change in eating and exercise habits."

It was *Dr. Atkins' Diet Revolution* that triggered the polarization of the emerging discipline of obesity medicine ever after. Atkins had been trained at Cornell University Medical College in New York City and claimed to have lost fifty pounds in the mid-1960s on a carbohydrate-restricted, high-fat diet. He would later say that he tried it "because that's what was being taught at the time."* After finding that the same approach worked for his patients and focusing on the ketosis created by a diet sufficiently restricted in carbohydrates and rich in fat—the first physician or diet book author to do so—he wrote his diet book, which became, as its publisher touted, the fastest selling book in history.

Atkins claimed that his patients lost "thirty, forty, one hundred,

---

* Among other similar discussions in the literature, Edward Gordon of the University of Wisconsin had published a 1963 article in *JAMA*—"A New Concept in the Treatment of Obesity"—suggesting carbohydrates were uniquely fattening: "An abundant supply of carbohydrate food exerts a powerful influence in directing the stream of glucose metabolism into lipogenesis [fat formation]," he had written, "whereas a relatively low carbohydrate intake tends to minimize the storage of fat. . . . The diet has been planned around the basic concept that its carbohydrate content should be low."

or more pounds" eating "lobster with butter sauce, steak with Bearnaise sauce . . . *bacon* cheeseburgers." "As long as you don't take in carbohydrates," he wrote, "you can eat any amount of this 'fattening' food and it won't put a single ounce of fat on you." That he was making these claims precisely as the AHA and ADA were widely promoting low-fat, high-carbohydrate diets to minimize the risk of heart disease and, supposedly, obesity itself—by avoiding the dense calories of dietary fat—was a problem that had to be resolved. One of these supposedly authoritative sources had to be wrong. The easy choice for the academic researchers and the medical organizations was that it was Atkins and his fellow diet-book doctors.

A quarter century later, when physicians found themselves confronted with the reality of the obesity epidemic and ever more patients struggling with weight control, they faced an obvious dilemma: either continue to give their patients the same questionable advice about weight loss—some variation on eat less and exercise more—or try something new. The only obvious alternatives, though, came from diet-book doctors like Atkins and what the nutrition and obesity researchers had taken to describing dismissively as "fad" diets.

For physicians to consider those options seriously required rejecting or relearning what they had been taught from medical school onward. Linda Stern, for instance, an internal medicine specialist at the Philadelphia Veterans Affairs Medical Center, described browsing the nutrition section at her local bookstore in the late 1990s, looking for ideas that might help her patients. "As a trained physician," she said, "I was trained to mock anything like the Atkins diet. It was quackery." Yet when she experimented herself with a similar approach (based on the 1996 book *Protein Power*, written by the physicians Michael and Mary Dan Eades), she found it made weight loss relatively easy. Stern organized a clinical trial to test the calorie-restricted, low-fat diet recommended by the American Heart Association against the kind of carbohydrate-restricted diet, unrestricted in calories, that popular diet books by Atkins, the Eadeses, and others were proposing. Stern's trial, carried out with Frederick Samaha and collaborators at the University of Pennsylvania and Drexel University, would eventually enroll more than 130 severely obese patients, averaging almost three hundred pounds, some weighing as much as four hundred pounds.

Albert Stunkard described the origin of a similar trial at the University of Pennsylvania as motivated by a chance meeting. The context was his assumption, based on his analysis from the 1950s, that patients who lost even forty pounds on a diet were akin to the medical equivalent of unicorns. As he told the story, though, he was walking down the street when he failed to recognize his hospital's chief of radiology. "The reason you don't know who I am is I've lost sixty pounds," the radiologist told Stunkard. When Stunkard replied, "How the hell did you do that?" the radiologist said, "The Atkins diet. And here's my associate, he lost thirty pounds on the Atkins diet." "Apparently it got to be hot stuff in the hospital," Stunkard later said. "All the young guys were doing it. So we decided to do a study."

Eric Westman, a physician and assistant professor of medicine at the Duke University Medical School in Durham, North Carolina, recalled a similar experience: two patients had lost more than fifty pounds each with relative ease using the Atkins diet. When one of them told Westman that "all he did is eat steak and eggs," Westman recalled, "I nearly choked. I was concerned about the cholesterol." And yet a checkup and blood tests revealed that the patient's risk factors for heart disease—including excess weight and cholesterol—had improved. By objective measures, the patient was healthier than he had been following Westman's conventional guidance. So Westman, too, decided to organize a clinical trial hoping to determine whether his patient's experience was a reproducible phenomenon. "I look back at the good fortune of these patients having told me what they had done," Westman said. "Most patients probably don't tell their doctors when they do something against their wishes. In this case, they did and it worked."

By the spring of 2002, these physicians-turned-diet-researchers were reporting at conferences on the results of five clinical trials comparing very-low-carbohydrate, high-fat, calorie-unrestricted diets to the kind of low-fat, calorie-restricted diet advice given then (and mostly still) by the AHA to prevent heart disease, by the ADA for patients with diabetes who struggle to maintain a healthy weight, by the U.S. Department of Agriculture in its Dietary Guidelines, and, by then, national health organizations worldwide. The participants in the five trials ranged from overweight adolescents in Long Island, who were followed for twelve weeks on their assigned diets, to the severely obese veterans of the trial conducted by Stern and her colleagues, who were followed for six months.

The results were again consistent and aligned with the expectations from Brillat-Savarin onward about the benefits of carbohydrate restriction and the evidence linking dietary carbohydrates, through their effect on insulin signaling, to fat accumulation. Stunkard put it simply when he described the results of the trial led by his University of Pennsylvania colleagues (collaborating with physician researchers at Thomas Jefferson University in Philadelphia, the Washington University School of Medicine in St. Louis, and the University of Colorado Health Sciences Center in Denver): "Atkins," he said, "beat[s] the standard diet all hollow." Patients assigned to eat the carbohydrate-restricted, high-fat diet, despite getting to eat as much other foods as they wanted—eating to satiety, as they were counseled—still lost considerably more weight than those counseled to follow the AHA calorie- and fat-restricted regime. Meanwhile, the metabolic abnormalities that so clearly associated with metabolic syndrome, with obesity and type 2 diabetes, consistently improved as well. "People always scorned the Atkins diet and paid no attention to it," Stunkard said in a 2002 interview, "and probably Atkins has hurt himself with all of his claims without solid evidence, but it really looks like the thing might work."*

The results of these early trials were particularly notable in that they challenged the validity of both of the dominant nutrition paradigms of the era: (1) that it's the saturated fat content that makes modern western diets atherogenic, that cause heart disease, and (2) that the factor responsible for causing obesity is the amount of calories consumed rather than the quality and quantity of the carbohydrates (and, presumably, their effect on insulin).

* One reason why the diet was scorned, Stunkard told me, was the personal animosity he and his colleagues in the obesity medicine world held for Atkins himself. "We thought he was a jerk," Stunkard said, "an idiot who just wants to make money."

The immediate effect of the publications, though, was to persuade yet more patients with obesity, physicians, and dietitians, and even journalists like myself, to take the carbohydrate-restricted, high-fat diets seriously as a means of reducing weight. "We were blazing a trail," Westman later said, "saying, yes, you can actually study these diets, and, yes, people can actually eat this way and it's not going to harm them."

It was at the Philadelphia Veterans Affairs Medical Center that the trials of a carbohydrate-restricted, high-fat diet for weight loss became de facto a trial of the same dietary therapy for type 2 diabetes, as close as researchers had come in eighty years to testing the Newburgh-Marsh prescription. By recruiting patients for their study who were severely obese, as Linda Stern and her colleagues had done, they had found themselves with a significant proportion of patients who also suffered either from type 2 diabetes (39 percent) or from metabolic syndrome (43 percent). "Although enrolling these subjects introduced confounding variables," Stern, Samaha, and their colleagues explained in 2003 in the article reporting their results in *The New England Journal of Medicine*, "it allowed the inclusion of subjects with the obesity-related medical disorders typically encountered in clinical practice."

In these patients, the carbohydrate-restricted, high-fat diet had improved insulin sensitivity independent of the amount of weight the subjects had lost, and had lowered the triglycerides as well, resolving what Joslin would have called the diabetic lipemia. Despite the diet's relatively high saturated fat content and the fact that these patients, suffering from severe obesity and, often, diabetes, had been allowed to eat *as much* food as they desired, they, too, had become both leaner and healthier while doing so. Stern, Samaha, and their colleagues wrote up their results with the appropriate caution. With all the major health organizations counseling that patients eat the opposite of what these trials suggested, these physicians would respect that guidance and suggest only that more research was necessary. "This study proves a principle," they wrote in *The New England Journal*, "and does not provide clinical guidance."

Meanwhile, other physicians continued to report on their clinical experience prescribing the diets for their own diabetic patients, confident now that the diets could be offered to their patients without fear of doing harm. Westman, his colleague Will Yancy at the Durham Veterans Affairs Medical Center, and the University of Kansas physician Mary Vernon reported in 2003 on their experience prescribing Atkins diets to fourteen diabetic patients in clinical practice and seven subjects in a "pilot trial" of the diet and then, two years later, their experience with twenty-one more. Not only did their patients get leaner, Westman, Yancy, and Vernon reported, but their triglyceride levels returned to normal, as did their blood sugar. This happened while they were weaned off, by necessity, most to all of their medications. By 2005, Westman, Yancy, and Vernon were no longer concerned that carbohydrate-restricted, high-fat diets were dangerous or ineffective, but that individuals who tried them would need "close medical supervision" to discontinue or reduce the doses of whatever blood pressure or glycemic medications they might be taking.

From the early 2000s onward, an ever-growing but still small minority of physicians treating obesity and type 2 diabetes took to siding with the diet-book doctors promoting these carbohydrate-restricted, high-fat (or higher-fat) diets, both studying the risks and benefits of the diets in clinical trials and prescribing them to patients. As Westman had said, the ongoing trials had revealed that the diets were safe and could be tested on a patient-by-patient basis as to their efficacy. It now became clear that these high-saturated-fat diets occasionally raised LDL cholesterol in those who adhered to them, but the elevation of LDL, as Westman and other physicians now argued, had to be weighed against all the other benefits.

When I interviewed more than 120 of these physicians between 2016 and 2018, this was a recurrent theme. Just as Stern, Westman, and the University of Pennsylvania physicians had initially tested the diets because of anecdotal evidence that they made their patients both leaner and healthier in a way that calorie-restricted, low-fat diets did not, so physicians now accepted the evidence primarily because these diets worked for them or their patients. "At the end of our clinic day," as Westman and Vernon wrote in a 2008 response to the failure of intensive insulin therapy reported in the

ACCORD trial, "we go home thinking, 'The clinical improvements are so large and obvious, why don't other doctors understand?'"

In 2015, twenty-six academic researchers and clinicians coauthored an article in the journal *Nutrition* to argue that the burden of proof had now shifted, and that dietary carbohydrate restriction should be—should *always* have been—the first approach in diabetes management. The authors still represented a minuscule minority of the relevant medical community, but it was the first time that such an argument had been made, collaboratively, in a well-respected journal in almost a century. Led by Richard Feinman, a biochemist at the State University of New York Downstate Medical Center, the authors included Richard Bernstein, Eric Westman, and many of the physicians-turned-researchers who had now published either the results of clinical trials or reports on their clinical experience. They enumerated twelve points of evidence, beginning with the simple propositions that high blood sugar is "the most salient feature of diabetes" and "dietary carbohydrate restriction has the greatest effect on decreasing blood glucose levels," neither of which could be reasonably contested. That "the best predictor of microvascular and, to a lesser extent, macrovascular complications . . . is glycemic control" was also incontestable, even if the randomized trials of drug therapy (and the AHA diet) had failed to confirm that improving glycemic control by these methods bestowed any benefit. "Logically," the authors wrote, "it is not the target but the method of trying to attain it" that matters.

There were still the assumed dangers of the saturated-fat content in these carbohydrate-restricted diets, and its effect on LDL cholesterol, but, as Feinman and his coauthors argued, the accumulated evidence suggested that whatever the benefits of cholesterol-lowering drugs, the clinical trials had consistently failed to confirm the idea that saturated fat itself was atherogenic. (The Look AHEAD trial, with its low-fat, low-saturated-fat diet prescription, was just the latest, and perhaps most relevant, example.) "To assess the disadvantages of carbohydrate restriction for individuals with diabetes," they concluded, "one has to ask what the standard is and where it came from. The idea that there is an effective diet of known macronutrient composition, one tested in long-term, or even short-term trials, that is beneficial in treating diabetes is implied by the question. To our knowledge, no such diet exists. The more dietary carbohydrate, the more medication will be required."

The most compelling evidence to support this position would come from a trial that began a year later, funded by a San Francisco start-up called Virta Health. The company had been founded in 2014 to capitalize on the proposition that the insurers and employers of individuals with type 2 diabetes could save significant money—five thousand dollars or more per patient annually on medical care—if the disease could be adequately controlled by diet.

Virta Health had been cofounded by two researchers, Steve Phinney and Jeff Volek, who had done early studies on Atkins-like ketogenic diets, working with the technology entrepreneur Sami Inkinen. Phinney had begun his research as a physician at the University of Vermont in the late 1970s, and then focused on ketogenic diets for his doctorate in nutritional biochemistry at the Massachusetts Institute of Technology. He had been surprised by how well the participants in his small studies tolerated the diets, whether lean, obese, or even well-trained endurance athletes, as Phinney reported in the third of the three trials he published. In the 1990s, Phinney had served as medical director of an obesity treatment center at the University of California, Davis, that prescribed a calorie-restricted ketogenic diet as therapy. According to Phinney, some two hundred patients a year averaged more than fifty pounds of weight loss on their program.*

Volek had been a competitive weightlifter who had first experimented with the diet himself to see how it influenced his perfor-

---

* The diet Phinney prescribed was modeled after the protein-sparing, modified-fast, a calorie-restricted ketogenic diet that had been developed in the 1970s by Bruce Bistrian and George Blackburn of Harvard Medical School. Bistrian and Blackburn had reported similar weight loss and success rates and had served as mentors to Phinney while he was doing his PhD. They were collaborators with him on his early trials. In 1976, Bistrian and Blackburn had reported that they had prescribed their protein-sparing modified fast to seven obese patients with diabetes and all had been able to discontinue their insulin therapy within three weeks. The problem with such a severely calorie-restricted approach, though, was always the question of what the patients should eat once they lost their excess fat. Bistrian and Blackburn didn't believe they could answer that in the 1980s. Now the evidence is more clear.

mance. When he found that it helped, he began to study it, first as a registered dietitian pursuing a doctorate in exercise science at Pennsylvania State University, followed by postdoctoral training at Ball State University. He published his first papers on the ketogenic diet in 2000 and 2002, describing the nature of the studies as "primarily putting people on the diet and drawing a lot of blood, measuring a lot of risk factors for cardiovascular disease." Simplistic as these studies were, Volek and his colleagues were among the very few researchers in the world at the time who had done even that. When Volek discovered that the ketogenic diet seemed beneficial for both body composition and heart disease risk, he collaborated with Eric Westman to publish reviews of the science. In 2003, he connected with Phinney.

A decade later, Volek and Phinney would meet and recruit Inkinen to be Virta's CEO. Inkinen was a graduate of Stanford business school and a cofounder of the real estate company Trulia. He was also a world-class triathlete who described himself as having spent his twenties eating essentially no added fats at all. ("I have one of the better willpowers in the world," Inkinen said. "If you tell me to eat no fat, I will eat no fat.") In 2011, he had finished first in his age group, thirty to thirty-nine, in the Ironman world triathlon championships. A year later, he was diagnosed as prediabetic (defined as hemoglobin A1C greater than 5.7 percent but less than the diabetic cutoff of 6.5 percent). By then Phinney and Volek had coauthored multiple papers and three books on ketogenic diets, and Inkinen sought them out for advice and guidance on using a ketogenic diet for both prediabetes and athletic endurance.

The concept of Virta Health emerged from their discussions: they would pioneer the use of twenty-first-century technology— smartphone apps and the internet providing patients with access at any time to health coaches, medical providers (physicians and nurse practitioners), and an "on-line peer community"—to help patients with type 2 diabetes follow what was, in effect, a nineteenth-century diet therapy. Phinney and Volek were confident that if they could get diabetic patients to understand the underlying principles of a well-formulated ketogenic diet—what the company would refer to as "nutritional ketosis," distinguishing it unambiguously from diabetic ketoacidosis—and then to live by those principles, they could put their diabetes into remission and that this could be sustained so long as the patients continued to eat appropriately. Virta Health would fund and carry out a clinical trial, Inkinen later said, because it would be the only way to demonstrate to potential clients—large employers and insurers, in their business plan—that they were right about what they were claiming. The diet world was full of radical ideas—every diet book, or so it seemed, offering up yet another one—so they had little choice. If Phinney and Volek were right about what they believed, the trial would confirm it. "Everyone who eats has an opinion about food," Inkinen said. "How do you sell the truth? How do you stand out? We really had to test the safety and effectiveness of what we were claiming."

The study itself would be run by Sarah Hallberg, a physician associated with Indiana University Health in Lafayette. Hallberg later said that she had been taught as both an undergraduate and a graduate student in exercise science to eat a low-fat diet and believe in its benefits. After getting her medical degree and working in internal medicine for IU Health—a statewide nonprofit health-care system that has a partnership with the university—she had been recruited in 2010 to create and direct a medically supervised weight loss clinic. To prepare herself for the job and to become certified in obesity medicine, she had set about reading the obesity literature, concluding that little of what she had been taught about the benefits of low-fat calorie-restricted diets had been confirmed in clinical trials; the more recent trials were suggesting very-low-carbohydrate, high-fat ketogenic diets would be effective. Hallberg spent time shadowing Westman at his Duke University clinic and was convinced of the potential benefits of the approach. After Phinney met Hallberg at a conference in early 2015, Hallberg was chosen as principal investigator of the trial and her Indiana clinic as the site. "It's not Stanford or Harvard," Inkinen later said, "but maybe it's better to run this kind of trial in the middle of the country where the problem is so severe."

Beginning in 2015, patients with type 2 diabetes in the IU Health system were offered a choice between the care they were already receiving—pharmaceutical therapy and conventional diet advice—or the Virta Health program, nutritional ketosis delivered by "continuous remote care." Those patients who opted for the Virta Health program—262 out of some 350 patients—would have their health status, medications, and health-care costs compared to the remainder who opted for the usual care.

In 2018, Hallberg and her colleagues published a report in *Diabetes Therapy* on the first year of patient experience and followed up in 2019 with the two-year data. Of the 262 patients who started on their program, 194 (74 percent) were still actively participating in the Virta Health continuous remote care program at the end of two years. These patients had achieved and maintained an average weight loss of thirty pounds—12 percent of their body weight—despite being encouraged to eat to satiety and eating to control their blood sugar rather than, specifically, their weight. As for their diabetes, the average hemoglobin A1C in the patients who had selected the Virta Health remote continuous care program had dropped by 1.3 percentage points, comparable to that observed in trials of drug therapy. In the ACCORD trial, for instance, intensive drug and insulin therapy had led to a reduction in hemoglobin A1C of 1.6 percentage points (only 0.6 with the standard therapeutic doses). In the Virta Health trial, this return to normoglycemia was achieved not by using more drugs and higher doses but significantly less of both: "discontinuing 67.0% of diabetes-specific prescriptions including most insulin and all sulfonylureas that engender risks for weight gain and hypoglycemia," as the 2019 paper reported.

Just over half of the Virta Health patients had achieved what Virta Health called "diabetes reversal" and the ADA had defined as "partial remission from diabetes," which still allowed for high blood sugar but below the diagnostic threshold for diabetes (a hemoglobin A1C below 6.5 percent) without diabetes-specific drug therapy. Seventeen percent had achieved "complete remission," defined as normoglycemia, normal healthy blood sugar (hemoglobin A1C less than 5.7 percent) maintained for at least a year without drug therapy. None of this had happened in the usual care group.*

---

* A 2014 analysis of more than 120,000 patients with type 2 diabetes in the United States reported that only 0.2 percent experienced complete remission each year, and 2.4 percent partial remission. "In community settings," the authors had said, "remission of type 2 diabetes does occur without bariatric surgery, but it is very rare."

The results of this kind of clinical trial are easy to overinterpret, and one has to be careful not to do so. The Virta Health trial was not a randomized trial, and its control group was not getting an equivalent intervention with a different dietary therapy. This means the benefit of the very-low-carbohydrate ketogenic diet alone cannot be assumed. One interpretation of such a nonrandomized trial is that any dietary therapy would induce the same or even greater benefits, provided it was delivered via the same continuous remote system.

Nonetheless, the results from the Virta Health trial represent yet another challenge to a core belief about type 2 diabetes: decades of observational studies and clinical trials of drug therapy for diabetes had consistently concluded that type 2 diabetes is a chronic, progressive, and degenerative condition—as had, most notably, the UKPDS trial. Diabetologists had concluded based on their drug trials that the primary challenge to better treatment—"the most frustrating barrier," as diabetes authorities had described it in 2017—is the reluctance of physicians to either add more medications or increase doses of those medications that patients were already taking. The Virta Health trial had demonstrated that this conclusion was wrong. It had demonstrated that the degeneration of the diabetic condition could be reversed, even that the disease itself could be put into a drug-free remission with continuous remote care and a diet that restricted carbohydrates and replaced those calories mostly with fat.

The belief that type 2 diabetes is necessarily a chronic, progressive disease—that it inevitably gets worse—can now be defended only in the context of the accepted standard of care for treatment. One way to look at the Virta Health results is that for many individuals with type 2 diabetes, the disease manifests itself only in the context of a carbohydrate-rich diet of the kind the ADA has been prescribing for more than half a century. It is a response to the carbohydrate content of the diet. For those with type 2 diabetes eating such a diet, the symptoms and the progression of the disease can *only* be managed by drug therapy. For those eating a mostly carbohydrate-free, fat-rich diet—marked by the state of nutritional ketosis—the disease ceases to progress and the symptoms essentially vanish.

As for the possibility that eating such a fat-rich diet, despite the benefit to glycemic control, increases heart disease risk—what Joslin and other physicians had worried about since the first years of the insulin therapy era—the Virta Health trial presents a significant challenge to that assumption as well. A year of intensive counseling to eat a ketogenic diet, as Hallberg and her colleagues reported, had resulted in the improvement of twenty-two of twenty-six well-established risk factors for heart disease, including, not surprisingly, HDL cholesterol, triglycerides, and blood pressure. Three of the twenty-six risk factors remained unchanged, and only one on average got worse—the amount of LDL cholesterol, albeit not the characteristics of the LDL particle (its size and density, in particular) that seem to make it potentially atherogenic.

To establish the overall benefit or risk—how the increase in LDL cholesterol should be weighed against the improvement of so many risk factors—Hallberg and her colleagues had calculated the change in what the American Heart Association calls the "aggregate atherosclerotic cardiovascular disease risk score." This is a measure of the ten-year risk of having a heart attack that had been developed jointly by the AHA and the American College of Cardiology. By this measure, the patients being counseled to eat the carbohydrate-restricted, high-fat, high-saturated-fat diet decreased their risk of having a heart attack by over 20 percent, compared to those patients following the standard of care prescribed by the ADA and the use of the drug therapies prescribed. In short, even with a rise in LDL cholesterol and assuming that the conventional thinking about the dangers of LDL cholesterol is true, these patients with type 2 diabetes got significantly healthier, as did their hearts.

The Virta Health clinical trial can be considered, ultimately, a technologically advanced version of the kind of clinical experience that led Joslin and other physicians, prior to the beginning of the clinical trial era in the 1950s, to formulate much of what had come to be considered conventional wisdom in diabetes therapy. As with what Westman had called the initial proof of principle in the trials in the early 2000s, the Virta Health observations demonstrated that the ketogenic diet prescribed was safe and effective in the context of continuous remote care; the patients thrived and did so in a way that patients following the usual care approach of a major health system did not. Those conclusions could not be avoided. In June 2022, the company announced the results of a five-year follow-up of their patients: a third of the patients who stayed with the program for five years showed no clinical signs of diabetes, while the diabetes medications prescribed to the remainder of this patient group had been reduced by half. Of the patients who started the program on insulin therapy, half of those had discontinued it. By then, Virta Health itself was thriving, with a $2 billion market valuation and more than three hundred major corporations and insurers—including Blue Shield of California and Humana*— offering the Virta system to their employees with type 2 diabetes and to those with type 2 diabetes whom they insure.

Both the American Diabetes Association and the American Heart Association have now come to accept that very-low-carbohydrate ketogenic diets can be beneficial for patients with type 2 diabetes and, with caveats, are willing to recommend them. The acceptance is evidence of how medical orthodoxy is still dependent on one or a few researchers who are either influential themselves or find themselves in a position of influence: in short, who is chosen to decide what is consensus.

In 2019, the ADA published two consensus reports on lifestyle therapy for diabetes. The first, published in January of that year, was the association's consensus report on the standard of care for patients with diabetes. The authors were anonymous, and that report repeated much of the conventional dietary wisdom about portion control and calorie-restricted diets for weight loss, and the desirability of eating carbohydrates that are high in fiber, "including vegetables, fruits, legumes, whole grains." It emphasized "healthful eating patterns" with "less focus on specific nutrients" and singled out Mediterranean diets, the Dietary Approaches to Stop Hypertension diet (known as the DASH diet), and plant-based diets as examples that could be offered to patients. This ADA report was still arguing for the benefits of a diet restricted to 25 to 30 percent fat, citing the recommendations from a fourteen-year-old National

* As of August 2022.

Academy of Medicine report, while suggesting that the "challenges with long-term sustainability" of carbohydrate-restricted eating plans made them of limited value.

Three months later, the ADA released a five-year update on nutrition therapy. This was also described as a consensus report and was authored by a fourteen-member committee of physicians, dietitians, and nutritionists, three of whom, including the chair of the committee, Will Yancy of Duke, had done research themselves on carbohydrate-restricted diets. The committee was charged with reviewing the literature since the last consensus report on nutrition therapy had been published five years earlier. Among their conclusions was that the diets recommended as examples of healthful eating patterns in the lifestyle management report—low-fat diets, Mediterranean diets, plant-based diets, and the DASH diet—were supported by surprisingly little evidence to suggest they were beneficial for patients with diabetes. In the clinical trials that had been done, the results had been inconsistent. "Lowering total fat intake," as the authors said about the ADA prescription since 1971 and the National Academy of Medicine report had advised in 2005, "did not consistently improve glycemia or [cardiovascular disease] risk factors in people with type 2 diabetes." As for low-carbohydrate or very-low-carbohydrate eating patterns, the latter of which would tend to be ketogenic diets, they were now "among the most studied eating patterns for type 2 diabetes" and the only ones in which the results and the benefits had been consistent. "Reducing overall carbohydrate intake for individuals with diabetes," the ADA report stated, "has demonstrated the most evidence for improving glycemia and may be applied in a variety of eating patterns that meet individual needs and preferences."

In 2022 when the American Heart Association published a "scientific statement" on the management of heart disease risk factors for adults with type 2 diabetes, the nine members of the AHA committee now accepted, as well, that the research was convincing that low-carbohydrate diets, high in fat and even saturated fat, were beneficial for diabetic patients, that the fewer the carbohydrates in the diet—"very low-carbohydrate versus moderate carbohydrate diets"—the greater the decrease in hemoglobin A1C, the greater the effect on weight loss, and the "fewer diabetes medications [used by] individuals with diabetes." The AHA committee suggested that it was the calorie-restricted diets that presented challenges to sustainability, reversing a century of dogma by doing so, and that "for those who are unable to adhere to a calorie-restricted diet, a low-carbohydrate diet reduces A1C and triglycerides."

In short, half a century after the American Medical Association had declared carbohydrate-restricted ketogenic diets to be "bizarre concepts of nutrition and dieting [that] should not be promoted to the public as if they were established scientific principles," that verdict had been reversed. Now enough trials had been performed, motivated by necessity and the rising tide of obesity and type 2 diabetes, and it had become ever more difficult to avoid the conclusion that the "bizarre concept of nutrition" has much to recommend it.

One consequence of this fifty-year lag in the medical and nutrition science is the enormous variation in dietary therapies that physicians will recommend to their patients with diabetes. The majority perspective remains what it has always been, but patients can now easily find alternatives. As of the spring of 2022, the ADA even publishes a "guide for health care providers" on low-carbohydrate and very-low-carbohydrate eating patterns for adults with type 2 diabetes. The pamphlet's guidance, though, conflicts with that given by the ADA in its most recent standard of care document for lifestyle management, and with all the numerous other guides that the ADA publishes and sells on its website, all variations on the conventional approach in which carbohydrates make up the bulk of the calories in the meal.

Physician awareness of the potential benefits of carbohydrate restriction and ketogenic diets for type 2 diabetes still often comes, as it did with Eric Westman, from patients who might try such diets first to lose weight. If the conventional approach has failed these patients, they are likely to try the unconventional, and these diets are an obvious choice. In the United Kingdom, for instance, David Unwin, a senior partner in what he calls a medium-sized practice—seven physicians looking after nine thousand patients—shifted his thinking on how to treat patients with type 2 diabetes essentially on the basis of observations made in one of his patients.

When he joined the practice in 1986, according to Unwin, they had fifty-seven patients with type 2 diabetes. By the 2010s, they had well over four hundred. Throughout that time, Unwin had counseled his patients to eat less and exercise more to lose weight and had seen not a single patient benefit from the advice. In 2011, one of his patients came in to have her prescriptions renewed and Unwin failed to recognize her because she had lost over fifty pounds. When results of her blood tests came back, Unwin realized that she was no longer suffering from diabetes. Both the weight loss and the reversal of diabetes were unique in his experience. His patient was embarrassed to tell him, although she did, that she had accomplished both by eating a very-low-carbohydrate, high-fat diet.

After reading up on the burgeoning literature on carbohydrate restriction and working with his wife, Jen, a clinical psychologist, Unwin began counseling his patients with type 2 diabetes to follow a very-low-carbohydrate, high-fat eating pattern. In 2020, he published a record of his clinical experience: over a quarter of the patients with type 2 diabetes in the group practice had been persuaded to try a carbohydrate-restricted diet, and 46 percent of those had achieved remission of their diabetes. Over sixty thousand dollars each year was saved in total by clinic patients on medications to control blood sugar and blood pressure. By the end of 2022, Unwin was reporting that 20 percent of the type 2 patients in his clinic—121 patients in total—had put their type 2 diabetes into remission.

In 2017, the United Kingdom's National Health Service had awarded Unwin its innovator of the year award for applying an approach to diabetes therapy that was two hundred years old and "was routine until 1923," as Unwin says. But the institutional resistance at the NHS to such a commonsensical approach nonetheless remains. The NHS website still advises physicians* to "encourage adults with type 2 diabetes to follow the same healthy eating advice as the general population" and to give the same advice to patients who are overweight and obese—eat less and exercise—that Unwin says had never worked in his practice and, as Albert Stunkard had

---

* As of this writing, in the fall of 2022.

reported more than sixty years ago, may never have had any clinical efficacy.

Meanwhile, a competing source of authoritative information on diabetes guidance has emerged in the United Kingdom, an online forum, Diabetes.co.uk, that has over 2 million participants, including one in every ten individuals in the United Kingdom living with diabetes. It describes itself as "a community of people with diabetes, family members, friends, supporters and carers, offering their own support and first hand knowledge." The site was founded in 2002 by Arjun Panesar, then a student studying artificial intelligence for his master's degree whose grandfather had just been diagnosed with type 2 diabetes. "My grandfather wasn't sure what he should be eating," Panesar says, "and my grandmother wasn't sure what she should be cooking for him." And so Panesar decided that he would create a website based on Facebook's model that would allow him to crowdsource information about how his grandfather should eat for optimal health. Panesar says that when Diabetes.co.uk was up and running, the overwhelming majority of comments came from individuals who said they avoid carbohydrate-rich foods. "It was all about reducing sugar, starch, carbohydrates and eating real food," Panesar says. In 2015, Diabetes.co.uk started working with Unwin to provide online counsel for those living with diabetes on how to follow low-carb diets. The organization is now working with the NHS, even as the two are providing very different, if not antithetical, guidance on how those with diabetes should eat.

Recommending a ketogenic or severely carbohydrate-restricted diet to a patient with type 1 diabetes comes with the threat of much greater harm than for those with type 2 diabetes. The benefits might be greater as well, but it is the risks that might be irreversible in the short term. "The patient who breaks over only in carbohydrate," who eats more than the insulin dose can cover, as Joslin wrote a century ago, "pays an immediate penalty and is warned by increased urination; the patient who breaks over only in fat and protein is not warned and dies." While insulin therapy mostly solved the problem of diabetic ketoacidosis, it had created the problem of hypoglycemia. Few if any physicians were willing to suggest to their patients who could so easily die from either of these disorders, that they should eat in a way that is characterized, in its essence, by its ability to put healthy individuals into a state of ketosis and lower their blood sugar.

This was one reason why it took a patient—Richard Bernstein— to be the first to promote such an approach publicly. By the time Bernstein had written his first book advocating for his program, he not only could draw on his own experience with type 1 diabetes but also that of the patients who participated in the small clinical trials at the local New York universities—Rockefeller, Cornell, and the State University of New York Downstate Medical Center—giving him the confidence that the same approach would help others. As he then accumulated his own clinical experience as a diabetologist, that further convinced him of the potential benefits of the approach and gave him ever more experience dealing with the risks. He counseled his patients and the readers of his books to identify the diet that best stabilized their blood sugar by carefully and safely experimenting on themselves. The diet Bernstein ultimately pre-scribed, that he believed his patients and readers would choose, was not very high in fat (not as high in fat as Newburgh and Marsh and Petrén's plans from the 1920s, or the kind of ketogenic diet pre-scribed in the Virta Health trial) but rather higher in protein and fat than more traditional diets and certainly the dietary prescrip-tion that the ADA was promoting. Patients who followed Bern-stein's thinking would have to learn how to dose their insulin to account for the glycemic effect of the protein. With carbohydrates coming only from non-starchy vegetables, the amino acids from the protein, converted into glucose, would now play the major role in raising blood sugar. Bernstein believed this trade-off was worth it, but with no clinical trials to test a competing approach, there was no way to know for sure.

Once Bernstein had initiated the process of making such a severely carbohydrate-restricted diet a possibility for patients with type 1 diabetes and then had provided the details in his books, oth-ers with type 1 diabetes could do so at their own risk. They, too, would likely be patients for whom the conventional approach was least satisfactory. If Bernstein's method and dietary prescription worked as Bernstein claimed, it too could begin to disseminate through the medical community. It would do so slowly because there would always be considerable and justifiable resistance—the great majority of physicians and diabetologists who thought it was too risky, if not quackery—but it would occur because at least some patients struggling with type 1 diabetes would have reason to try it, and at least some of them would find it beneficial.

The same technological forces that had helped disseminate information on carbohydrate-restricted diets to those who were overweight, obese, or suffered from type 2 diabetes now came into play with type 1 diabetes as well. The internet and social media made it possible for patients with type 1—those for whom the standard of care was not working to their satisfaction—to search out information about alternative approaches. They could find the equivalent of patient support groups providing the guidance that they wouldn't get from their physicians and diabetes educators. While these authority figures would remain hesitant to allow their type 1 patients to consider such an extreme way of eating, the internet would create, in effect, a record of the patients' clinical experience. This evidence, too, can easily be overinterpreted, but the clinical experience of these patients would nonetheless provide provisional answers to questions that the diabetology community had been unwilling to ask since the initial and very limited studies of Bernstein's approach in the late 1970s.

A necessary step in this progression would be the use of such an extreme dietary approach for a child or adolescent, when type 1 diabetes commonly appears. That adults might experiment on themselves is one thing; parents making such a decision for their child is always more difficult to accept. For researchers to test such a way of eating on children in a clinical trial, to establish whether the benefits outweigh the risks, might well be considered unethical. By accepting an alternative approach as standard of care for the better part of a century, covering the carbohydrates in the diet with however much insulin is necessary, the diabetes community had made it virtually impossible to deviate from it. The fact remains, though, that avoiding the foods that make all but the minimal insulin doses necessary might be the more beneficial approach.

The ethical and cognitive impasse would be overcome, again,

by anecdotal experience—in this case, that of a single child. Dave Dikeman was nine years old in 2013, living in Hawaii with his parents and younger brother, when he was diagnosed with type 1 diabetes. Dikeman had an advantage that few children have: both his parents had doctoral degrees and were willing to challenge authority if they believed that their son's health depended on it. His father, R. D. Dikeman, had been trained as a theoretical physicist and worked as a senior scientist for the aerospace company Lockheed Martin. His mother, Roxanne Uradomo-Dikeman, had a doctorate in psychology and worked professionally in that capacity.

As with many children with type 1 diabetes, Dave was initially misdiagnosed. By that time he had lost almost a quarter of his body weight. "At one point Dave vomited all over his shirt," his father R.D. said. "He was too weak to go to the bathroom and I had to carry him. When I picked him up under the blanket, I could tell how light he was. He felt like a bag of bones. When I helped him get his shirt off and into the shower, he looked totally emaciated. I called my wife and said, 'I think Dave is going to die.' We took him to the pediatrician, and she said we should feed him milkshakes, put some weight on him. But they did blood work and called the next morning to tell us to take him immediately to the emergency room." With the correct diabetes diagnosis and insulin therapy, Dave's energy returned. When they visited an endocrinologist and a dietitian, R.D. recalled, "just like everybody else, Dave was put on a very-low-fat, high-carb diet: lean meats, breads, pasta, margarine, potatoes, poultry. The endocrinologist gave us a lecture on how to use insulin, and he drew a time series of blood sugar on the whiteboard, showing how carbohydrate makes blood sugar go up and insulin makes it go down, and explained that we want to balance it so blood sugar stays in a certain range. When he drew a time series, I thought that's the kind of thing I do for a living as a physicist. I immediately grasped on to that and thought here's something I can work with."

Dikeman began preparing meals for his son just as he had been instructed. He would serve his son oatmeal in the morning, measure his blood sugar afterward, and then serve him lunch and dinner, weighing the foods and calculating the carbohydrates they contained, and do a blood sugar measurement. Then he would adjust the insulin dose as necessary to try to keep his son's blood sugar stable. Eventually, he was testing his son's blood sugar ten to twelve times each day. "I was writing algorithms and keeping data doing everything scientifically. I was staying home with Dave to do this and every day we'd get a little bit better. I would excitedly show Roxanne when she came home, and she would shrug because ultimately it was not that good. I felt I was learning. I wanted to be optimistic, [but] I would get closer one day and the next day would be a total failure. Dave's blood sugar would be up to four hundred. We would give too much insulin to bring it down, and it would drop to almost zero. We'd have to give him a bunch of juice."

After a month of failing to stabilize their son's blood sugar, to successfully cover the carbohydrates in his diet with insulin, they came to accept that it might not be possible. "The interesting thing," said Dikeman, "is that at diagnosis, when the endocrinologist was explaining how to control his blood sugar, Dave, this shriveled up dude in a hospital bed, said, 'I just won't eat carbs.' And you know what happened? I laughed at him and the endocrinologist laughed at him." Now Roxanne googled "low-carbohydrate diets" and "diabetes." Among the top listings was *Dr. Bernstein's Diabetes Solution*, and they ordered the book.

Bernstein's laws of low doses resonated with Dikeman's physics understanding, and they switched to Bernstein's very-low-carbohydrate approach immediately. Dikeman says he initially tried to compromise as necessary for Dave's athletics; he played quarterback on a youth flag-football team. He would have his son eat berries or drink chocolate milk before a game or practice, assuming Dave needed the carbohydrates for energy. But that, too, would cause excessive swings in his blood sugar. "The more closely we adopted the Bernstein approach, the better Dave's blood sugar average and the smaller the [deviations from it]. One day before a football game, when I was trying to 'load' with some chocolate milk and his blood sugars were spiking, Dave protested and said, 'I'm done with this . . . it's stupid.' After that we followed Bernstein's method precisely and still do. Dave's blood sugar has been about as good as his healthy nondiabetic little brother's ever since." When

Dikeman told me this story in the spring of 2020, Dave was then sixteen, a sophomore in high school, seven years out from his diagnosis. "Roxanne wants me to tell you," Dikeman added, "that Dave is six foot one, one hundred and seventy pounds, and now plays quarterback on his high school's junior varsity football team."

When the Dikemans discussed their son's experience on social media, thinking that others with type 1 diabetes would find it valuable, they were astonished by the responses. "No one else was doing what we were doing back then (with kids)," Dikeman later explained in an email. "People wrote crazy things like Dave was being abused. They even shared Dave's school name and someone hinted that he might be rescued and saved (inferring kidnapped)." By then, Dikeman had made adult friends with type 1 diabetes and, "despite the blowback, some parents were reaching out. We realized we needed to make a Bernstein-friendly low-carb group for people that were truly interested in learning . . . some place to be safe." A year after Dave's diagnosis, the Dikemans created a Facebook group called TypeOneGrit, where patients with type 1 diabetes and parents of children with type 1 could learn about Bernstein's approach, share their experiences, and provide guidance to others in the same situation.

By the fall of 2016, TypeOneGrit had accumulated 1,900 members and provided an opportunity that had never existed in diabetes research until then: a body of clinical experience on very-low-carbohydrate diets—specifically Bernstein's approach—that could be studied to begin to establish the risks and benefits. Two Harvard pediatric endocrinologists, David Ludwig, who had been studying variations on carbohydrate-restricted diets since the mid-1990s, and Belinda Lennerz, then took the opportunity to survey the TypeOneGrit members and document what they found. For a century, as Lennerz, Ludwig, and a team of coauthors* wrote in 2018 in the journal *Pediatrics*, very-low-carbohydrate diets had been tra-

---

* The coauthors on the study included, once again, many of the same physicians and researchers who had been arguing for the benefits of these carbohydrate-restricted diets for diabetes therapy: Eric Westman, Will Yancy, and Sarah Hallberg, most notably. R. D. Dikeman was also a coauthor as was Carrie Diulis, a spine surgeon in Ohio who uses a very-low-carbohydrate diet to help control her own type 1 diabetes.

ditionally "discouraged out of concern for potential diabetic keto-acidosis (DKA), hypoglycemia, dyslipidemia [high triglycerides, specifically], nutrient deficiency, growth failure in children, and sustainability."

Their survey of TypeOneGrit members had suggested that these concerns, while understandable, were unsubstantiated. More than three hundred members had responded to the survey request. Their average age at diagnosis had been sixteen and they had been living with the disease, on average, for more than a decade. The one undeniable observation was that these respondents had remarkable glycemic control for patients with type 1 diabetes. When they'd initially been diagnosed, their hemoglobin A1C had typically been over 11 percent. Eating as Bernstein advised and using minimal insulin, it averaged 5.7.

"Their blood sugar control seemed almost too good to be true," Lennerz was quoted as saying in *The New York Times*. The Type-OneGrit survey, Lennerz later said, had revealed "a finding that was thought to not exist. No one thought it possible that people with type one diabetes could have an A1C in the healthy range." The respondents had maintained this level of glycemic control with few of the side effects or complications that had been documented in the clinical trials of intensive drug therapy. They had (on average) maintained a healthy weight, and they experienced relatively few episodes of hypoglycemia and ketoacidosis. This was true of the children in the survey as well as the adults. "These findings are without precedent among people with T1DM [type 1 diabetes mellitus], revealing a novel approach to the prevention of long-term diabetes complications," Lennerz, Ludwig, and their coauthors wrote in their article in *Pediatrics*.

This observation, too, can be easily misinterpreted, as Lennerz, Ludwig, and their coauthors acknowledged and discussed. It was the relative likelihood of such misinterpretation that led to a follow-up debate in the journal that reflected all the challenges that had confronted the medical community on the diet-diabetes issue in the century since insulin had been discovered.

One critical problem with such a survey, as with all anecdotal evidence, is that what is not observed may be as important or more so than what is. Conceivably, thousands or tens of thousands of diabetic patients had tried the Bernstein protocol and only these three hundred had found it sufficiently beneficial that they had kept it up and were now testifying to its benefits. It's why physicians will disparage the use of "testimonials" in support of a particular therapy, as three leading authorities on diet and diabetes—a nutritionist and two diabetologists*—did in writing a letter to the editor in response to the paper. "Children and adults adhering to a VLCD [very-low-carbohydrate diet] and remaining in this online community may represent a special subpopulation with high levels of motivation and other health-related behaviors (e.g., physical activity), presenting another source of selection bias," Lennerz, Ludwig, and their coauthors wrote. "Therefore, the study sample may not be representative of all people with T1DM in the social media group and may differ from the general T1DM population in ways that could influence the safety, effectiveness, and practicality of a VLCD."

What the survey did establish, though, as the authors concluded and the critics did not dispute, was the fact that such a dietary therapy might be ideal for some patients. It "may allow for exceptional control of T1DM without increased risk of adverse event," Lennerz, Ludwig, and their coauthors wrote. "The results, if confirmed in clinical trials, indicate that the chronic complications of T1DM might be prevented by diet." The authorities who wrote in response worried that "promulgating such methodologically weak although enticing data broadly through the media" would create a risk that patients or physicians would pursue such a radical approach without being aware of the potential danger—of the need to adequately adjust insulin, if nothing else, as fewer carbohydrates were consumed—and that would cause irreparable harm.

Both sides argued, as I am, that what is now needed, after a century of reliance on insulin therapy and an increasing array of newer pharmaceuticals, is the research necessary to resolve these issues

* Elizabeth Mayer-Davis, chair of the Department of Nutrition at the University of North Carolina at Chapel Hill; Lori Laffel, chief of the Pediatric, Adolescent, and Young Adult Section at the Joslin Diabetes Center; and John Buse, chief of the Division of Endocrinology and Metabolism at the University of North Carolina at Chapel Hill.

in patients with both type 2 and type 1 diabetes. They should no longer be ignored by the diabetes community and, with luck, they won't be. We now know it's possible that a way of eating that minimizes the need for insulin and other medications can be ideal for those who suffer from type 1 diabetes. We don't know how often this will lead to harm, or how severe the harm, should patients attempt to eat this way and have it fail them. These questions can be answered definitively, but the community of diabetologists has to care enough about their patients that they commit to doing so.

Epilogue

The Conflicts of Evidence-Based Medicine

Why do physicians vary so much in the way they practice medicine?
At first view, there should be no problem. There are diseases—neatly
named and categorized by textbooks, journal articles, and medical
specialty societies. There are various procedures physicians can use to
diagnose and treat these diseases. It should be possible to determine the
value of any particular procedure by applying it to patients who have a
disease and observing the outcome. And the rest should be easy—if the
outcome is good, the procedure should be used for patients with that
disease; if the outcome is bad, it should not. . . .

The problem of course is that nothing is this simple. . . .

Uncertainty creeps into medical practice through every pore.
Whether a physician is defining a disease, making a diagnosis, selecting
a procedure, observing outcomes, assessing probabilities, assigning pref-
erences, or putting it all together, he is walking on very slippery terrain.
It is difficult for nonphysicians, and for many physicians, to appreciate
how complex these tasks are, how poorly we understand them, and how
easy it is for honest people to come to different conclusions.

—DAVID EDDY,
"Variations in Physician Practice: The Role of Uncertainty," 1984

Medical practice has always depended as much on what physicians do not know as on what they do. Physicians and medical authorities might not discuss that publicly, but it's this uncertainty that has always made medical practice as much an art as a science.

Among the revelations in medicine in the 1970s and 1980s was how uncertainty in medicine manifested itself and how it directly affected both the well-being of patients and the financial burden of health care. An emerging field of physician-researchers studying medical practices—led by David Eddy at Duke University, David Sackett and Gordon Guyatt at McMaster University in Ontario, Iain Chalmers at Oxford, and Archie Cochrane at the Medical Research Council in the United Kingdom—had concluded that the profession was facing a crisis. Not only were clinical trials demonstrating that at least some standard medical practices resulted in far more harm than good—the use of radical mastectomies, perhaps most famously, for the treatment of breast cancer—but research, led by John Wennberg at Dartmouth University, was revealing huge variations in medical practices from physician to physician, hospital to hospital, and state to state. As David Eddy would later write, they had realized that "medical decision making was not built on a bedrock of evidence or formal analysis, but was standing on Jell-O."

In the winter of 1990, *JAMA* introduced this research and its implications in a series of twenty-eight articles—"Clinical Decision Making: From Theory to Practice"—written by Eddy, whose language and conclusions were appropriately blunt. "There is a distinct possibility that many decisions made by practitioners are wrong," he wrote in the first of the series, subtitled "The Challenge"— "wrong in the sense that they are based on mistaken perceptions of the facts, and wrong in the sense that they are not in their patients' best interests." The same patient could "go to different physicians, be told different things, and receive different care. No doubt some of the differences will not be important. However, some will surely be important—leading to different chances of benefits, different harms, and different costs."

Moreover, if a single physician could be wrong, patients would have no guarantee that the physicians to whom they went for second opinions were any more likely to be right, nor that the collected wisdom was accurate. When physicians were confronted with the inherent uncertainty of treating a patient, Eddy explained, they would default to what was comfortable, to standard practices, to what they'd done in the past and what their colleagues were doing. Whether those practices were appropriate or not, whether they were working to the benefit or harm of the patient, would be assumed, but not on the basis of objective evidence. "The applicable maxim is 'safety in numbers,'" Eddy had written in 1984 in an issue of the journal *Health Affairs* dedicated to this emerging discipline of research. "A physician who follows the practices of his or her colleagues is safe from criticism, free from having to explain his or her actions, and defended by the concurrence of colleagues." Ultimately, the decisions made in the clinic would be checked not against any objective reality, but whether they agreed or not with a consensus that was also unreliable.

As Eddy argued and the *JAMA* editors had come to believe, the situation could not remain as it was: the medical establishment had to act, both to maximize the benefit to the patients and keep the rising tide of health-care costs—which had almost tripled since the 1950s as a percentage of gross national product—under control. If nothing else, the insurers, employers, and federal agencies that were now mostly paying for this health care wanted to assure that they and, by implication, patients, were getting their money's worth.

The result was what Arnold Relman, then editor of *The New England Journal of Medicine*, described as a "revolution" in medicine. It encompassed the emergence of two related movements—*outcomes management*, a term coined in 1988 in *The New England Journal of Medicine* by the Mayo Clinic pediatric neurologist and clinical epidemiologist Paul Ellwood, and what came to be called *evidence-based medicine*. Both of these concepts seem commonsensical, but they were not only new to medicine, they would be, and remain, controversial in their application.

Outcomes management works from the assumption that decisions in medical practice must ultimately serve the values and preferences of the patient. The outcomes that matter most to the patient's "function and well-being or quality of life," in Ellwood's words, have to be systematically studied, tracked, and measured and the results integrated into the decision-making process.

Evidence-based medicine, as Eddy would describe it, was in turn based on four further concerns, none of which could be assumed from either a poll of expert opinion or the fact that this is what physicians had done in the past:

> First, there must be good evidence that each test or procedure recommended is medically effective in reducing morbidity or mortality; second, the medical benefits must outweigh the risks; third, the cost of each test or procedure must be reasonable compared to its expected benefits; and finally, the recommended actions must be practical and feasible.

David Sackett would define evidence-based medicine as "the conscientious, explicit and judicious use of current best evidence in making decisions about the care of individual patients." Another way to phrase it, as Sackett himself occasionally would, is that it begins with the idea that half of what aspiring doctors are taught in medical school is "dead wrong" and then tries in the most thoughtful, critical way to establish what practices that half includes.

The movements and the attendant revolution implied a shift of emphasis on what physicians considered the nature of *the current best evidence*. Experimental evidence, the systematic analysis of the results from clinical trials, would take precedence over the kind of experiential evidence and intuition that would be garnered over years of medical practice. Physicians worried that "bean counters" and anonymous committees of biostatisticians and clinical epidemiologists, perhaps concerned as much about the financial cost of a particular intervention as the actual benefit to the patient, would dictate how they had to practice, rather than letting them rely on their own clinical judgment. Physicians and health-care analysts worried that randomized controlled trials could be as poorly done or inapplicable to many questions as any other experiment in medicine or science. Everyone agreed that better evidence was needed—"accurate, interpretable, applicable observations of the frequencies with which important outcomes occur with different practices," to quote Eddy again—but not on what to do when it was absent or when the existing evidence was equivocal.

Among the many complicating factors was the unavoidable fact that patients themselves are individuals. They will not only respond uniquely to different medical practices,* but they will weigh the risks and benefits and value of the ultimate outcomes differently, including the psychological and financial burdens. Randomized clinical trials could be done to answer some critical questions about the outcomes to be expected from a medical intervention, giving a probabilistic assessment of what was *likely* to happen, the *likely* risks and benefits, but certainly not answering every question, or even most. They would never tell physicians whether the results of the clinical trials were relevant to the individual patient who requires treatment.

In 1992, *JAMA* followed up on Eddy's articles with a report by the Evidence-Based Medicine Working Group, comprising thirty-one physicians and health-care researchers, many of whom, including the committee's chair, Gordon Guyatt, were colleagues or former students of David Sackett at McMaster University. Evidence-based medicine was now presented not as a new approach to medical policymaking or the drafting of standard of care documents, as Eddy had considered it, but as a new paradigm of teaching medical students and young physicians. Guyatt, Sackett, and their coauthors reiterated that the new philosophy was a challenge to the traditional practice of medicine that put a high value on "scientific authority and adherence to standard approaches." Evidence-based medicine, on the other hand, assumed the importance of clinical experience and the development of the physician's clinical instincts, combined it with the study and understanding of basic mechanisms of disease ("pathophysiological rationale"), the rules of evidence ("necessary to correctly interpret literature on causation, prognosis, diagnostic tests, and treatment strategy"), and the readiness to "accept and live with uncertainty." "The proof of the pudding of evidence-based medicine," Gordon, Sackett, and their coauthors suggested, would lie "in whether patients cared for in this fashion enjoy better health."

Four years later, Sackett, now directing the newly founded

---

* "Take two people who, to the best of our ability to define such things, are identical in all important respects," as Eddy had written in 1984, "submit them to the same operative procedure, and one will die on the operating table while the other will not."

National Health Service Research and Development Centre for Evidence-Based Medicine in Oxford, England, set about clarifying for British physicians what this new philosophy or movement was all about. Writing with four coauthors in a *BMJ* article, "Evidence Based Medicine: What It Is and What It Isn't," he reiterated that evidence-based medicine still depended, first and foremost, on the physician's intuition, proficiency, and judgment, and that could be acquired only through the actual practice of medicine. "Evidence based medicine is not 'cookbook' medicine," Sackett wrote with his coauthors. "Because it requires a bottom up approach that integrates the best external evidence with individual clinical expertise and patients' choice, it cannot result in slavish, cookbook approaches to individual patient care." While randomized controlled trials would always be the " 'gold standard' for judging whether a treatment does more good than harm," not all questions about therapy required them, and some could not wait for the trials to be conducted. "If no randomised trial has been carried out for our patient's predicament, we must follow the trail to the next best external evidence and work from there."

On issues related to diet and chronic disease, as we've seen, and particularly diabetes, this kind of thoughtful, judicious approach to therapy and clinical decision-making has not been embraced. The constraints of yet another emerging discipline, preventive medicine, coming of age in the same decades would assure this never happened. Once researchers decided that they could establish the dietary means (often incorrect) to prevent common chronic diseases, they rendered irrelevant the clinical observations and experience of the physicians themselves and set in motion waves of unintended consequences. While the ongoing epidemics of obesity and diabetes might well be one of them, a result of misidentifying the dietary triggers of these epidemics, there was a more subtle consequence, one that is profoundly harmful as well.

The shift from *curative medicine*—the physician treating the patient's disease or trying to minimize its symptoms—to the new public health paradigm of *preventive medicine* went along with the establishment of new sources of authority that took precedence over either physician or patient experience. Since neither physicians nor patients can see or experience a disease being prevented the way they can see or experience the symptoms of a disease being ameliorated or the disease itself being cured, they have to accept that this is happening because that's what they are being told. Prescriptions about which foods to eat and which to avoid are disseminated on the basis of what such an eating pattern might accomplish in theory, in the relatively distant future, not on any objective reality of how it improves the patient's condition.

The authoritative statements on diet and health came first from the American Heart Association, the American Diabetes Association, the U.S. Congress and Department of Agriculture, the National Institutes of Health, and related organizations worldwide, and then from the dietetics and nutrition associations and the armies of educators who accepted the authority of these organizations. Together they prescribed how patients should eat to remain healthy and prevent or delay chronic disease—the ultimate in "cookbook medicine" (pun acknowledged). Physicians and diabetes educators giving advice to their newly diagnosed patients deferred to the dietitians with whom they worked, and they, in turn, to these institutional authorities.

Nowhere in this process does the patients' experience on the diet or the diet's impact on the characteristic symptoms of the disease—high blood sugar—influence the nature of the advice given. If the patients' blood sugar remains high, or swings unpredictably from hyperglycemia to hypoglycemia, it does not suggest to the physician that the patients are eating the wrong foods, only that they aren't taking enough insulin or timing their insulin doses correctly or need other hypoglycemic agents as well or instead. For physicians to challenge this model and these authorities, to suggest to patients an alternative approach to treating their disease based on the physicians' own judgment, intuition, and experience—what they had learned, in effect from the anecdotal experiences of other patients or even their own, if they were living with diabetes—is perceived as quackery (and so unethical) or food faddism (which is to say, not the stuff of serious medicine).

Physicians and diabetologists, dietitians and diabetes educators had come to assume that their patients would rather not restrict their eating behavior if any drug or intervention existed to render it (apparently) unnecessary; fewer restrictions on food choices were always better than more. No clinical trials, however, were conducted to establish if that was true. It was assumed based on how patients acted on the dietary advice they had been given, which may have been the wrong advice. The preference for drug therapy over diet, even when the drugs themselves came with undeniable side effects and the burden of the disease remained high, might have been simply a patient's response to a dietary prescription that either didn't work or, perhaps, made the patient's symptoms and burden of disease worse. A different dietary prescription, as in Bernstein's case, might have solved the problem.

What makes diabetes so remarkable among common chronic diseases is that the symptomatic manifestation of the disease—the hyperglycemia—is undeniably a direct reaction to what the patient eats. Changes in response to diet and so different dietary therapies can be immediate and dramatic, and the patient's experience of the disease can and arguably should be a guide to therapy. Once patients gained the ability to monitor their own blood sugar, they could emulate Bernstein, whether they'd read him or not. They could correlate their own physical experience of the disease—their function and sense of well-being—with their blood sugar and their blood sugar to the foods they ate. They could make, in effect, their own clinical observations and decide for themselves their treatment.

The creation and dissemination of yet another new measurement technology, known as continuous glucose monitors, or CGMs, have had a profound effect here, as well. These devices first appeared in the early 2000s and the first truly practical, accurate model in 2011. They allow, as the name implies, for the continuous monitoring of blood sugar for two weeks, typically, with an unobtrusive device the size of a silver dollar. Patients can now see their blood sugar changing essentially in real time on their smartphones, responding to the foods they're eating, the exercise they are doing, and physiological variables over which they may have little control.

As with any new measurement technology, its use can cut both ways. For many patients who try a CGM, it can become yet another way to learn they are failing to control their blood sugar. This is why many will discontinue its use. If they think there's nothing they can do about their too-great glycemic variation, other than perhaps try yet again to adjust the timing and doses of insulin and other medications, they may prefer to live without still more evidence of the failure. But patients can also decide for themselves whether they want to change their diet as a result and then adjust their drug therapy to a way of eating that does not include those foods that drive the glycemic variations. In that sense, physicians like Westman consider CGMs to be gateways to carbohydrate-restricted diets, because foods that raise blood sugar are foods that are carbohydrate rich.

Diabetologists will freely acknowledge that, but their biases and preconceptions—based on how they've approached the problem throughout their careers—will influence, for better or worse, how they apply that knowledge. "We know how to lower blood glucose, and the best way to do that is by reducing the carbohydrate content," as Frank Nuttall, a coauthor of the ADA's 1979 report "Principles of Nutrition and Dietary Recommendations," told me. "But from my perspective, people should eat what they want to eat. I said to people, I think there are no bad foods, no good foods, just foods, and they should eat a balanced diet. That's about it." Nuttall acknowledges that a diet can be prescribed for patients with type 2 diabetes so that they can avoid taking most or all medications—as Virta Health has demonstrated—but he also believes that the long-term consequences are unknown, as are the psychological repercussions should the patient find the diet too difficult to sustain.

Other diabetologists I interviewed—Larry Deeb, for instance, who has over forty years' experience as a diabetologist and is director of the Diabetes Center at Tallahassee Memorial Hospital and clinical professor of pediatrics at the University of Florida, and Joseph Wolfsdorf, who was chief of pediatrics at the Joslin Diabetes Center from 1986 to 1997 and served from 1983 to 2018 as director of the diabetes program at Boston Children's Hospital—had seen in patients how remarkably beneficial Bernstein's approach or a ketogenic diet could be.

These diabetologists are aware of the apparent benefits of the approach, but still concerned about the issues surrounding it and the lack of clinical trial data to provide guidance. The complications of the standard approach are, in essence, the evils they know and have had decades of experience helping their patients to manage. Those of severe carbohydrate restriction, whether Bernstein's approach or a higher-fat ketogenic diet, are the evils that they don't. "There is no question in my mind at all that if you significantly curb carbohydrate consumption, it is much easier to control diabetes," Wolfsdorf told me. "There's no doubt about it. It will also reduce the insulin requirement, and so patients are likely to lose weight or at least not gain weight." Still, he questioned whether the long-term benefits are any greater than they would be on a more moderate diet that includes whole grains, legumes, and fruits. "I don't know the answer," he said. "I hope one day really good research will be done and will clarify it."

Among Deeb's concerns are the possible effect of such a diet on childhood growth. "There is some literature," he said, "suggesting kids don't grow as well, mostly because you can't get kids to eat as much protein and fat to make up the calories." He is also concerned that children whose parents insist they should eat a severely carbohydrate-restricted diet might "become the most rebellious teenager[s] on earth." And if they rebel, what happens then? Deeb considers his obligation as a physician to make the management of the disease fit the patient's preferences, not force the patient to do what he thinks might be best. "It's bad enough to have diabetes. At some point we really do want our patients to have a life."

Ultimately, the question becomes, as perhaps it always has been, whether a diabetic patient values the freedom of eating what others eat more than the freedom that comes with greater function, health, and well-being. That will depend on the individual, but it will also require that the patient struggling with a diagnosis, the attending physician, endocrinologist, dietitian, and diabetes educator fully understand the options. Consider Ross Wollen's perspective. Wollen, a chef turned journalist, was diagnosed with type 1 diabetes when he was thirty-six years old. Here's how he recalled first discussing this with his primary care physician in 2018 upon being informed of the diagnosis:

So I said, "Carbs are toxic to me now, right?"

My doctor said, "Don't think of it that way."

"But they are," I said. "It's like carbs are poison and insulin is the antidote that I have to take. So isn't it better that I take as little of the poison as I can; then I also need as little of the antidote?"

My doctor disagreed; he didn't want me to worry. I just wanted to be as healthy as possible.

When Wollen's diabetes educator suggested that he would find Bernstein's approach or a ketogenic diet unsustainable, a common criticism made of these strategies, implying that it was not worth the effort to try, Wollen responded, "Why is this a problem? If it doesn't work out, I'll try something else. If I told you I was going to start jogging ten miles a week, you wouldn't say that's not sustainable. You'd say, 'Oh, that's great.'"

If the patients desire to be as healthy as possible, as Wollen phrased it, then the choice of whether to comply with a diet is likely to be based on how healthy it makes them feel. If it indeed *works*, in the sense that it allows these individuals to maintain normal blood sugar and avoid the psychological and physical burden of the disease, it's a reasonable assumption that they will think of how they eat as sustaining their health, not sustaining a diet. This is the kind of shift in perspective that is required if it is indeed true, as I'm arguing, that the current dietary approaches to managing diabetes are inappropriate.

Consider the following experiences of physicians living with type 1 diabetes. Maryanne Quinn is a pediatric endocrinologist who treats patients with diabetes at Boston Children's Hospital and worked with Joseph Wolfsdorf for most of that period. Quinn was diagnosed with type 1 diabetes as a four-year-old living in Indiana in 1960. As Quinn tells her story, her father, a high school football coach, had taken responsibility for both her insulin injections and how she would eat. "In the old urine tests," Quinn said, "you put a pill in the test tube and if it turned blue that was good. My dad said it was more blue when I ate salads, the way he was telling me to, so that's what I ate." By the time Quinn read Bernstein's first book in the mid-1980s, the summer before she entered medical school, she had already spent a quarter of a century eating as Bernstein recommended. While reading the book, she recalled, she kept thinking, "This is interesting, but I already know this stuff. Here's somebody else doing this, too. People say he's too extreme. I say he's not."

As Quinn described it to me, she consistently eats the same foods at every meal and can't imagine why she would want to do it differently. Dinner, for instance, is chicken or fish with a green salad. One of her former colleagues at Boston Children's Hospital described sitting next to Quinn at an Italian restaurant where young physicians from the diabetes center had all gone to celebrate. This was a restaurant, he said, where you went to eat pizza, and so he was amazed when he saw Quinn had ordered a Caesar salad with chicken and then was picking out the croutons so that she didn't eat them by accident. "If I eat them," she told her colleague, "my blood sugar is going to go way up. And, yes, I might use some insulin to cover it, but that's going to be really hard to figure out the right dose. It's much easier just not to eat them." She doesn't change her approach and never has, she said, "because it works."

Craig Suchin, an interventional radiologist, was diagnosed with type 1 diabetes when he was nineteen years old. He has had what he calls "suboptimal blood sugar control" ever since. Because many of the patients he saw in his radiology practice had kidney disease, a consequence of years of poorly controlled diabetes, Suchin came to consider this reason enough to try to get better control of his own disease. After reading the available literature, both in the academic journals and that available now in books and online, and joining a Facebook group called "Keto for Type 1 Diabetes," he tried it himself. On February 25, 2019, he said, recalling the date exactly, he stopped eating carbohydrate-rich foods "cold turkey." "I was afraid I would die," he told me. "I had no idea what would happen. Within twenty-four hours I was one hundred percent convinced this was how I wanted to live the rest of my life. For the first time ever, I had normal blood sugars and I needed a fraction of the insulin I had been taking. My blood sugar curves, which had looked like ocean waves [on the CGM readout] were now flat." Nonetheless, when he saw his endocrinologist five weeks later and told him how he was eating, his endocrinologist responded with more anxiety about the potential harms—a serious hypoglycemic episode, in particular— than appreciation of the apparent benefits. He insisted Suchin find another endocrinologist to oversee his medical care.* Suchin did.

J. Daniel Jones, a radiologist who recently retired from his position working with Orlando Health in Florida, was diagnosed with type 1 diabetes in 1986 at age thirty-four. He "thought the world had come to an end," he recalled. While his physicians were satisfied with his blood sugar control on insulin therapy—they considered it "wonderful," he said, because his hemoglobin A1C tended to be under 7.0 percent, consistent with ADA recommendations—he was not. "It was all over the place," he added, describing the glycemic variations as "just insane." In the mid-2000s, as a radiologist at the peak of his career, Jones developed macular edema in one eye, an accumulation of fluid in the retina from the rupture and leakage of blood vessels that blurred his vision. Macular edema is typically treated first with injections into the retina of a medicine that inhibits blood vessel growth. "When the retina guy says, 'Okay, we're going to stick a needle in your eyeball,'" Jones recalled, "I said, 'Wait, let's back up. What's the root cause of the problem?' And the root cause of the problem is bad blood sugar, fluctuations, bad glycemic control. I had read some of Bernstein's stuff and I was on an insulin pump at the time and I felt, let's back off the needle in the eyeball thing and let's see if there's another way to approach it."

Jones began reading the literature on low-carbohydrate diets, an experience he described as "going to the dark side," since it went against "all medical dogma of the time" and his fellow physicians considered such diets quackery. He changed his diet and his approach to glycemic control. "I cut my carbohydrates to almost nothing," he told me. "Three months later my vision was fine. Fif-

---

* The endocrinologist sent Suchin a registered letter, which Suchin shared with me. "Due to irreconcilable differences between us regarding your medical care, I am notifying you in writing that I will no longer serve as your endocrinologist," it read. "I am unwilling to medically supervise your use of a very low carbohydrate, ketogenic diet, since I believe this to be a dangerous and untested experimental therapy for your type 1 diabetes mellitus." When I reached out to the endocrinologist to discuss his perspective, he politely declined.

teen years later, it's still fine. Even my retina doctor just scratches his head. I keep telling him that I can't do what the ADA tells me to do and expect to remain healthy. He looks at me like I'm some sort of alien. The easy way to go through diabetes is to run blood sugar of 150 all the time [an A1C of still less than 7 percent and the target of ADA guidelines] and wait twenty-five years until complications set in. I will not go back."

Laura Nally, a pediatric endocrinologist at the Yale School of Medicine, sees the trade-offs differently. Nally was diagnosed when she was six years old. Both her father and two aunts lived with type 1. Her aunts died young—in their thirties and forties—from complications of the diabetes. She said she has eaten the same diet throughout her childhood—roughly 100 grams of carbohydrates each day, far lower than the ADA recommendations, primarily from whole grains and legumes and, yes, the occasional ice cream or pizza. Her approach is based on her father's experience, "trying to figure out what worked and what didn't," which is what Bernstein did. She has maintained "remarkably good glycemic control," she said, hemoglobin A1C under 7 percent her whole life. She has patients who follow Bernstein's approach and do very well on it, and she experiments herself with different diets out of professional and personal curiosity. Nally had been running a small study at Yale, looking at twenty adolescent patients who spend two weeks each on a very-low-carbohydrate diet, similar to Bernstein's, and two weeks on a standard carbohydrate diet. She wanted to know how the diets affect insulin sensitivity and metabolism; how insulin doses might have to be adjusted, and the subjective experiences of her young patients.

Nally told me that she tried the very-low-carbohydrate diet herself, the diet she was using in her study, and for the first few days "was miserable." Then she got used to it. She had to adjust her insulin dose down by 25 percent but went back to eating as she had when her LDL cholesterol went up. "There's a significant amount of heart disease in my family," she explained. "I opted to go back to higher carb." In her practice, she said, she tries to encourage families "to continue as much the normal diet" when their children are first diagnosed, "so that they don't feel like they've changed everything in their diet all at once." But she does tell her patients to avoid sugar-containing beverages and syrups—"things we know that a shot of insulin is not able to match the blood sugar rise at all"—and eat healthier. If they notice, she said, that their blood sugar spikes every morning after eating a bowl of cereal, for instance, "we can have a discussion if we want to modify that meal."

A necessity, ultimately, for the acceptance of any radical new approach to medical therapy (even if two hundred years old) is the acceptance of the possibility, borrowing David Eddy's words, that what practitioners have been doing is perhaps not what they should be doing; that they may be causing harm, even as they no longer associate the harm with the intervention. Barring randomized clinical trials that can establish long-term risks and benefits unambiguously, the kind of trials that are likely never to be done because of expense and difficulty, policymakers and physicians will always default to the assumption that what they are doing entails the least risk, though that's not necessarily the case. As I've suggested here, it's quite possible that carbohydrate-rich diets for patients with diabetes, no matter how well covered by insulin therapy and any other hypoglycemic drugs used, cause harm in comparison to a diet absent those carbohydrates: in type 2 diabetes, the long-term chronic complications that associate with the disease may be an effect of the carbohydrate-rich diets prescribed. In type 1 diabetes, the carbohydrate-rich diets *and* the insulin necessary to cover those carbohydrates may be responsible for both the long-term complications *and* the short-term complications—the hyperglycemic swings and hypoglycemic reactions that make the disease both so difficult to live with and potentially dangerous. Here evidence-based medicine fails us. The evidence supporting the standard of care is weak, but diabetologists have never properly tested it against anything else.

When diabetologists assume that their patients living with diabetes would always prefer drug therapy to significant dietary restriction, they justify or perhaps rationalize their focus on new drug therapies, new surgical interventions, and even advances to insulin-delivery technology that will do a better job of treating the symptoms of the disease. These advances may be marvelously effective, but the history of diabetes therapies provides reason for pessimism.

These interventions, no matter how sophisticated, treat the symptoms of the disease, not the root cause. Patients with diabetes do not get healthier because they take them or use them; they do not regain the ability to metabolize the carbohydrates that are causing harm. Rather they mitigate the harm being done. A new class of drugs called SGLT2 (sodium-glucose co-transporter 2) inhibitors, for instance, work by inducing the kidney to excrete more glucose so that blood sugar stays lower even as urine glucose—what physicians used to consider a characteristic symptom of the disease—rises. Insulin signals cells to take up more glucose and use it for fuel, even as the cells themselves resist doing so. Other drugs slow the absorption of carbohydrate-rich foods in the gut to reduce the rise in blood sugar; they may stimulate insulin secretion and/or resolve insulin resistance—the science is uncertain—as do the newest drug therapies, known as GLP1 (glucagon-like peptide 1) and GIP (glucose-dependent insulinotropic polypeptide) agonists, which all have such a seemingly remarkable ability to induce weight loss, as well.

All of these drugs are intended to lower and stabilize blood sugar, but they are mitigating a symptom that might not exist (type 2 diabetes) or might be much easier to control (type 1) if the individual abstained from eating those carbohydrate-rich foods that are converted so quickly to blood sugar. It's hard to doubt that drug therapy will continue to improve, as will insulin delivery technology. The newest diabetes drugs, SGLT2 inhibitors and GLP-1 and GIP agonists, in particular, clearly reduce the burden of disease significantly in diabetic patients, although they are costly (thousands of dollars a year) as is insulin (several hundred dollars for a 10 milliliter vial containing 1,000 units of insulin, as of the fall of 2022), and no drug therapy comes free of side effects.

If the diabetes community is to solve the formidable problems confronting it—the ever-rising tide of both type 1 and type 2 diabetes and the failure to control the complications of the diseases, even as insulin delivery technology and drug therapy get ever more sophisticated—it is going to have to accept that some of its fundamental preconceptions may indeed be wrong. And as it does so, it will have to provide support for those living with diabetes who decide that what they have been doing is not working. Some patients, when confronted with the choice between following a restricted eating pattern that maximizes their health and well-being or liberalizing carbohydrate consumption and treating the symptoms with drug therapy, will prefer the former.

For those who do, who are willing to experiment with their own diet and health, the assistance and guidance of their physicians, endocrinologists, and diabetes educators will be critical. When I interviewed individuals living with type 1 diabetes, among the most poignant comments I heard was from a nutrition consultant diagnosed in 1977 when she was eight years old. She told me that she finally had faith she could manage her blood sugar and live with her disease when she finally met a physician who said to her, "What can I do to help you?" That's what changed her life, as much as any technology or medical intervention. In the context of the dietary therapies we're discussing, that requires practitioners who are themselves open-minded and willing to spend the necessary time and effort to truly understand an approach to controlling diabetes that is, by definition, unconventional and, in type 1 diabetes, still lacking clinical trials that test (or testify to) its safety and efficacy compared to more conventional diabetes therapy. Easy as it is for diabetologists and physicians to continue believing that what they should be doing is what they have been doing, they do not serve their patients best by doing so.

Acknowledgments

Early in my journalism career, I had the opportunity to research and write two books about researchers—physicists and chemists, specifically—who had reported the discovery of what turned out, regrettably, to be nonexistent phenomena. Both books documented how the relevant research communities responded to these claims and how these erroneous discoveries came to be acknowledged and corrected. In the first, *Nobel Dreams,* I was *embedded* (using the modern terminology) with the physicists while this correction process unfolded. In the second, *Bad Science,* about the late 1980s scientific fiasco known as *cold fusion,* I eventually interviewed over 300 researchers and administrators. My goal was to write what I hoped would be a definitive case study of "pathological science"—defined in 1953 by the Nobel laureate chemist, Irving Langmuir, as "the science of things that aren't so."

In the course of my research for both books, I had the good fortune of being mentored and counseled by some of the very best experimental physicists and chemists of that era. All of these scientists were accustomed to the reality that researchers and, indeed, entire research disciplines can simply go off the rails in the pursuit of what should be reliable knowledge. Getting things wrong in science may be at least as common as getting them right. To those scientists, I want to express my gratitude for their patience and their guidance and for teaching me to recognize the signs and symptoms of pathological science. I hope that I have done justice to their counsel.

As this early education led me to write critical articles on the research in far softer scientific disciplines than physics and chemistry—in public health, medicine, and nutrition—I entered into mostly uncharted territory. In the past few years, with the Covid epidemic and the rise of "fake news," the public dissemination of reasons to distrust what is often called the science, the consensus of opinion by the established researchers in the field, by anyone other than the researchers themselves is tolerated with little patience. In the 1990s and 2000s, only a very few of us in journalism had the temerity to do so. (In 2017, when *The Atlantic* reviewed my book *The Case Against Sugar,* the reviewer, Dan Engber, now an *Atlantic* editor, wrote, "It's extraordinary and refreshing to see a science journalist so wary of his sources, and so willing to present himself as someone who knows more than they do." Not everyone agrees with him.)

I have been able to do this for nearly forty years because of the support I received along the way from editors who were willing to trust their judgment in trusting mine. They were also willing to believe that researchers and research disciplines do get their science wrong and that science/health journalists have an obligation to critically assess the research on which they are reporting. These editors trusted my work even as I had to diverge from the standard procedures of reporting on science and health—i.e., quote the experts—to discussing the apparent flaws in the research and whether the experts themselves were credible sources. (That these articles then won major science journalism awards—including three Science in Society Journalism Awards from the National Association of Science Writers and inclusion in multiple "best of" anthologies of science and nature writing—provides some evidence, albeit not nearly proof, that this trust was warranted.) To these editors—at *Discover* in the 1980s, *Science* in the 1990s, and *The New York Times Magazine* in the decade that followed—I also owe a debt of gratitude.

Since 2002, none of this happens without the invaluable help and support of my editor at Knopf, Jon Segal. Jon has guided all of my books on nutrition, obesity, and chronic disease from the concept stage to publication and has been relentlessly critical throughout. He has been, for me, the ideal editor and has become a good friend. I'm also grateful at Knopf to the team that shepherded this book from manuscript to publication: Sarah Perrin, Victoria Pearson, Maggie Hinders, Peggy Samedi, and Linda Huang. And, once again, I could not be more grateful to my literary agent Kris Dahl at CAA, who has been in my corner since 1984 and has been integral to making all my books and my career as a journalist/author work.

Writing critically about diabetes therapy and science from the perspective of a journalist rather than a physician or a patient presents a unique set of challenges. The research for this book included not only studying a significant portion of the medical literature related to diabetes going back to the eighteenth century—made possible in the age of Covid shutdowns by Google books among other remarkably helpful internet repositories—but copious interviews with diabetes researchers, physicians, patients, and, in several cases, parents of patients. I am grateful to all of them for their time and their patience. I have to thank those, in particular, who took the time to read and critically assess significant portions of the draft, at least minimizing the errors I am likely to make: Amy Berger, R. D. Dikeman, Mark Friedman, Mariela Glandt, Bob Kaplan, David Ludwig, James McCarter, Charles Peterson, Eric Westman, Ross Wollen, and Will Yancy.

Finally, and simply, to my family, Sloane, Nick, and Harry, for making this all worthwhile.

Notes

INTRODUCTION

3 "The effect of any": Joslin et al. 1946: 353.
3 diagnosed by Hindu physicians: Tattersall 2009: 10.
3 "excessive formation": Harley 1866: 32.
3 "defective assimilation": Harley 1866: 33.
4 *le diabète gras*: Tattersall 2009: 26.
4 90 to 95 percent of all diagnoses: CDC 2021.
4 "Patients were always": Joslin 1933.
5 "integral" to diabetes therapy: ADA Professional Practice Committee 2022.
5 "lip service": Sawyer and Gale 2009.
6 "free diets": Tolstoi 1950.
6 Recent standard-of-care: ADA Professional Practice Committee 2022.
7 published their first report: Banting et al. 1922.
7 "therapeutic revolution": Marks 1997: 17.
7 "Were I a diabetic": Quoted in Collens 1954.
7 "set in motion": Feudtner 2003: 26.
8 "We have too much darkness": Brown 2017: 5.
9 "From an historical perspective": Sawyer and Gale 2009.
9 diabetes increased 600 percent: CDC 2017.
9 almost 30 million Americans: CDC 2022.
9 significant percentage of patients: See, for instance, Hartmann et al. 2021.
9 The prevalence of type 1 diabetes: Lawrence et al. 2021.
10 a century ago: Joslin et al. 1934.
10 one in nine: ADA 2022.
10 cognitive impairment and dementia: Gonzalez et al. 2018 and Luchsinger et al. 2018.
10 Their life expectancy: Rosenquist and Fox 2018.
10 accelerated aging: Gale 2009.
10 "Extending the average duration": Interview with James Foley, October 14, 2020.
11 "first and foremost": Zinman et al. 2017.
11 "The remarkable magnitude": Riddle and Herman 2018.
11 ninth leading cause of death: WHO 2021.
11 "slow-motion disaster": Chan 2017.
12 fewer than one in five: Foster et al. 2019.
12 "double diabetes": Schaffer 2017.
12 stunting of growth: Bonfig et al. 2012.
12 brain growth and development: Mazaika et al. 2016.
12 expectation of a premature death: Rawshani et al. 2018.

12   a 2021 review: Hill-Briggs et al. 2021.
13   "futility": Look AHEAD Research Group 2013.
13   ten-thousand-patient ACCORD trial: ACCORD 2008.
13   "Halted After Deaths": Kolata 2008.
13   twenty-country ADVANCE trial: Advance Collaborative Group 2008.
13   1,800-subject Veterans Affairs Diabetes Trial: Duckworth et al. 2009.
13   "still a serious shortage": Knowler et al. 2018.
14   "An Analysis of Failure": West 1973.
14   patients would prefer to take: Sawyer and Gale 2009.
15   cardiologist Robert Atkins: Atkins 1972.
15   In the early 1980s, Richard Bernstein: Bernstein 1981.
16   "laws of small numbers": Bernstein 1997: 102–8.
16   "Unquestionably large doses": Joslin 1928c: 74.
16   In 2019, the American Diabetes Association: Evert et al. 2019.
17   Researchers working with a San Francisco–based: Athinarayanan et al. 2019.
17   Almost half of the Virta patients: Virta Health 2022.
17   TypeOneGrit survey: Lennerz et al. 2018.
18   series of investigative articles: Taubes 1995, Taubes 1998a, Taubes 2001, and Taubes 2002.
18   *The Case for Keto*: Taubes 2020.
19   "the first principle": Feynman 1985: 343.
19   Bacon made this point: Bacon 1994.
20   "your best sense of the data": Interview with Carol Foreman, July 19, 1999.
21   "one of the very few": Popper 1963: 216.
21   "pathological physiology": Joslin 1916: 61.
23   "game-changers," "'the' transformative breakthrough," and "sustained applause": Prillaman 2023.
23   gain back much to all of the weight: Wilding et al. 2022.
24   the American Academy of Pediatrics published guidelines: Hampi et al. 2023.
24   "ripping up long-held beliefs": Petersen, Winkler and O'Brien 2023.

CHAPTER 1 The Nature of Medical Knowledge

26   "In diabetes . . . the chief difficulty": von Noorden 1912: 82.
26   "In the writer's experience": Rabinowitch 1944.
27   "A 32-year-old white male": Johnson and Rynearson 1951.
27   "the hinterland of Montana": Wilder 1958.
28   Restitution, in fact: See, for instance, Joslin 1923b: 160–62.
29   "appalling increase": Joslin 1950.
29   Joslin's patients in Boston: Joslin 1950.
29   "we shall learn": Joslin 1923b: 40.
29   reporting lesions of the retina: Wagener et al. 1934.
29   a new type of kidney disease: Kimmelstiel and Wilson 1936.
29   80 percent of their patients: Root et al. 1950.
29   "Of those who died": Tattersall 2009: 84.
29   a hundred times more common: Cited in Root 1950.
30   Joslin was suggesting: Joslin 1950.
30   "extraordinary frequency": Root 1950.
32   "When I was a student": Dunlop 1954.

CHAPTER 2 The Early History

33   "In the matter of diet": Smith 1889: 45–46.
33   "There is no cure": Lusk 1909.
33   "The ingenious author": Duncan and Duncan 1798.
34   "whatever was available": Presley 1991: 17.
34   Rollo had seen: Rollo 1797a.
35   "in perfect health": Rollo 1797b.
35   "in tolerable health": Anderson 1965.
36   Physicians were encouraged to write: Rollo 1797b and Rollo 1798.
36   "We have to lament": Rollo 1798: 405.
36   "a gradual return": Rollo 1798: 401–2.
37   a twelve-year-old girl: Rollo 1798: 271–82.
37   "The great point": Chambers 1866: 480–84. See also Chambers 1875: 274–75.
38   "Gauchos of South America": Pavy 1862: 144–45.
38   "a very insecure hold": Pavy 1862: 138–39.
38   "easily the most brilliant clinician": Allen et al. 1919: 25.
38   "*Manger le moins possible*": Bouchardat 1875: 205.
39   "is not in the drugstore": Quoted in Allen 1952: 31–32.
39   "followed with absolute rigor": Cantani 1876: 383.
39   "under lock and key": Allen et al. 1919: 31–32.
39   "the foremost diabetic authority": Allen et al. 1919: 37.
39   German physician Wilhelm Griesinger: Allen et al. 1919: 28.
39   "The introduction of fat": Allen et al. 1919: 40.
40   "The carbohydrates in the food": Osler 1893: 303.
41   "It is desirable": Saundby 1908.
41   "patients in France": Allen et al. 1919: 50.
41   list of compounds: Tattersall 2009: 23.
41   The only drug that appeared to help: Bliss 2007: 23.
42   "a man who loved": Allen et al. 1919: 27.
42   tried it on three patients: Tattersall 2009: 19.
42   One well-known physiologist: Allen et al. 1919: 27.
42   "produce complete relief": Donkin 1871: 174.
42   "It has not come into favor": Joslin 1916: 290.
42   "equal amount of butter": Falta 1909.
42   "when only bacon": von Noorden 1912: 93.
42   "the success can": Falta 1909.
42   a "most excellent" diet: Osler 1914: 436–37.
43   "We do not yet know": von Noorden 1912: 105.
43   "I have myself": von Noorden 1912: 102.
43   "The general direction": von Noorden 1912: 74.
43   "The diabetic's chance": Joslin 1905.

CHAPTER 3 Diabetes in Retrospect

44   "Each science confines": Whitehead 1938: 131.
44   "Diabetes, recognizable clinically": Embleton 1938.
45   typically assumed either the brain: For a good review, see Tattersall 2009: 32–35.
45   Bernard had famously discovered: Holmes 1986.

45 "more wear and tear": Saundby 1900.

45 "fat and ruddy": Harley 1866: 32. Summarized in Tattersall 2009: 35.

46 nestled in a bend of the small intestine: Tattersall 2009: 37.

46 Minkowski later said: Minkowski 1989.

46 before the days of ethics committees: Tattersall 2009: 35.

46 In a letter written in 1926: Minkowski 1989 and Houssay 1952.

47 "as soon as it comes back": Houssay 1952.

47 "existence of *a* pancreatic diabetes": Minkowski 1989.

47 "virtually ignored": Soskin 1941.

47 which dates back at least to 1919: Frederick Allen tells this version of the story in Allen et al. 1919, attributing it to a personal communication from A. E. Taylor.

47 This was the version preferred: McGarry 1992.

48 "a foundational pillar": Newgard 2018.

49 "grappling with the enormous complexity": McGarry 2002.

50 Medical historians date: For a comprehensive history of endocrinology, see Medvei 1982.

50 thanks to the suggestion: See Starling 1905 for Starling's first published use of the term "hormone."

50 The term "endocrine": Medvei 1982: 7–8.

51 "give to the blood": Brown-Séquard 1893.

51 Brown-Séquard gave a lecture: Tattersall 2009: 39.

51 "Though many jeered at him": Anon. 1893.

52 "Evidently the gland": Bliss 2007: 27.

52 coined in 1926: Cannon 1926.

52 the *milieu intérieur*: Bernard 1878: 113. An English translation is Bernard 1974 ("harmonious ensemble," "such a degree of perfection," 48). The Cooks translation of Bernard is occasionally awkward and so I have used Cannon's translation—"All the vital mechanisms . . ."—which is from Cannon 1939: 38. Footnote: "No more pregnant sentence": Cannon 1939: 38. Cannon also discussed homeostasis at length in Cannon 1929.

54 "common medium": Starling 1923.

54 "We really must learn": Bernard 1957: 89.

54 "the wholeness of the organism": Krebs 1971.

55 chapters in the textbooks: See, for instance, Williams 1968.

55 first demonstrated its efficacy: Yalow and Berson 1960c.

55 their earliest papers: Yalow and Berson 1960b and Yalow and Berson 1960a.

56 individuals with obesity: Yalow et al. 1965. See also Karam et al. 1963.

56 "brought about a revolution": Karolinska Institute 1977.

56 "a bewildering array": McGarry 1992.

56 suggested the name "insulin": Banting, Best, Collip, Macleod, and Noble 1922.

56 "a hormone produced in the pancreas": The Google definition of insulin comes from the Oxford Languages English dictionary. https://languages.oup.com/google-dictionary-en/.

58 "Council of Food Utilization": Feudtner 2003: xix–xx.

58 Its functions have to do: For a good review of the actions of insulin, see, for instance, Petersen and Shulman 2018.

59 It inhibits the excretion of salt: See, for instance, DeFronzo 1981.

59 "The striking exception": Tattersall 2009: 138.

CHAPTER 4 The Fear of Fat

61   "The history of diabetic therapy": Allen 1916.
61   "When nutrition is the paramount": Grafe 1933: 363.
62   "the staple food in diabetes": Grafe 1933: 345.
62   "a masterpiece": Lusk 1928: 663.
62   fat became a thing to be feared: Joslin 1916 ("poison": 59; "the chief source": 225).
63   "a jolly": Joslin 1946.
63   "a severe form": Feudtner 2003: 45.
63   "the heyday": Joslin 1946.
63   "A diabetic diet is really": Joslin 1905.
63   "vitamin-rich, low-carbohydrate": Joslin 1946.
63   Joslin discussed Case No. 8: Joslin 1923b (outliving: 456; "unusually strong and vigorous": 105).
64   "absolute adherence": Joslin 1946.
64   "On meticulous examination": Joslin 1950.
65   "the ones who have controlled": Joslin 1956.
65   "especially vigorous control": Ryan et al. 1970.
65   "too low in carbohydrates": Joslin 1946.
66   "the value of fat": Krall and Joslin 1971.
66   "world's most powerful": Proctor 1999: 15.
67   benefit of working together: See, for instance, Krebs 1967.
67   *Studies Concerning Glycosuria and Diabetes*: Allen 1913.
67   His influence rose: Bliss 2007: 239.
68   "the god of diabetes": Interview with Ronald Krauss, January 12, 2004.
68   "was the English book": Lawrence 2012: 3.
69   "since Rollo's time": Joslin 1915.
69   The blood and urine became more acidic: Allen 1916.
69   two-thirds of his patients: Joslin 1916: 292 and Joslin 1950.
71   "The subject of diabetes heretofore": Allen 1913: vi.
71   "patient scientific experimentation": Joslin 1915.
72   scrupulously note: Allen 1916.
72   "diabetes varying": Allen 1915.
72   "measured in weeks": Allen 1915.
72   "produces . . . an appearance": Allen 1917.
72   "Most of the diets": Allen et al. 1919: 45.
73   Allen was offered: For Allen's history, see Mazur 2011.
73   Allen's first opportunity: Allen et al. 1919: 177–85.
73   Allen discussed her treatment: Allen's unpublished memoir is quoted and discussed in Mazur 2011.
74   "the actual accomplishment": Allen et al. 1919: 178–85.
74   "uniformly beneficial": Allen 1915.
74   radical and revolutionary: Anon. 1916.
74   "Using Allen's program": Joslin 1916: 242.
74   "Fasting is never": Joslin 1916: 248.
74   "I have witnessed": Williams 1921.
75   "An adult diabetic": Bliss 2007: 35.
76   "contrary to my belief": In the discussion at the end of Woodyatt 1921a.
76   "disliked her diet": Bliss 2007: 44.
76   treatment failures: Allen et al. 1919: 567–72.

76 "If a healthy individual": Joslin 1923b: 303.
76 When the Viennese diabetologist Wilhelm Falta: Falta 1909.
77 "the incautious use of fat": Williams 1921.
78 published a series of four articles: Newburgh and Marsh 1920a, Newburgh and Marsh 1920b, Marsh, Newburgh, and Holly 1922, and Newburgh and Marsh 1923.
78 Louis Newburgh is a problematic figure: Davenport 1999: 153–62.
79 "Because of the low energy intake": Newburgh and Marsh 1920b.
80 "court of last appeal": Newburgh and Marsh 1920b.
80 "is entirely ungrounded": Newburgh and Marsh 1920b.
80 "the evils of fasting": Newburgh and Marsh 1923.
80 Joslin would later acknowledge: Joslin 1928c: 604.
81 The Montana farmer: Johnson and Rynearson 1951.
81 "respected intellectual giant": Campbell 1962.
81 "striking": Woodyatt 1921b.
81 "masterpiece": Lusk 1928: 663.
81 Woodyatt explained: Woodyatt's article on diet and diabetes was published both in the *Archives of Internal Medicine* (Woodyatt 1921b) and with a discussion section in the *Transactions of the Association of American Physicians* (Woodyatt 1921a).
82 "For diabetes itself": Woodyatt 1921a.
83 "with an increase": Ladd and Palmer 1923.
83 "extreme limitation": Wilder 1924.
83 His 1923 monograph: Petrén et al. 1924.
84 "much attention in Europe": Naunyn 1924.
84 "strictest limitation of protein": Petrén et al. 1924.
84 "grave diabetes": Petrén et al. 1924.
85 "martyrdom": Grafe 1933: 361.
85 "proved that the principle": Grafe 1933: 363.
85 "striking success": The discussion section of Woodyatt 1921a.
86 "The statement that severe cases": Leclercq 1922.
86 "In one acidosis": The discussion section of Woodyatt 1921a.
87 "unnecessary" deaths: Joslin 1916: 292 and Joslin 1950.
87 "Diabetic patients need fat": Joslin 1916: 59.
87 "fundamental principle upon which all treatment": Joslin 1916: 68.
87 "The whole aim": Joslin 1923b: 311.
87 "courageously demonstrated": Joslin 1923b: 459.
87 Their "diabetic creed": Joslin 1923b: 526.
87 "a portion of the fat": Joslin 1923b: 491–92.
87 the diet worked because: Joslin 1923b: 526.
88 "unalterably opposed": Joslin 1923b: 527.
88 "The patient who breaks": Joslin 1923b: 526.
88 "disease of the arteries": Joslin 1928a.
88 "in her early twenties": Joslin 1930.
89 "Arteriosclerosis today": Joslin 1928a.
89 "based upon the principles," "large golden yellow nodules": Gordon, Connor, and Rabinowitch 1928.
89 "very much like thickened cream": Joslin 1917a.
90 "did not live in vain": Joslin 1928a.
90 "ideal subject for treatment": Joslin 1928a.

90 "on account of his diabetes": Joslin 1928b.
90 "If our diets approach": Joslin 1930.

CHAPTER 5 Insulin

92 "Insulin does not cure": Joslin 1923b: 38.
92 "The diet in health": Joslin 1948: 84.
92 Leonard Thompson was thirteen: Banting, Best, Collip, Campbell, and Fletcher 1922, although the Toronto team incorrectly record Thompson's age as fourteen, when he was still thirteen.
92 "all knew that [Thompson]": Allen 1972.
93 "never seen a living creature": Bliss 2007: 112.
93 In 1908, a Berlin physician: Jörgens 2021.
93 little to lose: Campbell 1962.
93 Thompson's parents allowed the decision: Wrenshall et al. 1962: 74–75.
93 "The boy became brighter": Banting, Best, Collip, Campbell, and Fletcher 1922.
94 Thompson lived another thirteen years: Campbell 1962.
94 "markedly reduced even": Banting, Best, Collip, Campbell, and Fletcher 1922.
94 An informal committee: Allen 1922.
94 "weight of her bones": Bliss 2007: 150.
94 ramped up Mudge's dosage: Feudtner 2003: 89–90.
95 "walk with ease": Joslin 1928c: 69.
95 "marvelously good": Bliss 2007: 151.
95 "extremely emaciated": Bliss 2007: 152.
95 Hughes gained seven pounds: Cox 2009: 179–85.
95 "pint of heavy cream": Cox 2009: 182–84.
96 Their first patient was fifty-one years old: Tompkins 1977: 69–73.
96 "rapidly approaching death": Williams 1922.
96 The two conditions insulin definitively: See, for instance, Joslin 1929b.
96 diabetologist Nellis Foster: Foster 1923.
96 "spectacular": Joslin 1923a.
97 By then, Joslin's clinic: Joslin 1925.
97 the only drugs widely available: Pellegrino 1979.
97 "Insulin was totally different": Tattersall 2009: 65.
97 "warning train of symptoms": Joslin 1923b: 39.
97 Until the autumn of 1922 : Tattersall 2009: 67.
98 "prolonged undernutrition": Joslin 1923b: 174.
98 "The initial symptom": Fletcher and Campbell 1922.
99 "presence of diarrhea": Joslin 1928c: 72.
99 When two diabetologists: Korp and Zweymüller 1972.
99 "Tolerance for food": Geyelin 1926. See also Joslin 1928c: 52–53.
100 "severe hypoglycemic shocks": Geyelin 1926.
100 "oranges, glucose candies": Fletcher and Campbell 1922.
100 "The cost of one such test": Joslin 1923a.
101 "its use is restricted": Joslin 1923a.
101 1,500 patients: Joslin 1928c: 66.
101 "The diabetic taking insulin": Joslin 1928c: 20.
101 "If a diabetic is not happy": Joslin 1929a: 103 and Joslin 1948: 114.
101 "There are certain patients": Thomson 1924.

102 "In some cases, the proper adjustment": Geyelin 1926.
103 "You are out of your mind": Donaldson 1962: 103.

CHAPTER 6 Rise of the Carbohydrate-Rich Diet

104 "Insulin cannot act": Joslin, Root, White, et al. 1946: 375.
104 "The most important advantage": Forsyth, Kinnear, and Dunlop 1951.
105 this "free diet": Lichtenstein 1938.
106 "for evil and for good": Joslin 1923b: 38.
106 "The strongest argument against": Lawrence 1933.
107 "clear and I believe true": Lawrence 1933.
108 "Case No. 632": Joslin 1928c: 72.
108 "Diabetes mellitus is due": Banting 1924.
109 Banting was furious: Bliss 2007: 231–32.
109 At his Nobel lecture: Banting 1925.
110 "The condition of the pancreas": Woodyatt 1909.
110 Naunyn and von Noorden had both suspected: Joslin 1916: 48.
110 "The tendency of the diabetic patient": Joslin 1916: 68–69, Joslin 1923b: 106–7, Joslin 1928c: 109.
111 "whip up intrinsic": Schumacher and Schumacher 1956: 58.
111 The second principle, "With insulin," and "the treatment of diabetes": Joslin 1923b: 40–45.
111 the thinking of the pre-insulin era: Wilder 1940: 109.
112 "Thomas D., Case No. 1305": Joslin 1923b: 38.
112 Joslin was taking responsibility: Joslin 1928c: 74.
112 start the patients off on a single unit: Joslin 1923b: 49–50.
112 "quite theoretical": Joslin 1923b: 46.
112 starting patients with 5 units: Joslin 1928c: 73.
113 some parents either refused insulin therapy: Toverud 1932.
114 Lawrence, speaking from his personal experience: Tattersall 2009: 67.
114 "The common denominator": Gray and Sansum 1933.
114 "disagreeable and sometimes alarming": Geyelin et al. 1922.
114 By the autumn of 1925: Geyelin 1926.
115 "nutritionally adequate diets": Tolstoi 1950.
115 "striking" amounts: Geyelin 1926.
115 the children certainly appreciated: Geyelin 1926.
115 "gained as much": Geyelin 1934.
116 telling him of a fourteen-year-old patient: Rabinowitch 1930.
116 "normal human beings": Geyelin 1935.
116 "with delight": Stolte 1931.
116 "Psychologically it is better": Quoted in Lichtenstein 1938.
116 "It is necessary to witness": Lichtenstein 1938.
117 "luxury, which also ought": Lichtenstein 1938.
117 "whipped cream and sweets": Stolte 1933.
117 "Who thinks of changing the diet": Stolte 1933.
117 A single "discontented patient": Sansum et al. 1926.
118 Sansum reported in January 1926: Sansum et al. 1926.
118 In 1933, Sansum reported: Gray and Sansum 1933.
118 Joslin announced its arrival: Wilder and Wilbur 1936.

119   added a compound called protamine to insulin: Tattersall 2009: 80–82.
119   "extraordinary . . . less coma": Joslin et al. 1940: 16.
119   "too long without food": Wilder 1958.
119   "I don't have diabetes anymore": Both Tattersall 2009 and Lawrence 2012 attribute this quote to Wilder 1958, although it is not there. Hence "supposedly."
119   possible long-term effects: Sherrill and MacKay 1939.
120   "It is advantageous under such circumstances": Wilder and Wilbur 1938.
120   "socially intolerable": Fletcher 1980.
121   "Physical vigor": Gray and Sansum 1933.
121   the most notorious of the physicians: Tattersall 2009: 86.
121   "established and conventional criteria": Tolstoi 1943.
121   "Some of the patients": Tolstoi 1943.
122   "amazingly free": Tolstoi 1943.
122   "put the equipment away": Tolstoi 1943.
122   "impatient with all the fuss": Tolstoi 1943.
122   "We learned quite frequently": Tolstoi 1950.
123   "overindulgence": Tolstoi 1950.
123   relaxing their dietary restrictions: Joslin et al. 1946: 353.
123   "These patients were treated by calculated diets": Tolstoi 1950.
124   "it makes little difference": Dunlop 1954.
124   he had had "no doubt": Dunlop 1954.
124   "as their appetites dictated": Forsyth et al. 1951.
125   "Results were uniformly good": Forsyth et al. 1951.
125   "have been disastrous": Dunlop 1954.
125   "individuals whose diabetes has been under": Dunlop 1954.
126   "The problem, of course": Cahill, Jr., 1985.

CHAPTER 7 Good Science/Bad Science, Part I

127   "The use of insulin to fatten patients": Allen and Sherrill 1922.
127   "If these favorable reports": Metz 1932.
128   "cardiac decompensation": Blotner 1933.
128   like a sponge: Richardson et al. 1933. I am using the phrase to refer to the tissues of diabetic patients given insulin; Richardson et al. used it in the context of emaciated diabetics fed a high-fat diet as prescribed by Newburgh and Marsh.
128   "The fact that insulin increases": Haist and Best 1966.
129   "one very noticeable early effect": Allen and Sherrill 1922.
129   were starting patients off: Joslin 1928c: 75–77.
129   Patient 2476: Joslin 1928c: 76–77.
130   Sansum was starting his patients: Sansum et al. 1926.
130   "Sansum and his colleagues claimed": Lawrence 1933.
130   "it has not always been easy": Geyelin 1935.
131   he lacked "the courage": Bowen 1930.
131   "the exact data": Joslin 1923b: 137.
131   "Diabetes, therefore, is largely": Joslin 1923b: 140.
131   had suggested presciently: Joslin 1928c: 156.
131   forty times more likely: Joslin 1933.
131   "a splendid opportunity": Joslin 1923b: 474.
131   "the process of overeating": Joslin 1933.

131 "With an excess of fat": Ganda 1985.
132 "achieve and maintain body weight": ADA 2022.
132 "When diet failed": Tattersall 2009: 123.
133 a Pennsylvania physician named Robert Pitfield: Pitfield 1923.
133 In St. Louis, William Marriott: Marriott 1924.
133 Orville Barbour, a pediatrician from Peoria: Barbour 1924.
134 "indifferent" or "confusing": Fischer and Rogatz 1926.
134 could vary "tremendously": Tisdall et al. 1925.
134 "functionally intact pancreas": Falta 1913: 573.
134 "fattening cure": Falta 1925. Translated by DeepL.
135 "a wave of enthusiasm": Nichol 1932.
135 60 units of insulin a day: Nichol 1932.
135 "insulin hypernutrition": Appel et al. 1929.
135 In Boston, Harry Blotner: Blotner 1933.
136 "definitely obese": Blotner 1938.
136 "wonderfully well": Blotner 1933.
136 "Euphoric": Greco et al. 1942.
136 "became agreeable and cooperative": Appel et al. 1929.
136 "satisfied himself that hypoglycaemia": Jones 2000.
137 "learning analysis": Valenstein 1986: 47.
137 "the remarkable results": Wortis 1936.
137 "rule of thumb": Wortis 1936.
137 "An exception is allowed": Wortis 1936.
138 "Patients furthest along": Valenstein 1986: 56–57.
138 remission of schizophrenia: Wortis 1936.
138 "safer, easier to administer": Fink et al. 1958.
138 "is much more dangerous": Boardman et al. 1956.
139 "most patients emerged": Jones 2000.
139 A 1955 report: West, Bond, Shurley, and Meyers 1955.
139 Nobel laureate John Nash: Nasar 1998: 293.
139 The poet Sylvia Plath: Butscher 2003: 122.
139 "much more food": Anon. 1932b.
139 "richer and more varied food": Nasar 1998: 293.
139 "stimulant to the appetite": Nahum and Himwich 1932.
139 "The rats got so obese": Associated Press 1937.
139 suggested that the insulin caused hypoglycemia: Barnes and MacKay 1940.
140 patients on insulin therapy got fatter: See, for instance, DCCT Research Group 1988.
140 "imposes a situation": West et al. 1955.
140 "large and rapid gain": Dally and Sargant 1966.
140 "must play a role": This is Falta's paraphrasing of what he wrote in 1913; Falta 1925.
140 "a melting of body proteins": Falta 1925.
141 "For fattening": Falta 1913: 573.
141 "diabetogenous obesity": von Noorden 1905: 61.
141 He reported that it did: Falta 1925.
141 "halfway up the thighs": Eeg-Olofsson 1930. For another early report, see Rowe and Garrison 1932.
141 "Severe diabetes": Joslin 1928c: 235.
142 a blood sample had to be taken: Joslin 1933.
142 Joslin acknowledged: Joslin 1923b: 195–96.
142 In 1928, Israel Rabinowitch: Rabinowitch 1928.

143 "excellent fattening substance": Grafe 1933: 147.
143 "splendid builder of reserve": Grafe 1933: 75–76.
143 because of the way insulin had been perceived: Grafe 1933: 299.
144 "many stout persons": Osler and McCrae 1914: 450.
144 "The medical profession in general": Newburgh and Johnston 1930a.
144 "individual patient who *allows*": Joslin 1923b: 474.
145 "It's constitutional": Shaw 1914: 66.
145 "noted Vienna authority": Anon. 1930.
145 *Constitution and Disease*: Bauer 1917.
145 "the exaggerated tendency": Silver and Bauer 1931.
145 almost 90 percent: Silver and Bauer 1931.
146 "The genes responsible": Bauer 1940.
146 "perverted appetite": Newburgh and Johnston 1930a.
146 "no specific metabolic abnormality": Newburgh 1931.
147 "been deliberately trained to overeat": Newburgh and Johnston 1930a.
147 "various human weaknesses": Newburgh and Johnston 1930b.
147 "Just Gluttony": Anon. 1932a.
147 "undermined conclusively": Anon. 1939.
147 "an entirely preventable disease": Newburgh 1950.
147 "lame excuses": Rynearson 1963.
147 "avoidance of the necessary corrective measures": Rynearson and Gastineau 1949: 42.
148 "unresolved emotional conflicts": Wilson 1963.
148 "the nervous tensions": Gastineau 1963.
148 "comfort eating": Ulijaszek 2017: 104.
148 The two dominant hypotheses: See, for instance, Mayer 1955.
148 "remarkably poor": Stunkard and McLaren-Hume 1959.
149 "the reduction of energy intake": Van Gaal 1998.
149 "reduction of caloric intake": Maratos-Flier and Flier 2005.
150 "to present indeed": Wertheimer 1965.
150 "meager": Wertheimer 1965.
151 Woodyatt had directed attention: Woodyatt 1921a.
151 "The experimenter": Schoenheimer and Rittenberg 1940.
152 "indistinguishable as to their origin": Schoenheimer 1961: 56.
152 "Mobilization and deposition": Wertheimer and Shapiro 1948.
152 "must be controlled by a factor": Wertheimer and Shapiro 1948.
153 insulin's role in stimulating: Hausberger et al. 1954 and Anon. 1955.
153 McGarry would point out: McGarry 1992 and McGarry 2002.
154 National Institutes of Health: Gordon and Cherkes 1956.
154 Rockefeller Institute: Dole 1956.
154 University of Lund: Laurell 1956.
154 "great metabolic activity": Gordon and Cherkes 1956.
155 Bertram made a similar observation: Wrenshall et al. 1962: 167.
155 "encountered in rather large numbers": Wilder and Wilbur 1935. See also Himsworth 1936.
156 "6 and possibly 7": Joslin et al. 1940: 327–28.
156 nineteen thousand patients: Joslin et al. 1940: 5.
156 "Joslin, like all other": Gale 2013.
156 "the almost incredible potency": Levine 1967.
157 a millionth of a gram: Wrenshall et al. 1962: 112.

157 "interesting information": Yalow and Berson 1960c.
157 "as little as a fraction": Yalow and Berson 1960c.
157 "a revolution in biological": Karolinska Institute 1977.
157 "early or mild maturity-onset disease": Berson and Yalow 1961.
159 "is largely a penalty of the obesity": See, for instance, Joslin 1923b: 140 and Joslin, Root, White, Marble, and Bailey 1946: 73.
159 "We generally accept": Berson and Yalow 1965.
159 "Simply speaking": Cahill, Jr., 1971.
159 "the negative stimulus": Berson and Yalow 1965.
160 "A high insulin level": Cahill, Jr., 1971.
161 Cahill was among the prominent voices: See Cahill, Jr., et al. 1976.
161 the fatter the diabetic subjects: Rosenzweig 1994.
161 In patients with type 1 diabetes: Sinha, Formica, Tsalamandris et al. 1996.
162 the ACCORD trial: Action to Control Cardiovascular Risk in Diabetes Study Group 2008.
162 study called ADVANCE: Advance Collaborative Group 2008.
162 the Veterans Affairs Diabetes Trial: Duckworth et al. 2009.
162 "counterbalancing consequences": Skyler et al. 2009.
162 "because it caused weight gain": Tattersall 2009: 123 and 170.
162 "direct lipogenic effects": Rosenzweig 1994.
162 "The result of weight gain": Cheng and Zinman 2005.
163 "Diet therapy and weight loss": Rosenzweig 1994.
163 "Successful treatment of obesity": Flier 1994.
163 "The treatment of obesity should entail": Skyler 1978.

CHAPTER 8 Good Science/Bad Science, Part II

165 "Briefly summed up": Chamberlin 1890.
165 "Why the delay?": Gale 2013.
165 Joslin looked back: Joslin 1950.
166 "arteriosclerosis and cardiovascular-renal": Joslin 1950.
166 "One of the most striking facts": Millard and Root 1948.
167 "To a non-diabetic": Joslin et al. 1940: 428.
168 So maybe the diet made little difference: See, for instance, Gray and Sansum 1933 and Tolstoi 1950.
168 Joslin came to consider an obvious possibility: Joslin 1950.
169 the American Diabetes Association began prescribing: Anon. 1971.
171 For a more comprehensive review: Taubes 2007 and Teicholz 2014.
172 "The method of science": Popper 1979: 81.
172 first principle of science: Feynman 1985: 343.
172 "Most people are concerned": Alvarez 1987: 19.
172 "only trust what you can prove": Oft quoted by physicists in the 1980s, when I first heard it. One source is Stoll 1989: 105.
172 "that famous opening line": Merton 1973: 339.
173 "False conclusions": Sagan 2011: 20.
174 "Within a framework": Glueck et al. 1978.
174 at least when challenged: Mann 1977.
174 As Francis Bacon observed: Bacon 1994.
175 "The human understanding": Bacon 1994: 57.

175 In her 2010 book *Being Wrong*: Schulz 2010: 310.
175 "each new research adds detail": Keys 1957.
176 Chamberlin explained in 1890: Chamberlin 1890.
176 American Heart Association began promoting: Anon. 1961b.
177 "important question": Bishop 1961.
177 "because it is not known": Review Panel of the National Heart Institute 1969: 2.
177 "a definitive test": HEW 1971: 67.
177 "an unproved hypothesis": Dawber 1978.
178 "overwhelming evidence": Dawber 1980: 141.
178 "It's an imperfect world": Interview with Basil Rifkind, August 6, 1999.
178 In the 1971 edition: 11th edition: Krall and Joslin 1971, 12th edition: Flood et al. 1985, 13th edition: Crapo 1994, 14th edition: Chalmers 2005.
179 The great German pathologist Rudolf Virchow: Bruger and Oppenheim 1951 and Aschoff 1924: 131–33.
179 Windaus reported: Joslin 1928c: 687.
179 "There can be no atherosclerosis": Quoted in Hoeg and Klimov 1993.
179 very likely right: See, for instance, Aschoff 1924: 150 and Steinberg 2004.
180 "There is no doubt in my mind": Aschoff 1924: 150.
180 "may or may not": Joslin 1928c: 689.
180 "in the blood varies": Joslin 1933.
181 "Can it be": Joslin 1928c: 251.
181 "Old arteriosclerotic diabetics": Joslin 1933.
181 "The condition produced in the animal": Leary 1935.
181 Pathologists and biochemists: Landé and Sperry 1936.
181 "from persons who had died suddenly": Anon. 1936.
182 "a uniform failure": Wilder and Wilbur 1936.
182 Schoenheimer and Rittenberg had reported: Rittenberg and Schoenheimer 1937.
182 Serum cholesterol levels could be lowered: Keys, Mickelsen et al. 1950.
183 "Since many herbivorous animals": Lindsay and Chaikoff 1963. See also, for chickens: Katz and Pick 1961; naturally occurring atherosclerosis: Altschule 1966 and Lindsay and Chaikoff 1963; in primates: McGill, Jr., et al. 1960.
183 This was first reported in 1949: Duff et al. 1954.
183 more and more Americans were living long enough: See, for instance, HEW 1979: 1-1-1-5, CDC 2000.
183 one in every four hundred: Joslin et al. 1934.
184 "newer public health": Keys 1953a.
184 "We were at war": Interview with Fred Stare, March 19, 1999.
184 Congress had taken the first step: Harden undated.
185 the AHA transformed itself: Anon. 1947, Anon. 1948, Davies 1950, and Moore 1983.
185 as their critics would point out: See, for instance, Yerushalmy 1966, Feinstein and Horwitz 1982, Feinstein 1988, Taubes 1995, and Von Elm et al. 2007.
186 Keys, in fact, would be wrong: See, for instance, Keys 1951.
186 "I've come to think": Boffey 1987.
186 Keys was a remarkable man: Shurkin 1992: 132–36, 247–53.
187 "awfully provincial": Hoffman 1979.
187 *The Biology of Human Starvation*: Keys, Brozek et al. 1950.
187 "to find out why": Anon. 1961a.
187 "The news in the American public press": Keys 1994.
188 "with apparatus for measuring": Keys 1994.
188 Keys would acknowledge: Keys 1953b.

420 Notes

188 "fatty diet": Keys 1994.
188 "direct evidence": Keys 1952.
188 "public health programs": Keys 1953a.
189 "All doors of medical schools": Keys 1994.
189 "uncritically or even superficially": Yerushalmy and Hilleboe 1957.
190 "uncompromising stands": Page et al. 1957.
190 "the best scientific information": Central Committee for Medical and Community Program of the American Heart Association 1961.
191 "the highest medical": Laurence 1960.
191 given time and sufficient effort: See, for instance, Bishop 1961 and Anon. 1964b.
191 "including infants, children": Inter-Society Commission for Heart Disease Resources 1970.
192 A Hungarian trial: Korányi 1963.
193 "A low-fat diet has no place": Research Committee 1965.
193 the Anti-Coronary Club Trial: Christakis et al. 1966a.
193 the Los Angeles Veterans Administration Trial: Dayton et al. 1969. See also Brody 1970.
193 A Finnish trial: Miettinen et al. 1972.
193 Results from the Minnesota trial: Frantz et al. 1975 and Frantz et al. 1989.
194 $115 million: Kolata 1982.
194 "throwing the kitchen sink": Interview with Stephen Hulley, July 13, 1999.
194 The MRFIT researchers: MRFIT Research Group 1982.
194 "A Test Collapses": Bishop 1982.
194 "some aspect of the intervention": MRFIT Research Group 1982.
194 $150 million: Boffey 1984.
195 "little doubt of the benefit": Anon. 1984a.
195 "landmark": Boffey 1984.
195 "It is now indisputable": Anon. 1984c.
195 "Hold the Eggs and Butter": Wallis 1984.
196 "could and should be extended": Anon. 1984a.
196 "no whisper of benefit": Moore 1989: 60–61.
196 "an unconscionable exaggeration": Kolata 1985.
196 The NIH now declared: Consensus Conference 1985.
196 "massive health campaign": Levy's testimony in Select Committee 1977: 19.
196 National Cholesterol Education Program: Anon. 1988.
196 Surgeon General's Office: HEW 1988.
196 National Academy of Sciences: NRC 1989.
197 "the depth of the science": Koop 1988.
197 premature death in women: Jacobs et al. 1992.
197 "We are coming to realize": Hulley et al. 1992.
197 Women's Health Initiative: WHI results on breast cancer: Prentice et al. 2006. On heart disease and stroke: Howard, Van Horn, et al. 2006. On colorectal cancer: Beresford et al. 2006. On body weight: Howard, Manson et al. 2006.
198 released a statement: WHO 2006.
198 These included: Research Committee 1968 (soybean oil trial) and Leren 1966 (Oslo trial).
199 "Readers may wonder": Sacks et al. 2017.
199 "The totality of available evidence": Astrup et al. 2020.
199 Implicit in the Cochrane Collaboration's existence: Taubes 1996.
199 In the 2020 edition: Hooper et al. 2020.

201 The results were published: Look AHEAD Research Group 2013.
201 "Many important benefits": Wing et al. 2013.
201 "futility": Look AHEAD Research Group 2013.
203 dramatic decrease in the prevalence of heart disease: See, for instance, Strom and Jensen 1951.
203 "A major lesson gained": Keys 1975.
203 "would entirely disappear": Aschoff 1924: 143.

CHAPTER 9 Good Science/Bad Science, Part III

205 "How then can one with certainty determine": Virchow 1847: 15 (translated from the German).
206 "living more and more": Schweitzer 1957: ix.
207 George Campbell, director of a diabetes clinic: Campbell's testimony in Select Committee on Nutrition and Human Needs 1973: 208–18.
207 "Although atherosclerosis is still": Burkitt 1970.
208 In 1912, Richard Cabot: Cabot 1912.
208 Out of 48,000 patient reports: Fitz and Joslin 1898.
209 he recognized this ongoing trend: Joslin 1921.
209 one patient in every four: Anon. 2017.
209 "that diabetes has increased": Discussion in Emerson and Larimore 1924.
209 newspaper and magazine articles: See, for instance, Anon. 1923.
209 In 1924, two Columbia University researchers: Emerson and Larimore 1924.
210 conspicuous explosion in the amount of sugar: 1801–1868, Anon. 1873: 129; 1890–1923, Emerson and Larimore 1924; one hundred pounds per year in 1921, Anon. 1921.
210 at best informed guesses: Joslin 1917b: 59.
210 Children, in particular: Woloson 2002.
210 The chocolate, ice cream, and candy industries: Taubes 2016: 57–63.
210 "particularly striking": Emerson and Larimore 1924.
211 "diabetes in the tropics": Charles et al. 1907.
211 "The consumption of sugar": Allen 1913: 146–47.
212 "Sugar and candies": Duncan 1935: 59.
212 discussed the sugar hypothesis: See Joslin 1917b: 59, Joslin 1923b: 146–47, and Joslin 1928c: 165–67.
212 By the 1971 edition: Marble et al. 1971.
212 "upon a diet consisting largely": Joslin 1917b: 59.
213 "in the employees of candy factories": Joslin 1928c: 167.
213 Joslin was citing: Joslin et al. 1934.
213 "Of the thirteen countries": Mills 1930.
214 the arrival of insulin therapy: Joslin 1925.
214 "Sugar is what must be given": Himsworth 1931.
214 "A considerable number": Himsworth and Marshall 1935.
214 Researchers in the discipline: Richter and Schmidt 1941 and Richter, Schmidt, and Malone 1945.
215 a second lengthy review: Himsworth 1935.
215 "painstakingly accumulated": White and Joslin 1959: 70.
215 reason enough to eliminate that possibility: Himsworth 1935.
215 "the evidence as": Himsworth 1935.
215 a study of "fisherfolk": Mitchell 1930.

215 "settler dietary": Hutton Undated: 37.

215 The second study: Thomas 1927.

216 As late as the 1960s: Insull et al. 1968.

216 "The consumption of fat": Himsworth 1949.

216 "other, more important": Himsworth 1949.

216 another variable that tracks with fat: Yudkin 1964.

216 Israel was one of several natural laboratories: Dreyfuss 1953, Toor et al. 1960, and Ungar et al. 1963.

218 "the quantity of sugar": Cohen et al. 1961. See also Cohen 1963a and Cohen 1963b.

218 A similar observation: Campbell 1963.

218 "A veritable explosion": Cleave and Campbell 1966: 25.

218 one in one hundred estimated: Cleave and Campbell 1966: 21, 32, 39.

219 "a remarkably constant": Campbell 1963.

219 Cleave, a surgeon captain: See Wellcome Collection, "Cleave, Surgeon Captain 'Peter' Thomas Latimer." Online at https://wellcomecollection.org/works/aegabdcp.

219 His 1962 book on ulcers: Cleave 1962.

220 "refined-carbohydrate disease": Cleave and Campbell 1966: iv.

220 "the Law of Adaptation": Cleave and Campbell 1966: 1.

220 "Such processes": Cleave 1956.

221 Native American tribes: West 1974 and West 1978: 134.

221 Cleave was writing: Cleave 1940.

221 "The Neglect of Natural Principles": Cleave 1956.

221 Campbell had written: Campbell 1961.

221 "radical dietary change" and Cleave reaching out to Campbell: Campbell 1996.

221 "a bigger contribution": Doll 1966.

222 "generally agreed": Dunlop 1966.

222 His doctoral research had been instrumental: Monod 1965.

222 criticized Keys's interpretation: Yudkin 1957.

222 "anatomically, physiologically": Yudkin 1963.

223 Yudkin conducted experiments: For Yudkin's review of his research, see Yudkin 1972 and Yudkin 1988.

223 were doing the same: See, for instance, Cohen and Shafrir 1965 and Cohen et al. 1966.

223 Cohen, Campbell, Cleave, and Yudkin testified: Select Committee 1973 ("The only question" and "Of course, one": 256; "They die": 155).

224 "For a modern disease": Select Committee 1973: 246.

224 "high blood cholesterol in itself": Select Committee 1973: 228–29.

224 It hired pollsters: National Analysts 1974: 33.

225 "plays an etiological role": McGandy and Mayer 1973.

225 the sugar industry hosted a pair of conferences: The Washington, DC, meeting: Hillebrand 1974. The Montreal meeting: ISRF 1975.

225 "establish definitively": ISRF 1976.

225 "Sugar in the Diet of Man": Stare 1975.

226 "enemies of sugar": Tatem 1976.

226 The chapter on heart disease: Grande 1975.

226 "The causes of primary diabetes": Bierman and Nelson 1975.

226 "We eat three times": Kellock 1985: 180–81.

227 "one of the world's": Auerbach 1974.

227 returning to England: Galton 1976: 6 and Cummings and Engineer 2018.

227 "perceptive genius": Burkitt 1979: 12.

227 "I knew from my experience": Burkitt 1991a.
228 "Before the cause of syphilis": Burkitt 1970.
228 "Changes made in carbohydrate food": Burkitt 1971.
229 "the greatest cause of dental caries": Cleave 1956.
229 *Nutrition and Physical Degeneration*: Price 1939.
229 In 1968, Burkitt had met Neil Painter: Cummings and Engineer 2018.
229 "do you remember the story in Sherlock Holmes": Galton 1976: 21.
229 Burkitt was suggesting a chain of causality: Burkitt et al. 1972.
230 Hugh Trowell finished the job: Cummings and Engineer 2018.
230 "three million men": Trowell 1981.
230 the first clinical diagnosis: Trowell and Singh 1956.
230 *Non-Infective Disease in Africa*: Trowell 1960: 465–66.
231 "the towns were full": Trowell 1975a.
231 took up the fiber hypothesis himself: Cummings and Engineer 2018.
231 suggested that they add the subtitle: Trowell 1978; the first book: Burkitt and Trowell 1975.
231 he began publishing articles: Trowell 1972.
231 overeat and become obese: Trowell 1975a and Trowell 1978.
231 Diabetes would be prevented: Trowell 1973 and Trowell 1975b.
232 "the tonic of our time": Auerbach 1974.
232 reporting on a review of the science: Burkitt et al. 1974.
232 "furor over fiber": Mayer and Dwyer 1977.
232 yearly doubling in sales: Kellock 1985: 166–67.
232 "A good diet": Mayer and Dwyer 1977.
232 nutrition researchers worldwide: See, for instance, Kelsay 1978 and NRC 1989: 291–310.
233 "at the expense": Koop 1988.
233 "particularly vulnerable": HEW 1988: 3.
233 "avoid eating too much sugar": USDA and HEW 1980.
233 "sugar does not cause diabetes": USDA and HEW 1985.
233 "generally recognized as safe": Glinsmann et al. 1986: S15.
233 "foods and dietary components": NRC 1989: 703; benefits of fiber remained unknown: NRC 1989: 303.
233 "no direct causal role": COMA 1989: 43.
234 two major trials in the 1990s: See Schatzkin et al. 2000 and Alberts et al. 2000.
234 working under the auspices: Asano and Mcleod 2002.
234 "does not support": Yao et al. 2017.
234 provided no apparent benefits: Beresford et al. 2006, Howard, Manson et al. 2006, Howard, Van Horn et al. 2006, and Prentice et al. 2006.
234 "plenty of reasons": Burros 2000.
234 "Burkitt's hypothesis": Richard Doll interview, April 24, 2003.

CHAPTER 10 The End of Carbohydrate Restriction

236 "In the end": Meinert 2015: xv–xvi, 5.
236 "tyranny of the insulin syringe": Quoted in Silverman 1957.
236 these drugs would do just that: The history of BZ-55 and tolbutamide is told in Silverman 1957, considered by Meinert as the best contemporary source.
237 unacceptable side effects: Silverman 1957.

237 it would be tested and marketed: Laurence 1957.

237 "With the advent": Silverman 1957.

238 "The consumption, day after day": Wrenshall et al. 1962: 197.

238 "to know if blood sugar control": Meinert 2015: 11.

239 leaked first to *The Wall Street Journal*: Leger 1970.

239 "If insulin—the diabetic's": Ricketts 1970.

239 "most forms of treatment": Tattersall 2009: 133.

239 "assailed": Anon. 1970.

239 "severely symptomatic": Arky 1971.

240 another *New England Journal* article: Brunzell et al. 1971.

240 The three were friends: Interview with Ronald Arky, February 27, 2004.

240 "In today's world": Interview with Ronald Arky, February 27, 2004.

241 "the enormous gaps": Arky 1971.

241 "may have implications": Brunzell et al. 1971.

241 "the common thread": Wood and Bierman 1972.

241 A handful of physicians: Singh 1955, Kempner et al. 1958, Stone and Connor 1963, and Ernest et al. 1965.

241 The report cited most frequently: Stone and Connor 1963.

242 "principles of nutrition" and "in some patients": Kinsell et al. 1967.

242 Kinsell had been the first to report: Kinsell et al. 1953.

242 "whether or not control": Kinsell et al. 1967.

243 "There no longer appears": Bierman et al. 1971.

243 "ingestion of simple sugars": Bierman et al. 1971.

244 "It is not yet possible": Bierman et al. 1971.

244 "important gaps in our knowledge": Bierman et al. 1971.

244 "Medical Group": Altman 1971.

245 "the field of nutrition": Nuttall and Brunzell 1979.

245 "ideally up to 55–60%": Anon. 1987b.

246 "The long-term effectiveness": Kolata 1987a.

246 "his data speak for themselves": Kolata 1987b.

247 The existence of the conflict: Anon. 1987b.

CHAPTER 11 Diabetes and Heart Disease

248 "Science is helplessly": Judson 1979: 95.

248 "When unproved hypotheses": Ahrens et al. 1957.

250 "research in the prevention": Joslin 1950.

251 WYSIATI: Kahneman 2011: 82–88.

252 "Human blood is": Cohn et al. 1950.

253 cost $30,000: Grundy 2004.

254 The very first reports in 1951: Barr et al. 1951 and Russ et al. 1951.

254 "simply ran against": Gordon 1988.

255 The "striking" revelation: Gordon et al. 1977. See also Castelli et al. 1977.

255 the only three lifestyle interventions: Noted in Castelli et al. 1977, citing Hulley et al. 1972, and Wilson and Lees 1972.

257 "At a particular cholesterol level": Gofman et al. 1950.

257 "does not account": Gofman et al. 1950.

257 In the late 1950s: For Gofman on LDL, VLDL, diet, and heart disease, see Gofman 1958.

258 "false and highly dangerous": Gofman 1958.
258 "revolutionize the field": Quoted in Bardossi and Schwartz 1984.
258 Ahrens's first contribution: Ahrens et al. 1957.
259 In 1961, Ahrens reported: Ahrens et al. 1961.
259 "This fat is carried": Olson 1998; "deficient in lipoprotein": Ahrens et al. 1961.
259 "This phenomenon": Ahrens et al. 1961.
260 "a common phenomenon": Ahrens et al. 1961.
260 "as something of a surprise": Osmundsen 1961.
260 "quite as impossible": Man and Peters 1935.
261 Albrink had joined Peters's laboratory: Interview with Margaret Albrink, June 18, 2002.
261 "suggests that an error": Albrink and Man 1959.
261 they reported similar numbers: Albrink et al. 1961.
261 "just about brought": Interview with Margaret Albrink, June 18, 2002.
262 heart attack survivors in the Seattle area: Goldstein et al. 1973.
262 a seminal five-part, fifty-page series: Fredrickson et al. 1967a–e.
262 four of the five disorders: Fredrickson et al. 1967c: 149 (table 2).
262 warned about the dangers: See, for instance, Fredrickson et al. 1967b: 219.
262 "sometimes considered": Fredrickson et al. 1967a.
262 "Patients with this syndrome": Lees and Wilson 1971.
262 Albrink and Man would report: Albrink et al. 1963.
263 suggested that all these disorders: Davidson and Albrink 1965.
263 Yudkin would cite Albrink's research: See Yudkin 1963 and Yudkin 1972.
263 often had poor glycemic control: Reaven et al. 1963.
264 insulin response to the carbohydrates: Reaven et al. 1967.
264 "within a group": Olefsky et al. 1974.
264 "In this proposal": Farquhar et al. 1966.
265 "price paid": Reaven et al. 1967.
265 Aharon Cohen: Cohen 1963b.
265 Yudkin might have predicted: Yudkin 1972.
265 carbohydrate-rich diets for diabetes was a mistake: Reaven et al. 1979 and Reaven 1980.
265 "the appropriate amounts": Reaven 1986.
266 researchers had demonstrated: For a review of this science, see DeFronzo 1981.
266 "Hypertension may thus": Landsberg 1986.
266 "the central role": Reaven 1988.
266 renamed it "metabolic syndrome": WHO 1999.
266 In 2002, the NCEP published: NCEP 2002 ("and its associated": II-26; "the primary target": II-36; "the greatest potential": II-26; "therapeutic lifestyle" and "maximum reduction of": V-2).
267 "emerging risk factor": Grundy, Brewer et al. 2004 and Grundy, Hansen et al. 2004.
267 "mass elevations": NCEP 2002: II-28.
267 "What you're faced with": Interview with Scott Grundy, May 4, 2004.
269 "nature has perfected": Lawrence 1933.
269 "the rising star": Gebel 2012.
269 he detected a novel form: Rahbar 1968 and Gebel 2012.
269 they confirmed the observation: Rahbar et al. 1969.
270 "clandestinely": Bunn 2013.
270 Put simply: Koenig et al. 1976.
270 it embraced hemoglobin A1C: ADA 2010.

271   true not just for patients with diabetes: Selvin et al. 2010.
271   "conform to a tightly regulated": Bunn and Higgins 1981.
271   is dependent only: Bunn et al. 1976.
271   a "conceptual framework": Cerami et al. 1978.
272   The process starts: For a good review, see Brownlee et al. 1988.
272   "If you remove": Interview with Anthony Cerami, July 15, 2004.
272   By the mid-1980s: Cerami et al. 1985 and Brownlee et al. 1984.
273   Brownlee's hypothesis: Brownlee 2005.
274   it's this oxidized LDL: Steinberg 1997; AGEs, glycation, and oxidized LDL: Bucala et al. 1993.
274   "markedly elevated": Stitt et al. 1997.
274   "Current evidence points": Peppa et al. 2003.
276   20 to 60 units a day: See, for instance, Eaton et al. 1980 and Kruszynska et al. 1987.
276   "at a rate": Stout 1982: 70, citing Werther et al. 1980.
277   Duff had reported: Duff et al. 1954.
277   insulin, unsurprisingly, has similar effects: Stout 1968.
277   *anabolic* effects on the arterial walls: Stout et al. 1975.
278   a series of clinical trials: ACCORD 2008, Advance Collaborative Group 2008, and Duckworth et al. 2009.
278   more deaths related to hypoglycemia: ACCORD 2008 and Taubes 2008.
278   appeared not to be true: Riddle et al. 2010.
279   almost twenty pounds: Henry et al. 1993.
279   "demonstrated that insulin": Di Pino and DeFronzo 2019.
279   the American Diabetes Association was wrong: DeFronzo 2009.
279   lip service to the role of diet: Sawyer and Gale 2009.

CHAPTER 12 What You See Is All There Is

280   "Over the past thirty years": Bernstein 1981: 222–24.
281   "The patient with diabetes": Tattersall et al. 1980.
281   "beware of the educated diabetic": Joslin 1923b: 682.
282   "Good Diabetic Control": Malone et al. 1976.
282   "there was no real alternative": Peacock and Tattersall 1982.
282   Chemists had known: Waymouth Reid 1896.
283   "at intervals of months": Joslin et al. 1937: 207 and Joslin et al. 1940: 206.
283   "frequent or severe reactions": Colwell 1970.
283   "The diabetic taking insulin": Joslin 1928c: 20.
284   fanciful, if not dangerous: Heller 2019, interview with Charles Peterson, February 12, 2021, and interview with Jay Skyler, October 11, 2021.
284   what had been learned from "the English experience" and "How . . . can you expect": Tattersall and Clarke 1982.
285   obvious limitation to hemoglobin A1C: Walford et al. 1978.
285   "important effects" and "It has shown": Tattersall and Clarke 1982.
286   averaging one a week: Cryer et al. 1989.
286   "explosive rebellion": Tattersall 1981.
287   medical profession lost fifteen years: Interview with Edwin Gale, December 7, 2021.
287   "otherwise be impractical": Keen and Knight 1962.
288   "It's not that I had": Interview with Richard Bernstein, January 3, 2020.
288   "conversion narrative": Gladwell 1998.

289 The Ames Reflectance Meter: ACS 2010.

289 "'ordinary' diabetic": Bernstein 1997: xii.

290 "was clearly becoming untenable": Bernstein 1997: xv.

290 Bernstein had seen an ad: Bernstein 1981: xxi–xxii.

291 "The very high and low": Bernstein 1997: xvi.

291 "either in a hypoglycemic state": Bernstein 1981: xiii.

291 "If a change": Bernstein 1997: xvii.

291 "laid the groundwork": Mendosa Undated a.

292 the first principle of nutritional guidance: See, for instance, Skyler 1978 and Colwell 1970.

292 "Big inputs": Bernstein 1997: 102.

293 "You're driving down": Bernstein 1997: 105.

293 no way to do it with large doses: Galloway et al. 1981.

293 By minimizing the amount of insulin: Bernstein 1997: 106–8.

294 "go way up": Bernstein 1997: xvi–xvii.

295 Charles Suther at Ames: Mendosa Undated a.

295 "fine control": Letter from Franz Ingelfinger to Richard Bernstein, December 28, 1976 (provided by Bernstein).

295 "use the electric device": Letter from Henry Ricketts to Richard Bernstein, November 2, 1976 (provided by Bernstein).

295 Bernstein responded by networking: Interview with Richard Bernstein, December 23, 2019.

296 the very first paper: Danowski and Sunder 1978.

296 used the Ames devices at a summer camp: Interview with Jay Skyler, October 11, 2021, email from Charles Peterson, September 26, 2021, and acknowledgments in published papers to Ames for providing the devices.

297 Bouchardat saying in 1885: Joslin 1959: 230.

297 By the 1970s: See, for instance, Carrington 1970, Mintz et al. 1978, Lowy 1998, White 1971, and Joslin 1959: 230–32.

297 "often waterlogged": Joslin 1959: 230.

297 "Provided [the newborns]": Mintz et al. 1978.

298 "So at 26 weeks": Lowy 1998.

298 "major or minor congenital": Jovanovic et al. 1981.

298 "By so doing": Sönksen et al. 1978.

299 "abolish completely": Mintz et al. 1978.

299 "Several patients": Walford et al. 1978.

299 covered the travel expenses: Mendosa Undated b.

299 "Children accepted": Silink 1982.

300 their patients at Rockefeller: Interview with Charles Peterson, February 12, 2021.

300 "aggressive insulin therapy": Bleicher et al. 1980.

300 "severe depression": Bogdonoff 1982.

300 "a diet that optimized": Peterson et al. 1979.

300 "an empirical approach": Peterson et al. 1980.

301 The results of this experiment: Peterson et al. 1979 and Peterson et al. 1980.

301 "the beginning of": Bleicher 1980.

301 the ADA estimated that a million patients: Anon. 1987a.

302 "choos[ing] a diet": Peterson et al. 1979.

302 "jet delivery": Danowski and Sunder 1978.

302 that the AHA and the ADA were correct: See, for instance, Schade et al. 1983: 154.

303 "the carbohydrate wars": Charles Peterson email, September 26, 2021.

303   "We certainly thought": Interview with Edwin Gale, December 7, 2021.
303   "many of the insulin regimens": Tattersall and Clarke 1982.
304   "Some physicians are led": West 1978.
304   "insulin wars": Sun 1980.
305   Eight thousand pounds of pancreas: Fraser 2016.
305   "into factories": McElheny 1977.
305   Considered even more important: Sun 1980.
306   The first publications: Pickup et al. 1978 and Tamborlane et al. 1979.
306   reports of thirty-five deaths: CDC 1982 and Unger 1982.
306   "meticulous control": Unger 1982.
307   "intensive insulin therapy": DCCT 1993a.
307   "methods of insulin delivery": DCCT 1986.
307   investigators amended the protocol: DCCT 1993b.
307   "For the DCCT protocol": Interview with Charles Peterson, January 14, 2022.
307   "the most important discovery": Blakeslee 1993.
307   intensive insulin therapy worked: DCCT 1993a.
308   "Joslin was right": Interview with Jake Kushner, January 30, 2022.
308   In many ways, the study had failed: DCCT 1993a.
308   "generate a windfall": Blakeslee 1993.
308   "an expert team": DCCT 1993a.
308   Diabetologists had difficulty imagining: Anon. 1993.
308   gained, on average, more than eleven pounds: In the first year, DCCT 1988.
308   Both groups continued to gain weight: After five years, DCCT 1993a.
309   "further study": DCCT 1988.
309   "two fatal motor vehicle accidents": DCCT 1993a.
309   "despite the added demands": DCCT 1993a.
310   "bringing blood sugar down": Kaufman 2005: 86.
310   "Hyperinsulinemia and insulin resistance": Lasker 1993.
311   "'Vindication' for a Diabetes Expert": Singer 1993.
311   had written letters of recommendation: Interview with Jay Skyler, January 27, 2021, and interview with Charles Peterson, February 12, 2021.
311   "was the only way": Singer 1988.
311   more than twenty-five thousand copies: Singer 1988.
311   "The GlucograF method": Turkington 1981.
311   "Those in the study": Singer 1993.
312   "avoid behaviors": DCCT 1991.
313   "explore the effects": Bernstein 1992.
313   The DCCT investigators agreed: Lorenz 1993.

CHAPTER 13 Low Blood Sugar

314   "The patient should be trained": Unger 1982.
314   "When a factor is known": Cannon 1945: 114.
314   "meticulous control": Unger 1982.
315   "appear warranted": Anon. 1993.
316   "The 'cost' to the patient": Tattersall 2009: 166.
316   "Even patients with optimally controlled": Lee et al. 2016.
316   "dark side of insulin": Porcellati et al. 2021.
317   "What If Minkowski": McGarry 1992.

317 is "that nature has perfected": Lawrence 1933.

318 "Whereas a single hormone": Unger 1966.

318 "fiercely independent": Wascovich 2020.

319 "sugar exists constantly": Bernard 1848: 6.

319 Bernard had controversially demonstrated: Bernard 1848.

319 The high blood sugar of diabetes: See, for instance, Soskin 1941.

319 "Only in this way": Unger 1966.

320 "the means by which": Holmes 1986.

320 "and the animal": Mann and Magath 1921, expanded on at length in Mann and Magath 1922.

320 Banting and Best in Toronto were aware: Bliss 2007: 109 and Visscher 1965.

320 comes from the kidney: Gerich 2000.

321 "no longer responds": Soskin 1941.

321 How this process is regulated: See, for instance, Levine and Fritz 1956.

321 "two schools confront[ing]": de Duve 2004.

321 "in accordance with": Soskin 1941.

322 That's how this is described: NIDDK 2016.

322 "the liver was": de Duve 2004.

323 damage to specific cell clusters: Opie 1901a and Opie 1901b.

323 "in all probability": Lane 1907.

323 "failed to investigate": Quoted in Unger 1976.

324 they had isolated a second substance: Kimball and Murlin 1923.

324 contaminated with glucagon: de Duve 2004.

324 Chemists at Eli Lilly: Kirtley et al. 1953.

324 "One need only inject": Blackman 1961.

325 extracted by the liver: Bolli et al. 2021.

325 "physiologic balance": Madison and Unger 1958.

326 Unger published his first papers: See, for instance, Unger et al. 1959.

326 "a single bihormonal": Unger 1976.

327 "the hormone of glucose production": In the discussion of McGarry et al. 1976.

327 "remarkably similar": Müller et al. 2017.

328 "adequate glucose delivery": Unger 1976.

328 "double-trouble hypothesis": Unger 1976.

328 "inappropriate" secretion: Holst et al. 2017.

328 "maintain[ing] glucose utilization": Unger 1982.

328 "massive doses of insulin": Unger 1976.

329 In one trial of pregnant women: Jovanovic et al. 1981.

330 "carbohydrate wars": Charles Peterson email, September 26, 2021.

CHAPTER 14 High-Fat Diets

331 "It is noteworthy": Richter et al. 1945.

332 Arnold Durig, for instance: Burtscher et al. 2012.

332 "mild diabetic": Richter et al. 1945.

333 "chief physiological means": Richter and Schmidt 1941. Animals that had their adrenal glands removed: Richter 1936.

333 "the physiological mechanisms": Richter and Schmidt 1941. Animals that had their thyroid glands removed: Richter and Eckert 1939.

333 "we were interested": Richter and Schmidt 1941.

334 "It was not possible": Richter, Schmidt, and Malone 1945.
334 made diabetic by injection: Young 1937.
334 rabbits and laboratory rodents: Shaw Dunn et al. 1943.
334 "We have been able": Burn et al. 1944. This was confirmed in the United States (Janes and Prosser 1947) and in Australia (Bornstein and Nelson 1949).
335 "It is unfortunate": Marks and Young 1939.
335 "a very definite change": Best et al. 1939.
335 "resting" the pancreas: Haist et al. 1940.
336 Such children, Best suggested: Anon. 1941.
337 discovered in the urine: Van Itallie and Nufert 2003.
337 "noxious substances": MacKay 1943.
337 "The clinician thinks": Cahill, Jr., et al. 1973.
337 "disappear as if by magic": Joslin 1923b: 303–4.
338 "with too much oil": Woodyatt 1916.
338 a dominant source of energy: Barnes et al. 1940.
338 healthy young men *always* have measurable amounts: Johnson et al. 1958.
338 more than a third of all the energy: Van Itallie and Nufert 2003 and Reichard et al. 1974.
339 "If the nutrition scientists": Van Itallie and Nufert 2003.
339 "intellectual capacity and performance": Cahill, Jr., et al. 1973.
240 "not the inflexible": Sokoloff 1973.
240 The role of glucagon: See, for instance, McGarry 1979.
240 "gets into trouble": Cahill, Jr., et al. 1973.
341 "prolonged starvation": Drenick 1976.
341 "normal young individuals": Cahill, Jr., and Aoki 1980.
341 "Brittle diabetics": Drenick et al. 1972.
341 "Both the potential danger": Cahill, Jr., and Aoki 1980.
342 "If known [blood sugar]": Cannon 1929.
342 Tattersall had made this point: Tattersall 2009: 138.
342 "Diabetes is defined by hyperglycemia": Riddle et al. 2021.
345 "low in saturated fat": UKPDS 1998.
345 "The UKPDS is": ADA 2000.
346 "The increasing failure": Turner et al. 1999.
347 "Few substantial improvements": National Commission on Diabetes 1995: 1–2.
348 The first three trials: ACCORD 2008, Advance Collaborative Group 2008, and Duckworth et al. 2009.
349 "one of the most contentious": Taubes 2008.
349 $200 million Look AHEAD: Look AHEAD Research Group 2013.
350 "put 'meaningful' in quotes": Kolata 2012.
350 apparent benefits: Wing et al. 2013.
350 "specific lifestyle intervention": Look AHEAD Research Group 2013.

CHAPTER 15 Very-Low-Carbohydrate Diets

352 "Overriding all rules": Micks 1943.
352 "There are two cosmic truths": Roth et al. 2004.
353 "at best, unlikely": Reaven 1986.
354 In 1985, he reported: Mattson and Grundy 1985.
354 These trials were small: Coulston et al. 1987, Garg et al. 1988, Coulston et al. 1989, Garg, Grundy, and Koffler 1992, Garg, Grundy, and Unger 1992, and Garg et al. 1994.

355   "caused persistent deterioration": Garg et al. 1994.

355   "more moderate carbohydrate intake": ADA 1998.

356   rates of obesity and overweight: Taubes 1998b.

356   "If this was about tuberculosis": Burros 1994.

357   "remarkably ineffective": Stunkard and McLaren-Hume 1959.

357   "Obesity is a chronic condition": Stunkard 1973.

357   "The Fattening of America": Pi-Sunyer 1994.

358   "simple sobriety in eating": Tanner 1870: 217.

358   "more or less rigid abstinence": Brillat-Savarin 1986: 251.

358   he listed three versions: Osler 1893: 1020.

359   seven such conferences: Proceedings of the Harvard conference: Barr et al. 1953; the Swedish conference: Blix 1964; the UCSF conference: Wilson 1969; the 1968 London conference: McLean Baird and Howard 1969; the Paris conference: Apfelbaum 1973; the NIH conference: Bray 1976a; and the 1973 London conference: Burland et al. 1974.

360   "Weight loss, fat loss": Young 1976.

362   "nutrition nonsense": White 1962.

362   "bizarre concepts of nutrition": Anon. 1973.

362   "unrestricted in calories": Anon. 1973.

362   "because that's what was being taught": Interview with Robert Atkins, March 20, 2002.

362   "An abundant supply": Gordon et al. 1963.

362   Atkins claimed: Atkins 1972 ("thirty, forty, one hundred": 3; "lobster with butter," and "As long as you": 15).

363   "As a trained physician": Interview with Linda Stern, June 11, 2002.

363   Stern organized a clinical trial: Samaha et al. 2003.

364   "The reason you don't know": Interview with Albert Stunkard, March 11, 2022.

364   "all he did is eat steak": Interview with Eric Westman, February 20, 2002.

364   five clinical trials: Philadelphia VA Center: Samaha et al. 2003; University of Pennsylvania (with Thomas Jefferson University, Washington University, and the University of Colorado Health Sciences Center in Denver): Foster et al. 2003; Duke University: Yancy et al. 2004; University of Cincinnati: Brehm et al. 2003; and Schneider Children's Hospital in Long Island: Sondike et al. 2003.

365   "beat[s] the standard diet": Interview with Albert Stunkard, May 22, 2002.

366   "We were blazing a trail": Interview with Eric Westman, February 20, 2002.

366   "Although enrolling these subjects": Samaha et al. 2003.

366   other physicians continued to report: Fourteen diabetic patients: Vernon et al. 2003; seven in a "pilot trial": Yancy, Jr., Vernon, and Westman 2003; twenty-one more: Yancy, Jr., Foy et al. 2005.

367   no longer concerned: Yancy, Jr., Foy et al. 2005.

367   "At the end of our clinic day": Westman and Vernon 2008.

368   twenty-six academic researchers: Feinman et al. 2015 (published online in July 2014).

368   "To assess the disadvantages": Feinman et al. 2015.

369   the three trials he published: Phinney et al. 1980, Phinney, Bistrian, Wolfe, and Blackburn 1983, and Phinney, Bistrian, Evans et al. 1983.

369   two hundred patients a year: Email from Stephen Phinney, April 11, 2022.

369   their protein-sparing modified fast: Bistrian et al. 1976.

370   He published his first papers: Volek et al. 2000, Volek et al. 2001, Sharman et al. 2002, and Volek et al. 2002.

370   "primarily putting people": Interview with Jeff Volek, April 29, 2002.

370   "I have one of the better willpowers": Interview with Sami Inkinen, December 15, 2020.

371 Hallberg later said: Interview with Sarah Hallberg, August 5, 2020.

371 "It's not Stanford": Interview with Sami Inkinen, December 15, 2020.

372 the first year of patient experience: Hallberg et al. 2018.

372 the two-year data: Athinarayanan et al. 2019.

372 "partial remission": Buse et al. 2009.

372 A 2014 analysis: Karter et al. 2014.

373 most notably, the UKPDS trial: Turner et al. 1999.

373 "the most frustrating barrier": Zinman et al. 2017.

373 The Virta Health trial had demonstrated: Athinarayanan et al. 2019.

374 A year of intensive counseling: Bhanpuri et al. 2018. Two-year follow-up on heart disease risk: Athinarayanan et al. 2020.

375 the results of a five-year follow-up: Virta Health 2022.

375 $2 billion market valuation: Jennings 2021.

375 three hundred major corporations: Email from Sami Inkinen, September 30, 2022.

375 "including vegetables, fruits," "healthful eating patterns," and "less focus on specific nutrients": ADA 2019.

375 fourteen-year-old National Academy of Medicine report: IOM 2005.

376 "Lowering total fat intake": Evert et al. 2019.

375 "very low-carbohydrate versus": Joseph et al. 2022.

377 "bizarre concepts of nutrition": Anon. 1973.

377 "guide for health care providers": Silverhus 2022.

377 the numerous other guides: Online at https://shopdiabetes.org/collections/nutrition.

377 David Unwin, a senior partner: Interview with David Unwin, September 1, 2017.

378 In 2020, he published: Unwin et al. 2020.

378 20 percent of the type 2 patients: Email from David Unwin, October 4, 2022.

378 "was routine until 1923": Interview with David Unwin, September 1, 2017.

378 "encourage adults with type 2": Online at https://www.nice.org.uk/guidance/ng28/chapter/Recommendations#dietary-advice-and-bariatric-surgery.

378 eat less and exercise: Online at https://www.nice.org.uk/guidance/cg189/chapter/Recommendations#behavioural-interventions.

379 "My grandfather wasn't sure": Interview with Arjun Panesar, December 4, 2020.

379 "The patient who breaks over": Joslin 1923b: 526.

382 Dave Dikeman was nine years old: Interview with R. D. Dikeman, March 26 and March 31, 2020, interview with Dave Dikeman, March 31, 2020.

385 Their survey of TypeOneGrit members: Lennerz et al. 2018.

385 "Their blood sugar control": O'Connor 2018.

385 "a finding that was thought": Interview with Belinda Lennerz, August 20, 2020.

385 "These findings are without precedent": Lennerz et al. 2018.

386 disparage the use of "testimonials": Mayer-Davis et al. 2018.

386 "Children and adults adhering": Lennerz et al. 2018.

386 "promulgating such methodologically": Mayer-Davis et al. 2018.

EPILOGUE The Conflicts of Evidence-Based Medicine

388 "Why do physicians": Eddy 1984.

389 the use of radical mastectomies: Fisher et al. 1977.

389 research, led by John Wennberg: Wennberg 1984.

389 "medical decision making": Eddy 2011.

389 "There is a distinct possibility": Eddy 1990.

390 "The applicable maxim": Eddy 1984.

390 checked not against any objective reality: Eddy 1990.

390 almost tripled since the 1950s: Relman 1988.

390 a "revolution" in medicine: Relman 1988.

390 "function and well-being": Ellwood 1988.

391 "First, there must be good evidence": Eddy 2011.

391 "the conscientious, explicit": Sackett et al. 1996.

391 "dead wrong": Online at https://en.wikipedia.org/wiki/David_Sackett.

391 "accurate, interpretable, applicable": Eddy 1990.

392 "Take two people": Eddy 1984.

392 "scientific authority": EBM Working Group 1992.

393 "Evidence based medicine is not": Sackett et al. 1996.

396 "We know how to lower blood glucose": Interview with Frank Nuttall, November 23, 2020.

397 "There is no question": Interview with Joseph Wolfsdorf, June 24, 2020.

397 "There is some literature": Interview with Larry Deeb, September 14, 2020.

398 "So I said, 'Carbs are toxic'": Interview with Ross Wollen, December 17, 2020.

398 Maryanne Quinn is a pediatric endocrinologist: Interview with Maryanne Quinn, January 20, 2020.

399 "If I eat them": Interview with Jake Kushner, December 9, 2019.

399 Craig Suchin, an interventional radiologist: Interview with Craig Suchin, January 14, 2020.

400 J. Daniel Jones, a radiologist: Interview with J. Daniel Jones, October 13, 2020.

401 Laura Nally, a pediatric endocrinologist: Interview with Laura Nally, August 21, 2020.

404 "What can I do to help you?": Interview with Lisa-Marie Newton, November 10, 2020.

Bibliography

Abraira, C., W. C. Duckworth, T. Moritz, et al. 2009. "Glycaemic Separation and Risk Factor Control in the Veterans Affairs Diabetes Trial: An Interim Report." *Diabetes, Obesity and Metabolism* 11, no. 2 (Feb.): 150–56.

Action to Control Cardiovascular Risk in Diabetes (ACCORD) Study Group. 2008. "Effects of Intensive Glucose Lowering in Type 2 Diabetes." *The New England Journal of Medicine* 358, no. 24 (June 12): 2545–59.

Advance Collaborative Group. 2008. "Intensive Blood Glucose Control and Vascular Outcomes in Patients with Type 2 Diabetes." *The New England Journal of Medicine* 358, no. 24 (June 12): 2560–72.

Ahrens, E. H., Jr., J. Hirsch, W. Insull, Jr., et al. 1957. "Dietary Control of Serum Lipids in Relation to Atherosclerosis." *The Journal of the American Medical Association* 164, no. 17 (Aug. 24): 1905–11.

Ahrens, E. H., Jr., J. Hirsch, K. Oette, J. W. Farquhar, and Y. Stein. 1961. "Carbohydrate-Induced and Fat-Induced Lipemia." *Transactions of the Association of American Physicians* 74: 134–46.

Alberts, D. S., M. E. Martinez, D. J. Roe, et al. 2000. "Lack of Effect of a High-Fiber Cereal Supplement on the Recurrence of Colorectal Adenomas. Phoenix Colon Cancer Prevention Physicians' Network." *The New England Journal of Medicine* 342, no. 16 (Apr. 20): 1156–62.

Albrink, M. J., P. H. Lavietes, and E. B. Man. 1963. "Vascular Disease and Serum Lipids in Diabetes Mellitus: Observations over Thirty Years (1931–1961)." *Annals of Internal Medicine* 58, no. 2 (Feb.): 305–23.

Albrink, M. J., and E. B. Man. 1959. "Serum Triglycerides in Coronary Artery Disease." *Archives of Internal Medicine* 103, no. 1 (Jan.): 4–8.

Albrink, M. J., J. W. Meigs, and E. B. Man. 1961. "Serum Lipids, Hypertension and Coronary Artery Disease." *The American Journal of Medicine* 31 (July): 4–23.

Allen, F. N. 1972. "Diabetes Before and After Insulin." *Medical History* 16, no. 3 (July): 266–73.

Allen, F. M. 1952. "Arnoldo Cantani, Pioneer of Modern Diabetes Treatment." *Diabetes* 1, no. 1 (Jan.–Feb.): 63–64.

Allen, F. M. 1922. "Preface." *Journal of Metabolic Research* 2, nos. 5–6 (Nov.–Dec.): no page numbers.

Allen, F. M. 1917. "The Role of Fat in Diabetes." *The American Journal of the Medical Sciences* 153, no. 3 (Mar.): 313–71.

Allen, F. M. 1916. "Investigative and Scientific Phases of the Diabetic Question." *The Journal of the American Medical Association* 66, no. 20 (May 13): 1525–32.

Allen, F. M. 1915. "Prolonged Fasting in Diabetes." *The American Journal of the Medical Sciences* 150, no. 4 (Oct.): 480–85.

Allen, F. M. 1913. *Studies Concerning Glycosuria and Diabetes.* Cambridge, MA: Harvard University Press.

Allen, F. M., and J. W. Sherrill. 1922. "Clinical Observations with Insulin. 1. The Use of Insulin in Diabetic Treatment." *Journal of Metabolic Research* 2, 804–985.

Allen, F. M., E. Stillman, and R. Fitz. 1919. *Total Dietary Regulation in the Treatment of Diabetes.* New York: Rockefeller Institute for Medical Research.

Altman, L. K. 1971. "Medical Group, in a Major Change, Urges a Normal Carbohydrate Diet for Diabetics." *The New York Times* (Oct. 3): 63.

Altschule, M. D. 1966. "The Uselessness of Diet in the Treatment of Atherosclerosis." In *Controversy in Internal Medicine,* ed. F. J. Ingelfinger, A. S. Relman, and M. Finland. Philadelphia: W. B. Saunders, 69–78.

Alvarez, L. W. 1987. *Adventures of a Physicist.* New York: Basic Books.

American Chemical Society (ACS). 2010. "The Development of Diagnostic Test Strips." Online: http://www.acs.org/content/acs/en/education/whatischemistry/landmarks/diagnosticteststrips.html.

American Diabetes Association (ADA). 2022. "Statistics About Diabetes." Online: https://www.diabetes.org/about-us/statistics/about-diabetes#.

American Diabetes Association (ADA). 2019. "5. Lifestyle Management: Standards of Medical Care in Diabetes—2019." *Diabetes Care* 42, suppl. 1 (Jan.): 546–60.

American Diabetes Association (ADA). 2010. "Diagnosis and Classification of Diabetes Mellitus." *Diabetes Care* 33, suppl. 1 (Jan.): 562–69.

American Diabetes Association (ADA). 2000. "Implications of the United Kingdom Prospective Diabetes Study." *Diabetes Care* 23, suppl. 1 (Jan.): S27–31.

American Diabetes Association (ADA). 1998. "Nutrition Recommendations and Principles for People with Diabetes Mellitus." *Diabetes Care* 21, suppl. 1 (Jan.): 532–33.

American Diabetes Association (ADA) Professional Practice Committee. 2022. "5. Facilitating Behavior Change and Well-Being to Improve Health Outcomes: Standards of Medical Care in Diabetes—2022." *Diabetes Care* 45, suppl. 1 (Jan.): 560–82.

Anderson, F. J. 1965. *Journal of the History of Medicine and Allied Sciences* 20, no. 2 (Apr.): 163–64.

Anon. 2017. "Diabetes Mellitus Federal Health Data Trends." *Federal Practitioner* (July): S20–S21. Online: https://www.mdedge.com/fedprac/article/152642/diabetes/diabetes-mellitus-federal-health-data-trends-full.

Anon. 1993. "Implications of the Diabetes Control and Complications Trial. American Diabetes Association." *Diabetes Care* 16, no. 11 (Nov.): 1517–20.

Anon. 1988. "Report of the National Cholesterol Education Program Expert Panel on Detection, Evaluation, and Treatment of High Blood Cholesterol in Adults. The Expert Panel." *Archives of Internal Medicine* 148, no. 1: 36–69.

Anon. 1987a. "Consensus Statement on Self-Monitoring of Blood Glucose." *Diabetes Care* 10, no. 1 (Jan.–Feb.): 95–99.

Anon. 1987b. "Nutritional Recommendations and Principles for Individuals with Diabetes Mellitus: 1986. American Diabetes Association." *Diabetes Care* 10, no. 1 (Jan.–Feb.): 126–32.

Anon. 1984a. "The Lipid Research Clinics Coronary Primary Prevention Trial Results. I. Reduction in Incidence of Coronary Heart Disease." *JAMA* 251, no. 3 (Jan. 20): 351–64.

Anon. 1984b. "The Lipid Research Clinics Coronary Primary Prevention Trial Results. II. The Relationship of Reduction in Incidence of Coronary Heart Disease to Cholesterol Lowering." *JAMA* 251, no. 3 (Jan. 20): 365–74.

Anon. 1984c. "Sorry, It's True. Cholesterol Really Is a Killer." *Time* (Jan. 23): 30.

Anon. 1973. "A Critique of Low-Carbohydrate Ketogenic Weight Reduction Regimens. A Review of *Dr. Atkins' Diet Revolution.*" *JAMA* 224, no. 10 (June 4): 1415–19.

Anon. 1971. "Principles of Nutrition and Dietary Recommendations for Patients with Diabetes Mellitus." *Diabetes* 20, no. 9 (Sept.): 633–34.

Anon. 1970. "FDA's Antidiabetic-Drug Caution Assailed by 40 Specialists Attending AMA Session." *The Wall Street Journal* (Dec. 2): 10.

Anon. 1968. "The National Diet-Heart Study Final Report." *Circulation* 37, no 3. suppl. (Mar.): I1–428.

Anon. 1964a. "Heart Group Urges All to Eat Less Fatty Food." *The New York Times* (June 9): 71.

Anon. 1964b. "Heart Association Stirs Up a Controversy by Urging Public to Alter Intake of Fats." *The Wall Street Journal* (June 10): 6.

Anon. 1961a. "The Fat of the Land." *Time* 73, no. 3 (Jan. 13): 48–52.

Anon. 1961b. "Dietary Fat and Its Relation to Heart Attacks and Strokes. Report by the Central Committee for Medical and Community Program of the American Heart Association." *JAMA* 175, no. 5 (Feb. 4): 389–91.

Anon. 1955. "Insulin and Fat Metabolism." *Nutrition Reviews* 13, no. 4 (Apr.): 115–17.

Anon. 1948. "Reports of Local Heart Association Activities." *American Heart Journal* 36: 158–59.

Anon. 1947. "National Heart Week." *The New England Journal of Medicine* 236, no. 5 (Jan. 30): 185–86.

Anon. 1941. "Prevention of Diabetes." *British Medical Journal* 2, no. 4,206 (Aug. 16): 234–35.

Anon. 1939. "Professor of Medicine Augments Teaching with Research." *The Michigan Alumnus* 45, no. 22 (June 10): 415.

Anon. 1936. "Blood Cholesterol and Atherosclerosis." *The Journal of the American Medical Association* 107, no. 24 (Dec. 12): 1970–71.

Anon. 1932a. "Just Gluttony Makes Obesity." *Los Angeles Times* (Jan. 5): 3.

Anon. 1932b. "Insulin and Hunger." *The Journal of the American Medical Association* 99, no. 20 (Nov. 12): 1692.

Anon. 1930. "Vienna Specialist Blames 'Mass Suggestion' for Parrot Fever Scare, Which He Holds Baseless." *The New York Times* (Jan. 16): 3.

Anon. 1923. "War on Diabetes." *Time* (Apr. 21): 20.

Anon. 1921. "To Be Record Year in Use of Sugar." *The New York Times* (June 19): 24.

Anon. 1916. "Radical New Method of Treating Diabetes." *The New York Times* (Feb. 13): SM19.

Anon. 1893. "Animal Extracts as Therapeutic Agents." *British Medical Journal* 1, no. 1,694 (June 17): 1279.

Anon. 1873. *One Hundred Years' Progress of the United States.* Hartford, CT: L. Stebbins.

Apfelbaum, M., ed. 1973. *Régulation de l'Équilibre Énergetique Chez l'Homme. Energy Balance in Man.* Paris: Masson et Cie.

Appel, K. E., C. B. Farr, and H. K. Marshall. 1929. "Insulin in Undernutrition in the Psychoses." *Archives of Neurology and Psychiatry* 21, no. 1 (Jan.): 149–64.

Arky, R. A. 1971. "Treating, Eating and Impeding." *The New England Journal of Medicine* 284, no. 10 (Mar. 11): 553–54.

Asano, T., and R. S. Mcleod. 2002. "Dietary Fibre for the Prevention of Colorectal Adenomas and Carcinomas." *Cochrane Database of Systematic Reviews.* 2: CD003430.

Aschoff, L. 1924. *Lectures on Pathology (Delivered in the United States. 1924).* New York: Paul B. Hoeber.

Associated Press. 1937. "Obesity Is Linked to Blood Control." *The New York Times* (June 22): 25.

Astrup, A., F. Magkos, D. M. Bier, et al. 2020. "Saturated Fats and Health: A Reassessment and Proposal for Food-Based Recommendations, *JACC* State-of-the-Art Review." *Journal of the American College of Cardiology* 76, no. 7 (Aug. 18): 844–57.

Athinarayanan, S. J., R. N. Adams, S. J. Hallberg, et al. 2019. "Long-Term Effects of a Novel Continuous Remote Care Intervention Including Nutritional Ketosis for the Management of Type 2 Diabetes: A 2-Year Non-randomized Clinical Trial." *Frontiers in Endocrinology* 10 (June 5): 348.

Athinarayanan, S. J., S. J. Hallberg, A. L. McKenzie, et al. 2020. "Impact of a 2-Year Trial of Nutritional Ketosis on Indices of Cardiovascular Disease Risk in Patients with Type 2 Diabetes." *Cardiovascular Diabetology* 19, no. 1 (Dec. 8): 208.

Atkins, R. C. 1972. *Dr. Atkins' Diet Revolution: The High Calorie Way to Stay Thin Forever.* New York: David McKay.

Auerbach, S. 1974. "Roughing It—Tonic for Our Time." *The Washington Post* (Aug. 19): B1.

Bacon, F. 1994. *Novum Organum* [1620]. Ed. and trans. P. Urbach and J. Gibson. Chicago: Open Court.

Banting, F. G. 1925. "Nobel Lecture: Diabetes and Insulin." Online: https://www.nobelprize.org/prizes/medicine/1923/banting/lecture.

Banting, F. G. 1924. "Insulin." In *Diabetes: A Medical Odyssey*. Tuckahoe, NY: USV Pharmaceutical Corp., 1971: 151–61.

Banting, F. G., C. H. Best, J. B. Collip, W. R. Campbell, and A. A. Fletcher. 1922. "Pancreatic Extracts in the Treatment of Diabetes Mellitus." *Canadian Medical Association Journal* 12, no. 3 (Mar.): 141–46.

Banting, F. G., C. H. Best, J. B. Collip, J. J. R. Macleod, and E. C. Noble. 1922. "The Effect of Pancreatic Extract (Insulin) on Normal Rabbits." *American Journal of Physiology* 62, no. 1 (Sept.): 162–176.

Barbour, O. 1924. "The Use of Insulin in Undernourished Non-Diabetic Children." *Archives of Pediatrics* 41, no. 10 (Oct.): 707–12.

Bardossi, F., and J. N. Schwartz. 1984. "Cholesterol-Watching: [Dr. Edward H. Ahrens, Jr.]." *Rockefeller University Research Profiles* 26, no. 18 (Fall): 1–6.

Barnes, R. H., D. R. Drury, P. O. Greeley, and A. N. Wick. 1940. "Utilization of the Ketone Bodies in Normal Animals and in Those with Ketosis." *American Journal of Physiology* 130, no. 1 (June): 144–50.

Barnes, R. H., and E. M. MacKay. 1940. "Influence of Protamine Zinc Insulin upon Appetite During Anorexia of Vitamin B1 Deficiency." *Proceedings of the Society for Experimental Biology and Medicine* 45, no. 3 (Dec.): 759–62.

Barr, D. P., J. R. Brobeck, H. W. Brosin, et al. 1953. *Overeating, Overweight and Obesity.* Nutrition Symposium Series, no. 6. New York: National Vitamin Foundation.

Barr, D. P., E. M. Russ, and H. A. Eder. 1951. "Protein-Lipid Relationships in Human Plasma. II. In atherosclerosis and related conditions." *The American Journal of Medicine* 11, no. 4 (Oct.): 480–93.

Bauer, J. 1940. "Some Conclusions from Observations on Obese Children." *Archives of Pediatrics* 57: 631–40.

Bauer, J. 1917. *Die Konstitutionelle Disposition zu Inneren Krankheiten.* Berlin: Springer.

Beresford, S. A., K. C. Johnson, C. Ritenbaugh, et al. 2006. "Low-Fat Dietary Pattern and Risk of Colorectal Cancer: The Women's Health Initiative Randomized Controlled Dietary Modification Trial." *JAMA* 295, no. 6 (Feb. 8): 643–54.

Bernard, C. 1974. *Phenomena of Life Common to Animals and Vegetables.* Trans. R. P. Cook and M. A. Cook. Dundee: R. P. and M. A. Cook.

Bernard, C. 1957. *An Introduction to the Study of Experimental Medicine.* Trans. H. C. Green. New York: Dover Publications.

Bernard, C. 1878. *Leçons sur les Phénomènes de la Vie Communs aux Animaux et aux Végétaux.* Paris: Librairie J.-B. Baillière et fils.

Bernard, C. 1848. *De l'Origine du Sucre dans l'Économie Animale.* Paris: Rignoux.

Bernstein, R. K. 1997. *Dr. Bernstein's Diabetes Solution, Newly Revised and Updated. The Complete Guide to Achieving Normal Blood Sugars.* New York: Little, Brown.

Bernstein, R. K. 1992. "Effects of Low Insulin and Low Carbohydrate on Frequency and Severity of Hypoglycemia." *The American Journal of Medicine* 92, no. 3 (Mar.): 339–40.

Bernstein, R. K. 1981. *Diabetes: The GlucograF Method for Normalizing Blood Sugar.* Los Angeles: Jeremy P. Tarcher.

Berson, S. A., and R. S. Yalow. 1965. "The Banting Memorial Lecture 1965: Some Current Controversies in Diabetes." *Diabetes* 14, no. 9 (Sept.): 549–72.

Berson, S. A., and R. S. Yalow. 1961. "Plasma Insulin in Health and Disease." *The American Journal of Medicine* 31 (Dec.): 874–81.

Best, C. H., R. E. Haist, and J. H. Ridout. 1939. "Diet and the Insulin Content of Pancreas." *Journal of Physiology* 97, no. 1 (Nov. 14): 107–19.

Bhanpuri, N. H., S. J. Hallberg, P. T. William, et al. 2018. "Cardiovascular Disease Risk Factor Responses to a Type 2 Diabetes Care Model Including Nutritional Ketosis Induced by Sustained Carbohydrate Restriction at 1 Yr: An Open Label, Non-Randomized, Controlled Study. *Cardiovascular Diabetology* 17, no. 1 (May 1): 56.

Bierman, E. L., M. J. Albrink, R. A. Arky, et al. 1971. "Principles of Nutrition and Dietary Recommendations for Patients with Diabetes Mellitus: 1971." *Diabetes* 20, no. 9 (Sept.): 633–34.

Bierman, E. L., and R. Nelson. 1975. "Carbohydrates, Diabetes, and Blood Lipids." In *World Review of Nutrition and Dietetics, volume* 22, ed. G. H. Bourne. Basel: S. Karger, 280–87.

Bishop, J. E. 1982. "Heart Attacks: A Test Collapses." *The Wall Street Journal* (Oct. 6): 32.

Bishop, J. E. 1961. "Helping the Heart: Major Research Effort Started to See if Diet Can Prevent Attacks." *The Wall Street Journal* (Oct. 27): 1.

Bistrian, B. R., G. L. Blackburn, J.-P. Flatt, et al. 1976. "Nitrogen Metabolism and Insulin Requirements in Obese Diabetic Adults on a Protein-Sparing Modified Fast." *Diabetes* 25, no. 6 (June): 494–504.

Blackman, B. 1961. "The Use of Glucagon in Insulin Coma Therapy." *Psychiatric Quarterly* 35, no. 3 (Sept.): 387–89.

Blakeslee, S. 1993. "Doctors Announce Way to Forestall Effect of Diabetes." *The New York Times* (June 14): 1, 12.

Bleicher, S. J. 1980. "Symposium on Home Blood Glucose Monitoring. Chairman's Introduction." *Diabetes Care* 3, no. 1 (Jan.–Feb.): 57.

Bleicher, S. J., T. Y. Lee, R. Bernstein, et al. 1980. "Effect of Blood Glucose Control on Retinal Vascular Permeability in Insulin-Dependent Diabetes Mellitus." *Diabetes Care* 3, no. 1 (Jan.–Feb.): 184–86.

Bliss, M. 2007. *The Discovery of Insulin.* 25th anniversary ed. Chicago: University of Chicago Press.

Blix, G. 1964. *Occurrence, Causes and Prevention of Overnutrition.* Upsala: Almqvist & Wiksell.

Blotner, H. 1938. "Late Results Following the Use of Insulin in One Hundred Cases of Malnutrition." *The New England Journal of Medicine* 218, no. 9 (Mar. 3): 371–74.

Blotner, H. 1933. "Observations on the Effect of Insulin in Thin Persons." *The Journal of the American Medical Association* 100, no. 2 (Jan. 14): 88–92.

Boardman, R. H., J. Lomas, and M. Markowe. 1956. "Insulin and Chlorpromazine in Schizophrenia: A Comparative Study in Previously Untreated Cases." *The Lancet* 271, no. 6941 (Sept. 8): 487–91.

Boffey, P. M. 1987. "Cholesterol: Debate Flares over Wisdom in Widespread Reductions." *The New York Times* (July 14): C1.

Boffey, P. M. 1984. "Study Backs Cutting Cholesterol to Curb Heart Disease Risk." *The New York Times* (Jan. 13): A1.

Bogdonoff, M. D. 1982. "Psychological Aspects of Home Blood Glucose Monitoring." In *Diabetes Management in the '80s*, ed. C. M. Peterson. New York: Praeger Publishers, 168–72.

Bolli, G. B., F. Porcellati, P. Lucidi, and C. G. Fanelli. 2021. "The Physiological Basis of Insulin Therapy in People with Diabetes Mellitus." *Diabetes Research and Clinical Practice* 175 (May): 108839.

Bonfig, W., T. Kapellen, A. Dost, et al. 2012. "Growth in Children and Adolescents with Type 1 Diabetes." *Journal of Pediatrics* 160, no. 6 (June): 900–903.

Bornstein, J., and J. F. Nelson. 1949. "Observations on the Effect of High Fat Diet in Alloxan Diabetic Rats." *Medical Journal of Australia* 1, no. 5 (Jan. 29): 121–26.

Bouchardat, A. 1875. *De la Glycosurie ou Diabète Sucré, son Traitement Hygiénique*. Paris: Librarie Germer Baillière.

Bowen, B. D. 1930. "Economic Results of the Modern Treatment of Diabetes. Status of Patients Treated with Insulin Compared with Their Prediabetic Condition." *The Journal of the American Medical Association* 95, no. 8 (Aug. 23): 565–68.

Bray, G. A., ed. 1976. *Obesity in Perspective*. DHEW pub. no. (NIH) 76-852. Washington, DC: U.S. Government Printing Office.

Brehm, B. J., R. J. Seeley, S. R. Daniels, and D. A. D'Alessio. 2003. "A Randomized Trial Comparing a Very Low Carbohydrate Diet and a Calorie-Restricted Low Fat Diet on Body Weight and Cardiovascular Risk Factors in Healthy Women." *Journal of Clinical Endocrinology and Metabolism* 88, no. 4 (Apr.): 1617–23.

Brillat-Savarin, J. A. 1986. *The Physiology of Taste* [1825]. Trans. M. F. Fisher. San Francisco: North Point Press.

Brody, J. 1970. "Heart Experts Disagree on Desirability of a Diet That Is Low in Saturated Fats." *The New York Times* (Nov. 13): 14.

Brown, A. 2017. *Bright Spots & Landmines. The Diabetes Guide I Wish Someone Had Handed Me*. San Francisco: The diaTribe Foundation.

Brownlee, M. 2005. "The Pathobiology of Diabetic Complications: A Unifying Mechanism." *Diabetes* 54, no. 6 (June): 1615–25.

Brownlee, M., A. Cerami, and H. Vlassara. 1988. "Advanced Glycosylation End Products in Tissue and the Biochemical Basis of Diabetic Complications." *The New England Journal of Medicine* 318, no. 20 (May 19): 1315–21.

Brownlee, M., H. Vlassara, and A. Cerami. 1984. "Nonenzymatic Glycosylation and the Pathogenesis of Diabetic Complications." *Annals of Internal Medicine* 101, no. 4 (Oct.): 527–37.

Brown-Séquard, C.-É. 1893. "On a New Therapeutic Method Consisting in the Use of Organic Liquids Extracted from Glands and Other Organs." *British Medical Journal* 1, no. 1693 (June 3): 1212–14.

Bruger, M., and E. Oppenheim. 1951. "Experimental and Human Atherosclerosis: Possible Relationship and Present Status." *Bulletin of the New York Academy of Medicine* 27, no. 9 (Sept.): 539–59.

Brunzell, J. D., R. I. Lerner, W. R. Hazzard, D. Porte, Jr., and E. L. Bierman. 1971. "Improved Glucose Tolerance with High Carbohydrate Feeding in Mild Diabetes." *The New England Journal of Medicine* 284, no. 10 (Mar. 11): 521–24.

Bucala, R., Z. Makita, T. Koschinsky, A. Cerami, and H. Vlassara. 1993. "Lipid Advanced Glycosylation: Pathway for Lipid Oxidation in Vivo." *Proceedings of the National Academy of Sciences* 90, no. 14 (July 15): 6434–38.

Bunn, H. F. 2013. "Embedded in the Red Cell." *Hematologist* 10, no. 2 (Mar.–Apr.).

Online: https://ashpublications.org/thehematologist/article/doi/10.1182/hem.V10.2 .1072/462555/Embedded-in-the-Red-Cell.

Bunn, H. F., D. N. Haney, S. Kamin, K. H. Gabbay, and P. M. Gallop. 1976. "The Biosynthesis of Human Hemoglobin A1c: Slow Glycosylation of Hemoglobin in Vivo." *The Journal of Clinical Investigation* 57, no. 6 (June): 1652–59.

Bunn, H. F., and P. J. Higgins. 1981. "Reaction of Monosaccharides with Proteins: Possible Evolutionary Significance." *Science* 213, no. 4504 (July 10): 222–24.

Burkitt, D. P. 1991a. In interview with Max Blythe, Gloucestershire, February 26, 1991, Interview III. Royal College of Physicians and Oxford Brookes Medical Sciences Video Archive (MSVA 64).

Burkitt, D. P. 1991b. In interview with Max Blythe, Gloucestershire, October 29, 1991, Interview IV. Royal College of Physicians and Oxford Brookes Medical Sciences Video Archive (MSVA 64).

Burkitt, D. P. 1979. *Eat Right—To Keep Healthy and Enjoy Life More: How Simple Diet Changes Can Prevent Many Common Diseases.* New York: Arco Publishing.

Burkitt, D. P. 1971. "Some Neglected Leads to Cancer Causation." *Journal of the National Cancer Institute* 47, no. 5 (Nov.): 913–19.

Burkitt, D. P. 1970. "Relationship as a Clue to Causation." *The Lancet* 296, no. 7685 (Dec. 12): 1237–40.

Burkitt, D. P., and H. C. Trowell, eds. 1975. *Refined Carbohydrate Foods and Disease: Some Implications of Dietary Fibre.* New York: Academic Press.

Burkitt, D. P., A. R. Walker, and N. S. Painter. 1974. "Dietary Fiber and Disease." *JAMA* 229, no. 8 (Aug. 19): 1068–74.

Burkitt, D. P., A. R. Walker, and N. S. Painter. 1972. "Effect of Dietary Fibre on Stools and the Transit-Times, and Its Role in the Causation of Disease." *The Lancet* 300, no. 7792 (Dec. 30): 1408–12.

Burland, W. L., P. D. Samuel, and J. Yudkin, eds. 1974. *Obesity.* New York: Churchill Livingstone.

Burn, J. H., T. H. C. Lewis, and F. D. Kelsey. 1944. "The Dietary Control of Alloxan Diabetes in Rats." *British Medical Journal* 2, no. 4379 (Dec. 9): 7522–24.

Burros, M. 2000. "Plenty of Reasons to Say, 'Please Pass the Fiber.'" *The New York Times* (Apr. 26): F5.

Burros, M. 1994. "Despite Awareness of Risks, More in U.S. Are Getting Fat." *The New York Times* (July 17): 1.

Burtscher, M., E. Gnaiger, J. Burtscher, W. Nachbauer, and A. Brugger. 2012. "Arnold Durig (1872–1961): Life and Work. An Austrian Pioneer in Exercise and High Altitude Physiology." *High Altitude Medicine and Biology* 13, no. 3 (Sept.): 224–31.

Buse, J. B., S. Caprio, W. T. Cefalu, et al. 2009. "Consensus Statement: How Do We Define Cure of Diabetes?" *Diabetes Care* 32, no. 11 (Nov.): 2133–35.

Butscher, E. 2003. *Sylvia Plath: Method and Madness.* Tucson, AZ: Schaffner Press.

Cabot, R. C. 1912. "Diagnostic Pitfalls Identified During a Study of Three Thousand Autopsies." *The Journal of the American Medical Association* 59, no. 26 (Dec. 28): 2295–98.

Cahill, G. F., Jr. 1985. "Current Concepts of Diabetes." In *Joslin's Diabetes Mellitus,* 11th edition, ed. A. Marble, L. P. Krall, R. F. Bradley, A. R. Christlieb, and J. S. Soeldner. Philadelphia: Lea & Febiger, 1–11.

Cahill, G. F., Jr. 1971. "Physiology of Insulin in Man." *Diabetes* 20, no. 12 (Dec.): 785–99.

Cahill, G. F., Jr., and T. T. Aoki. 1980. "Alternate Fuel Utilization by Brain." In *Cerebral Metabolism and Neural Function,* ed. J. V. Passonneau, R. A. Hawkins, W. D. Lust, and F. A. Welsh. Baltimore: Williams & Wilkins, 234–42.

Cahill, G. F., Jr., T. T. Aoki, and N. B. Ruderman. 1973. "Ketosis." *Transactions of the American Clinical and Climatological Association* 84: 184–202.

Cahill, G. F., Jr., D. D. Etzwiler, and N. Freinkel. 1976. " 'Control' and Diabetes." *The New England Journal of Medicine* 294, no. 18 (Apr. 29): 1004–5.

Campbell, G. D. 1996. "Cleave the Colossus and the History of the 'Saccharine Disease' Concept." *Nutrition and Health* 11, no. 1 (Apr.): 1–11.

Campbell, G. D. 1963. "Diabetes in Asians and Africans in and Around Durban." *South African Medical Journal* 37 (Nov. 30): 1195–1208.

Campbell, G. D. 1961. "Connubial Diabetes and the Possible Role of 'Oral Diabetogens.' " *British Medical Journal* 1, no. 5238 (May 27): 1538–39.

Campbell, W. R. 1962. "Anabasis." *Canadian Medical Association Journal* 87, no. 20 (Nov. 17): 1055–61.

Campbell, W. R. 1946. "The First Clinical Trials of Insulin." *Proceedings of the American Diabetes Association: Annual Meeting* 6: 97–106.

Cannon, W. B. 1945. *The Way of an Investigator: A Scientist's Experiences in Medical Research.* New York: W. W. Norton.

Cannon, W. B. 1939. *The Wisdom of the Body.* New York: W. W. Norton.

Cannon, W. B. 1929. "Organization for Physiological Homeostasis." *Physiological Reviews* 9, no. 3 (July): 399–431.

Cannon, W. B. 1926. "Physiological Regulation of Normal States: Some Tentative Postulates Concerning Biological Homeostatics." In *À Charles Richet: Ses amis, ses collègues, ses élèves*, ed. A. Pettit (in French). Paris: Les Éditions Médicales, 91.

Cantani, A. 1876. *Le Diabète Sucré et son Traitement Diététique.* Trans. H. Charvet. Paris: V. Adrien Delahaye.

Carrington, E. R. 1970. "Diabetes in Pregnancy." In *Diabetes Mellitus: Theory and Practice*, ed. M. Ellenberg and H. Rifkin. New York: McGraw-Hill, 710–23.

Castelli, W. P., J. T. Doyle, T. Gordon, et al. 1977. "HDL Cholesterol and Other Lipids in Coronary Heart Disease. The Cooperative Lipoprotein Phenotyping Study." *Circulation* 55, no. 5 (May): 767–72.

Centers for Disease Control and Prevention (CDC). 2022. "National Diabetes Statistics Report." Online: https://www.cdc.gov/diabetes/data/statistics-report/index.html.

Centers for Disease Control and Prevention (CDC). 2021. "Diabetes Basics." Online: https://www.cdc.gov/diabetes/basics/diabetes.html.

Centers for Disease Control and Prevention (CDC). 2000. "Achievements in Public Health, 1900–1999: Changes in the Public Health System." *JAMA* 283, no. 6 (Feb. 9): 735–38.

Centers for Disease Control and Prevention (CDC). 1982. "Deaths Among Patients Using Continuous Subcutaneous Insulin Infusion Pumps—United States." *Morbidity and Mortality Weekly Report* 31, no. 46 (Nov. 26): 625–26. Online: https://www.cdc.gov/mmwr/preview/mmwrhtml/00001195.htm.

Centers for Disease Control and Prevention (CDC) Division of Diabetes Translation. 2017. "Long-Term Trends in Diabetes." Online: https://www.cdc.gov/diabetes/statistics/slides/long_term_trends.pdf.

Central Committee for Medical and Community Program of the American Heart Association. 1961. "Dietary Fat and Its Relation to Heart Attacks and Strokes." *JAMA* 175, no. 5 (Feb. 4): 389–91.

Cerami, A., R. Koenig, and C. M. Peterson. 1978. "Annotation: Haemoglobin A1C and Diabetes Mellitus." *British Journal of Haematology* 38, no. 1 (Jan.): 1–4.

Cerami, A., H. Vlassara, and M. Brownlee. 1985. "Protein Glycosylation and the Pathogenesis of Atherosclerosis." *Metabolism* 34, no. 12, suppl. 1 (Dec.): 37–44.

Chalmers, K. H. 2005. "Medical Nutrition Therapy." In *Joslin's Diabetes Mellitus*, 14th edition, ed. C. R. Kahn, G. C. Weir, G. L. King, A. M. Jacobson, A. C. Moses, and R. J. Smith. Philadelphia: Lippincott Williams & Wilkins, 611–32.

Chamberlin, T. C. 1890. "The Method of Multiple Working Hypotheses." *Science* 15, no. 366 (Feb. 7): 92–96.

Chambers, T. K. 1875. *A Manual of Diet in Health and Disease.* Philadelphia: Henry C. Lea.

Chambers, T. K. 1866. *The Renewal of Life. Lectures, Chiefly Clinical.* 2nd American edition. Philadelphia: Lindsay & Blakiston.

Chan, M. 2017. "Obesity and Diabetes: The Slow-Motion Disaster." *The Millbank Quarterly* 95, no. 1 (Mar.): 11–14.

Charles, R. H., R. K. Chunder Bose, C. L. Bose, et al. 1907. "Discussion on Diabetes in the Tropics." *British Medical Journal* 2, no. 2442 (Oct. 19): 1051–64.

Cheng, A. Y. Y., and B. Zinman. 2005. "Principles of Insulin Therapy." In *Joslin's Diabetes Mellitus,* 14th edition, ed. C. R. Kahn, G. C. Weir, G. L. King, A. M. Jacobson, A. C. Moses, and R. J. Smith. Philadelphia: Lippincott Williams & Wilkins, 659–70.

Christakis, G., S. H. Rinzler, M. Archer, and A. Kraus. 1966. "Effect of the Anti-Coronary Club Program on Coronary Heart Disease. Risk-Factor Status." *JAMA* 198, no. 6 (Nov. 7): 597–604.

Christakis, G., S. H. Rinzler, M. Archer, et al. 1966. "The Anti-Coronary Club. A Dietary Approach to the Prevention of Coronary Heart Disease—A Seven-Year Report." *American Journal of Public Health and the Nation's Health* 56, no. 2 (Feb.): 299–314.

Clarkson, T. B., and H. B. Lofland. 1961. "Effects of Cholesterol-Fat Diets on Pigeons Susceptible and Resistant to Atherosclerosis." *Circulation Research* 9 (Jan.): 106–9.

Cleave, T. L. 1962. *Peptic Ulcer, a New Approach to Its Causation, Prevention, and Arrest, Based on Human Evolution.* Bristol, U.K.: John Wright & Sons.

Cleave, T. L. 1956. "The Neglect of Natural Principles in Current Medical Practice." *Journal of the Royal Naval Medical Service* 42, no. 2 (Spring): 55–82.

Cleave, T. L. 1940. "Instincts and Diet." *The Lancet* 235, no. 6087 (Apr. 27): 809.

Cleave, T. L., and G. D. Campbell. 1966. *Diabetes, Coronary Thrombosis, and the Saccharine Disease.* Bristol, U.K.: John Wright & Sons.

Cohen, A. M. 1963a. "Effect of Environmental Changes on Prevalence of Diabetes and of Atherosclerosis in Various Ethnic Groups in Israel." In *The Genetics of Migrant and Isolate Populations,* ed. E. Goldschmidt. New York: Williams & Wilkins, 127–30.

Cohen, A. M. 1963b. "Fats and Carbohydrates as Factors in Atherosclerosis and Diabetes in Yemenite Jews." *American Heart Journal* 65, no. 3 (Mar.): 291–93.

Cohen, A. M., S. Bavly, and R. Poznanski. 1961. "Change of Diet of Yemenite Jews in Relation to Diabetes and Ischaemic Heart-Disease." *The Lancet* 278, no. 7217 (Dec. 23): 1399–1401.

Cohen, A. M., N. A. Kaufmann, R. Poznanski, S. H. Blondheim, and Y. Stein. 1966. "Effect of Starch and Sucrose on Carbohydrate-Induced Hyperlipaemia." *British Medical Journal* 1, no. 5483 (Feb. 5): 339–40.

Cohen, A. M., and E. Shafrir. 1965. "Carbohydrate Metabolism in Myocardial Infarction, Behavior of Blood Glucose and Free Fatty Acids After Glucose Loading." *Diabetes* 14, no. 2 (Feb.): 84–86.

Cohn, E. J., F. R. N. Gurd, D. M. Surgenor, et al. 1950. "A System for the Separation of the Components of Human Blood: Quantitative Procedures for the Separation of the Protein Components of Human Plasma." *Journal of the American Chemical Society* 72, no. 1 (Jan. 1): 465–74.

Collens, W. S. 1954. "Regulated Versus Free Diet in the Treatment of Diabetes Mellitus." *Journal of Clinical Nutrition* 2, no. 3 (May–June): 195–202.

Colwell, A. R., Sr. 1970. "Clinical Use of Insulin." In *Diabetes Mellitus: Theory and Practice*, ed. M. Ellenberg and H. Rifkin. New York: McGraw-Hill, 624–37.

Committee on Medical Aspects (COMA) of Food Policy. 1989. *Report on Health and Social Subjects: No. 37, Dietary Sugars and Human Disease*. London: Her Majesty's Stationery Office.

Consensus Conference. 1985. "Lowering Blood Cholesterol to Prevent Heart Disease." *JAMA* 253, no. 14 (Apr. 12): 2080–86.

Coulston, A. M., C. B. Hollenbeck, A. L. Swislocki, Y. D. Chen, and G. M. Reaven. 1987. "Deleterious Metabolic Effects of High-Carbohydrate, Sucrose-Containing Diets in Patients with Non-Insulin-Dependent Diabetes Mellitus." *The American Journal of Medicine* 82, no. 2 (Feb.): 213–20.

Coulston, A. M., C. B. Hollenbeck, A. L. Swislocki, and G. M. Reaven. 1989. "Persistence of Hypertriglyceridemic Effect of Low-Fat High-Carbohydrate Diets in NIDDM Patients." *Diabetes Care* 12, no. 2 (Feb.): 94–101.

Cox, C. 2009. *The Fight to Survive. A Young Girl, Diabetes, and the Discovery of Insulin*. New York: Kaplan Publishing.

Crapo, P. A. 1994. "Dietary Management." In *Joslin's Diabetes Mellitus*, 13th edition, ed. C. R. Kahn and G. C. Weir. Philadelphia: Lea & Febiger, 415–30.

Cryer, P. E., C. Binder, G. B. Bolli, et al. 1989. "Hypoglycemia in IDDM." *Diabetes* 38, no. 9 (Sept.): 1193–99.

Cummings, J. H., and A. Engineer. 2018. "Denis Burkitt and the Origins of the Dietary Fibre Hypothesis." *Nutrition Research Reviews* 31, no. 1 (June): 1–15.

Dally, P., and W. Sargant. 1966. "Treatment and Outcome of Anorexia Nervosa." *British Medical Journal* 2, no. 5517 (Oct. 1): 793–95.

Danowski, T. S., and J. H. Sunder. 1978. "Jet Injection of Insulin During Self-Monitoring of Blood Glucose." *Diabetes Care* 1, no. 1 (Jan.–Feb.): 27–33.

Davenport, H. W. 1999. "Not Just Any Medical School: The Science, Practice, and Teaching of Medicine at the University of Michigan 1850–1941." Ann Arbor: University of Michigan Press.

Davidson, P. C., and M. J. Albrink. 1965. "Insulin Resistance in Hyperglyceridemia." *Metabolism* 14, no. 10 (Oct.): 1059–70.

Davies, L. E. 1950. "$4,000,000 Is Raised in Heart Campaign." *The New York Times* (June 25): 37.

Dawber, T. R. 1980. *The Framingham Study: The Epidemiology of Atherosclerotic Disease*. Cambridge, MA: Harvard University Press.

Dawber, T. R. 1978. "Annual Discourse—Unproved Hypotheses." *The New England Journal of Medicine* 299, no. 9 (Aug. 31): 452–58.

Dayton, S. D., M. L. Pearce, S. Hashimoto, W. J. Dixon, and U. Tomiyasu. 1969. "A Controlled Clinical Trial of a Diet High in Unsaturated Fat in Preventing Complications of Atherosclerosis." *Circulation* 40, no. 1s2 (July): II-1–II-63.

de Duve, C. 2004. "My Love Affair with Insulin." *Journal of Biological Chemistry* 279, no. 21 (May 21): 1679–88.

DeFronzo, R. A. 2009. "Banting Lecture. From the Triumvirate to the Ominous Octet: A New Paradigm for the Treatment of Type 2 Diabetes Mellitus." *Diabetes* 58, no. 4 (Apr.): 773–95.

DeFronzo, R. A. 1981. "The Effect of Insulin on Renal Sodium Metabolism. A Review with Clinical Implications." *Diabetologia* 21, no. 3 (Sept.): 165–71.

Diabetes Control and Complications Trial Research Group (DCCT). 1993a. "The Effect of Intensive Treatment of Diabetes on the Development and Progression of Long-Term Complications in Insulin-Dependent Diabetes Mellitus." *The New England Journal of Medicine* 329, no. 14 (Sept. 30): 977–86.

Diabetes Control and Complications Trial Research Group (DCCT). 1993b. "Nutrition Interventions for Intensive Therapy in the Diabetes Control and Complications Trial." *Journal of the American Dietetic Association* 93, no. 7 (July): 768–72.

Diabetes Control and Complications Trial Research Group (DCCT). 1991. "Epidemiology of Severe Hypoglycemia in the Diabetes Control and Complications Trial." *The American Journal of Medicine* 90, no. 4 (Apr.): 450–59.

Diabetes Control and Complications Trial Research Group (DCCT). 1988. "Weight Gain Associated with Intensive Therapy in the Diabetes Control and Complications Trial." *Diabetes Care* 11, no. 7 (July–Aug.): 567–73.

Diabetes Control and Complications Trial Research Group (DCCT). 1986. "The Diabetes Control and Complications Trial (DCCT): Design and Methodologic Considerations for the Feasibility Phase." *Diabetes* 35, no. 5 (May): 530–45.

Di Pino, A., and R. A. DeFronzo. 2019. "Insulin Resistance and Atherosclerosis: Implications for Insulin-Sensitizing Agents." *Endocrine Reviews* 40, no. 6 (Dec.): 1447–67.

Dole, V. P. 1956. "A Relation Between Non-Esterified Fatty Acids in Plasma and the Metabolism of Glucose." *The Journal of Clinical Investigation* 35, no. 2 (Feb.): 150–54.

Doll, R. 1966. "Foreword." In Cleave, T. L., and G. D. Campbell, *Diabetes, Coronary Thrombosis, and the Saccharine Disease*. Bristol, U.K.: John Wright & Sons, xi.

Donaldson, B. F. 1962. *Strong Medicine*. Garden City, NY: Doubleday.

Donkin, S. A. 1871. *The Skim-Milk Treatment of Diabetes and Bright's Disease*. London: Longmans, Green.

Doroshow, D. B. 2007. "Performing a Cure for Schizophrenia: Insulin Coma Therapy on the Wards." *Journal of the History of Medicine and Allied Sciences* 62, no. 2 (Apr.): 213–43.

Drenick, E. J. 1976. "Weight Reduction by Prolonged Fasting." In *Obesity in Perspective*, DHEW pub. no. (NIH) 76-852, ed. G. A. Bray. Washington DC: U.S. Government Printing Office, 341–60.

Drenick, E. J., L. C. Alvarez, G. C. Tamasi, and A. S. Brickman. 1972. "Resistance to Symptomatic Insulin Reactions After Fasting." *The Journal of Clinical Investigation* 51, no. 10 (Oct.): 2757–62.

Dreyfuss, F. 1953. "The Incidence of Myocardial Infarctions in Various Communities in Israel." *American Heart Journal* 45, no. 5 (May): 749–55.

Duckworth, W., C. Abraira, T. Moritz, et al. 2009. "Glucose Control and Vascular Complications in Veterans with Type 2 Diabetes." *The New England Journal of Medicine* 360, no. 2 (Jan. 8): 129–39.

Duff, G. L., D. J. Brechin, and W. E. Finkelstein. 1954. "The Effect of Alloxan Diabetes on Experimental Cholesterol Atherosclerosis in the Rabbit. IV. The Effect of Insulin Therapy on the Inhibition of Atherosclerosis in the Alloxan-Diabetic Rabbit." *Journal of Experimental Medicine* 100, no. 4 (Oct.): 371–80.

Duncan, A., Sr., and A. Duncan, Jr. 1798. "Account of Two Cases of Diabetes Mellitus, with Remarks." *Annals of Medicine for the Year 1797*, vol. 2. Edinburgh: G. Mudie & Son, 85–105.

Duncan, G. G. 1935. *Diabetes Mellitus and Obesity*. Philadelphia: Lea & Febiger.

Dunlop, D. 1966. "The Saccharine Disease." *British Medical Journal* 2, no. 5506 (July 16): 163.

Dunlop, D. M. 1954. "Are Diabetic Degenerative Complications Preventable?" *British Medical Journal* 2, no. 4884 (Aug. 14): 383–85.

Eaton, R. P., R. C. Allen, D. S. Schade, and J. C. Standefer. 1980. " 'Normal' " Insulin Secretion: The Goal of Artificial Insulin Delivery Systems?" *Diabetes Care* 3, no. 2 (Mar.–Apr.): 270–73.

Eddy, D. M. 2011. "The Origins of Evidence-Based Medicine—A Personal Perspective." *Virtual Mentor* 13, no. 1 (Jan.): 55–60.

Eddy, D. M. 1990. "The Challenge." *JAMA* 263, no. 2 (Jan. 12): 287–90.

Eddy, D. M. 1984. "Variations in Physician Practice: The Role of Uncertainty." *Health Affairs* 3, no. 2 (Summer): 74–89.

Eeg-Olofsson, R. 1930. "Local Changes in the Subcutaneous Tissue Following Injections of Insulin." *Acta Medica Scandinavica* 73, no. 1 (Jan.–Dec.): 89–98.

Ellwood, P. M. 1988. "Shattuck lecture—Outcomes management. A technology of patient experience." *The New England Journal of Medicine* 318, no. 23 (June 9): 1549–56.

Embleton, D. 1938. "Dietetic Treatment of Diabetes Mellitus with Special Reference to High Blood-Pressure." *Proceedings of the Royal Society of Medicine* 31, no. 10 (Aug.): 1183–1204.

Emerson, H., and L. D. Larimore. 1924. "Diabetes Mellitus—A Contribution to Its Epidemiology Based Chiefly on Mortality Statistics." *Archives of Internal Medicine* 34, no. 5 (Nov.): 585–630.

Ernest, I., E. Linnér, and A. Svanborg. 1965. "Carbohydrate-Rich, Fat-Poor Diet in Diabetes." *The American Journal of Medicine* 39, no. 4 (Oct.): 594–600.

Evert, A. B., M. Dennison, C. D. Gardner, et al. 2019. "Nutrition Therapy for Adults with Diabetes or Prediabetes: A Consensus Report." *Diabetes Care* 42, no. 5 (Apr. 15): 731–54.

Evidence-Based Medicine (EBM) Working Group. 1992. "Evidence-Based Medicine, a New Approach to Teaching the Practice of Medicine." *JAMA* 268, no. 17 (Nov. 4): 2420–25.

Falta, W. 1925. "Ueber Mastkuren mit Insulin und über Insuläre Fettsucht." *Wiener Klinische Wochenschrift* 38 (July 2): 757–58.

Falta, W. 1913. *The Ductless Glandular Diseases.* Philadelphia: P. Blakiston's Son.

Falta, W. 1909. "The Therapy of Diabetes Mellitus." *Archives of Internal Medicine* 3, no. 2 (Mar.): 159–74.

Farquhar, J. W., A. Frank, R. C. Gross, and G. M. Reaven. 1966. "Glucose, Insulin and Triglyceride Responses to High and Low Carbohydrate Diets in Man." *The Journal of Clinical Investigation* 45, no. 10 (Oct.): 1648–56.

Feinman, R. D., W. K. Pogozelski, A. Astrup, et al. 2015. "Dietary Carbohydrate Restriction as the First Approach in Diabetes Management: Critical Review and Evidence Base." *Nutrition* 31, no. 1 (Jan.): 1–13.

Feinstein, A. R. 1988. "Scientific Standards in Epidemiologic Studies of the Menace of Daily Life." *Science* 242, no. 4883 (Dec. 2): 1257–63.

Feinstein, A. R., and R. I. Horwitz. 1982. "Double Standards, Scientific Methods, and Epidemiologic Research." *The New England Journal of Medicine* 307, no. 26 (Dec. 23): 1611–17.

Feudtner, C. 2003. *Bittersweet: Diabetes, Insulin, and the Transformation of Illness.* Chapel Hill: University of North Carolina Press.

Feynman, R. P. 1985. *Surely You're Joking, Mr. Feynman: Adventures of a Curious Character.* New York: W. W. Norton.

Fink, M., R. Shaw, G. E. Gross, and F. S. Coleman. 1958. "Comparative Study of Chlorpromazine and Insulin Coma in Therapy of Psychosis." *The Journal of the American Medical Association* 166, no. 15 (Apr. 12): 1846–50.

Fischer, L., and J. L. Rogatz. 1926. "Insulin in Malnutrition." *American Journal of Diseases of Children* 31, no. 3 (Mar.): 363–72.

Fisher, B., E. Montague, C. Redmond, et al. 1977. "Comparison of Radical Mastectomy with Alternative Treatments for Primary Breast Cancer. A First Report of Results from a Prospective Randomized Clinical Trial." *Cancer* 39, suppl. 6 (June): 2827–39.

Fitz, R. H., and E. P. Joslin. 1898. "Diabetes Mellitus at the Massachusetts General Hospital from 1824 to 1898. A Study of the Medical Records." *The Journal of the American Medical Association* 31, no. 4 (July 23): 165–71.

Fletcher, A. A., and W. R. Campbell. 1922. "The Blood Sugar Following Insulin Administration and the Symptom Complex—Hypoglycemia." *Journal of Metabolic Research* 2, nos. 5–6 (Nov.–Dec.): 637–49.

Fletcher, C. 1980. "One Way of Coping with Diabetes." *British Medical Journal* 1, no. 6222 (Apr. 26): 1115–16.

Flier, J. S. 1994. "Obesity." In *Joslin's Diabetes Mellitus*, 13th edition, ed. C. R. Kahn and G. C. Weir. Philadelphia: Lea & Febiger, 351–62.

Flood, T. M., B. N. Halford, R. Cooppan, and A. Marble. 1985. "Dietary Management of Diabetes." In *Joslin's Diabetes Mellitus*, 12th edition, ed. A. P. Marble, L. P. Krall, R. F. Bradley, A. R. Christlieb, and J. S. Soeldner. Philadelphia: Lea & Febiger, 357–72.

Forsyth, C. C., T. W. G. Kinnear, and D. M. Dunlop. 1951. "Diet in Diabetes." *British Medical Journal* 1, no. 4715 (May 19): 1095–101.

Foster, G. D., H. R. Wyatt, J. O. Hill, et al. 2003. "A Randomized Trial of a Low-Carbohydrate Diet for Obesity." *The New England Journal of Medicine* 348, no. 21 (May 22): 2082–90.

Foster, N. B. 1923. "The Treatment of Diabetic Coma with Insulin." *The American Journal of the Medical Sciences* 166, no. 5 (Nov.): 699.

Foster, N. C., R. W. Beck, K. M. Miller, et al. 2019. "State of Type 1 Diabetes Management and Outcomes from the T1D Exchange in 2016–2018." *Diabetes Technology and Therapeutics* 21, no. 2 (Feb.): 66–72.

Frantz, I. D., Jr., E. A. Dawson, P. L. Ashman, et al. 1989. "Test of Effect of Lipid Lowering by Diet on Cardiovascular Risk. The Minnesota Coronary Survey." *Arteriosclerosis* 9, no. 1 (Jan.–Feb.): 129–35.

Frantz, I. D., Jr., E. A. Dawson, K. Kuba, et al. 1975. "The Minnesota Coronary Survey: Effect of Diet on Cardiovascular Events and Deaths." *American Heart Association Scientific Proceedings* 51 and 52, suppl. 2 (Oct.): II-4.

Fraser, L. 2016. "Cloning Insulin." Online: https://www.gene.com/stories/cloning-insulin.

Fredrickson, D. S., R. I. Levy, and R. S. Lees. 1967a. "Fat Transport in Lipoproteins—An Integrated Approach to Mechanisms and Disorders." *The New England Journal of Medicine* 276, no. 5 (Feb. 2): 273–81.

Fredrickson, D. S., R. I. Levy, and R. S. Lees. 1967b. "Fat Transport in Lipoproteins—An Integrated Approach to Mechanisms and Disorders." *The New England Journal of Medicine* 276, no. 4 (Jan. 26): 215–25.

Fredrickson, D. S., R. I. Levy, and R. S. Lees. 1967c. "Fat Transport in Lipoproteins—An Integrated Approach to Mechanisms and Disorders." *The New England Journal of Medicine* 276, no. 3 (Jan. 19): 148–56.

Fredrickson, D. S., R. I. Levy, and R. S. Lees. 1967d. "Fat Transport in Lipoproteins—An Integrated Approach to Mechanisms and Disorders." *The New England Journal of Medicine* 276, no. 2 (Jan. 12): 94–103.

Fredrickson, D. S., R. I. Levy, and R. S. Lees. 1967e. "Fat Transport in Lipoproteins—An Integrated Approach to Mechanisms and Disorders." *The New England Journal of Medicine* 276, no. 1 (Jan. 5): 34–42.

Gale, E. A. M. 2013. "Commentary: The Hedgehog and the Fox: Sir Harold Himsworth (1905–93)." *International Journal of Epidemiology* 42, no. 6 (Dec.): 1602–7.

Gale, E. A. M. 2009. "How to Survive Diabetes." *Diabetologia* 52, no. 4 (Apr.): 559–67.

Galloway, J. A., C. T. Spradlin, R. L. Nelson, et al. 1981. "Factors Influencing the Absorption, Serum Insulin Concentration, and Blood Glucose Responses After Injections of Regular Insulin and Various Insulin Mixtures." *Diabetes Care* 4, no. 3 (May–June): 366–76.

Galton, L. 1976. *The Truth About Fiber in Your Food.* New York: Crown.

Ganda, O. P. 1985. "Pathogenesis of Macrovascular Disease Including the Influence of Lipids." In *Joslin's Diabetes Mellitus*, 12th edition, ed. A. Marble, L. P. Krall, R. F. Bradley, A. R. Christlieb, and J. S. Soeldner. Philadelphia: Lea & Febiger, 217–50.

Gardner, H. W. 1901. "The Dietetic Value of Sugar." *British Medical Journal* 1, no. 2104 (Apr. 27): 1010–13.

Garg, A., J. P. Bantle, R. R. Henry, et al. 1994. "Effects of Varying Carbohydrate Content of Diet in Patients with Non-Insulin-Dependent Diabetes Mellitus." *JAMA* 271, no. 18 (May 11): 1421–28.

Garg, A., A. Bonanome, S. M. Grundy, Z.-J. Zhang, and R. H. Unger. 1988. "Comparison of a High-Carbohydrate Diet with a High-Monounsaturated-Fat Diet in Patients with Non-Insulin-Dependent Diabetes Mellitus." *The New England Journal of Medicine* 319, no. 3 (Sept. 29): 829–34.

Garg, A., S. M. Grundy, and M. Koffler. 1992. "Effect of High Carbohydrate Intake on Hyperglycemia, Islet Function, and Plasma Lipoproteins in NIDDM." *Diabetes Care* 15, no. 11 (Nov.): 1572–80.

Garg, A., S. M. Grundy, and R. H. Unger. 1992. "Comparison of Effects of High and Low Carbohydrate Diets on Plasma Lipoproteins and Insulin Sensitivity in Patients with Mild NIDDM." *Diabetes* 41, no. 10 (Oct.): 1278–85.

Gebel, E. 2012. "The Start of Something Good: The Discovery of HbA1c and the American Diabetes Association Samuel Rahbar Outstanding Discovery Award." *Diabetes Care* 35, no. 12 (Dec.): 2429–31.

Gerich, J. E. 2000. "Physiology of Glucose Homeostasis." *Diabetes, Obesity and Metabolism* 2, no. 6 (Dec.): 345–50.

Geyelin, H. R. 1935. "The Treatment of Diabetes with Insulin (After Ten Years)." *The Journal of the American Medical Association* 104, no. 14 (Apr. 6): 1203–8.

Geyelin, H. R. 1934. "The Treatment of Diabetes with Diets Normal in Carbohydrate and Low in Fat." *Bulletin of the New York Academy of Medicine* 10, no. 6 (June): 369–76.

Geyelin, H. R. 1926. "Recent Studies of Diabetes in Children." *Atlantic Medical Journal* 29, no. 12 (Sept.): 825–35.

Geyelin, H. R., G. Harrop, M. F. Murray, and E. Corwin. 1922. "The Use of Insulin in Juvenile Diabetes." *Journal of Metabolic Research* 2, nos. 5–6 (Nov.–Dec.): 767–91.

Gladwell, M. 1998. "The Pima Paradox." *The New Yorker* (Feb. 2): 44–57.

Glinsmann, W. H., H. Irausquin, and Y. K. Park. 1986. *Report from FDA's Sugars Task Force, 1986: Evaluation of Health Aspects of Sugars Contained in Carbohydrate Sweeteners.* Washington, DC: Food and Drug Administration.

Glueck, C. J., F. Mattson, and E. L. Bierman. 1978. "Diet and Coronary Heart Disease: Another View." *The New England Journal of Medicine* 298, no. 26 (June 29): 1471–74.

Gofman, J. W. 1958. "Diet in the Prevention and Treatment of Myocardial Infarction." *American Journal of Cardiology* 1, no. 2 (Feb.): 271–83.

Gofman, J. W., F. P. Lindgren, H. Elliott, et al. 1950. "The Role of Lipids and Lipoproteins in Atherosclerosis." *Science* 111, no. 2877 (Feb. 17): 166–86.

Goldstein, J. L., W. R. Hazzard, H. G. Schrott, E. L. Bierman, and A. G. Motulsky. 1973. "Hyperlipidemia in Coronary Heart Disease. I. Lipid Levels in 500 Survivors of Myocardial Infarction." *The Journal of Clinical Investigation* 52, no. 7 (July): 1533–43.

Gonzalez, J. S., K. K. Hood, S. A. Esbitt, et al. 2018. "Psychiatric and Psychosocial Issues Among Individuals Living with Diabetes." In *Diabetes in America*, 3rd edition. Bethesda, MD: National Institutes of Health, 33-1-33-34.

Gordon, A. H., C. L. Connor, and I. M. Rabinowitch. 1928. "An Unusual Case of Diabetes

Mellitus: Death After Thirteen Years' Observation; Necropsy." *The American Journal of the Medical Sciences* 175, no. 1 (Jan.): 22–31.

Gordon, E. S., M. Goldberg, and G. J. Chosy. 1963. "A New Concept in the Treatment of Obesity." *JAMA* 186, no. 1 (Oct. 5): 50–60.

Gordon, R. S., Jr., and A. Cherkes. 1956. "Unesterified Fatty Acid in Human Blood Plasma." *The Journal of Clinical Investigation* 35, no. 2 (Feb.): 206–12.

Gordon, T. 1988. "The Diet-Heart Idea: Outline of a History." *American Journal of Epidemiology* 127, no. 2 (Feb.): 220–25.

Gordon, T., W. P. Castelli, M. C. Hjortland, W. B. Kannel, and T. R. Dawber. 1977. "High Density Lipoprotein as a Protective Factor Against Coronary Heart Disease. The Framingham Study." *The American Journal of Medicine* 62, no. 5 (May): 707–14.

Grafe, E. 1933. *Metabolic Diseases and Their Treatment.* Trans. M. G. Boise. Philadelphia: Lea & Febiger.

Grande, F. 1975. "Sugar and Cardiovascular Disease." In *World Review of Nutrition and Dietetics, volume 22,* ed. G. H. Bourne. Basel: S. Karger, 248–69.

Gray, P. A., and W. D. Sansum. 1933. "The Higher Carbohydrate Diet Method in Diabetes Mellitus. Analysis of One Thousand and Five Cases." *The Journal of the American Medical Association* 100, no. 20 (May 20): 1580–84.

Greco, J. B., A. O. Lima, and J. R. Cançado. 1942. "The Treatment of Underweight with Insulin." *The American Journal of the Medical Sciences* 204, no. 2 (Aug.): 258–61.

Grundy, S. M. 2004. "Richard Havel, Howard Eder, and the Evolution of Lipoprotein Analysis." *The Journal of Clinical Investigation* 114, no. 8 (Oct.): 1034–37.

Grundy, S. M., H. B. Brewer, Jr., J. I. Cleeman, S. C. Smith, Jr., and C. Lenfant. 2004. "Definition of Metabolic Syndrome: Report of the National Heart, Lung, and Blood Institute/American Heart Association Conference on Scientific Issues Related to Definition." *Circulation* 109, no. 3 (Jan. 27): 433–38.

Grundy, S. M., B. Hansen, S. C. Smith, Jr., J. I. Cleeman, and R. A. Kahn. 2004. "Clinical Management of Metabolic Syndrome: Report of the American Heart Association /National Heart, Lung, and Blood Institute/American Diabetes Association Conference on Scientific Issues Related to Management." *Circulation* 109, no. 4 (Feb. 3): 551–56.

Haist, R. E., and C. H. Best. 1966. "Carbohydrate Metabolism and Insulin." In *The Physiological Basis of Medical Practice,* 8th edition, ed. C. H. Best and N. M. Taylor. Baltimore: Williams & Wilkins, 1329–67.

Haist, R. E., J. Campbell, and C. H. Best. 1940. "The Prevention of Diabetes." *The New England Journal of Medicine* 223, no. 16 (Oct. 17): 607–15.

Hallberg, S. J., A. L. McKenzie, P. T. Williams, et al. 2018. "Effectiveness and Safety of a Novel Care Model for the Management of Type 2 Diabetes at 1 Year: An Open-Label, Non-Randomized, Controlled Study." *Diabetes Therapy* 9, no. 2 (Apr.): 583–612.

Hampl, S. E., S. G. Hassink, A. C. Skinner, et al. 2023. "Clinical Practice Guideline for the Evaluation and Treatment of Children and Adolescents with Obesity." *Pediatrics* 151, no. 2 (Feb.): e2022060640.

Harden, V. A. Undated. "A Short History of the National Institutes of Health." Online at https://history.nih.gov/display/history/WWII+Research+and+the+Grants+Program.

Harley, G. 1866. *Diabetes: Its Various Forms and Different Treatments.* London: Walton & Maberly.

Hartmann, B., S. Lanzinger, P. Bramlage, et al. 2021. "Lean Diabetes in Middle-Aged Adults: A Joint Analysis of the German DIVE and DPV Registries." *PLOS ONE* 12, no. 8 (Aug. 21): 1–14.

Hausberger, F. X., S. W. Milstein, and R. J. Rutman. 1954. "The Influence of Insulin on Glucose Utilization in Adipose and Hepatic Tissues in Vitro." *Journal of Biological Chemistry* 208, no. 1 (May): 431–38.

Heller, S. R. 2019. "Robert Tattersall, a Diabetes Physician Ahead of His Time." *Diabetes Care* 42, no. 6 (May): 1005–8.

Henry, R. R., B. Gumbiner, T. Ditzler, et al. 1993. "Intensive Conventional Insulin Therapy for Type II Diabetes. Metabolic Effects During a 6-Mo Outpatient Trial." *Diabetes Care* 16, no. 1 (Jan.): 21–31.

Higgins, H. L. 1916. "The Rapidity with Which Alcohol and Some Sugars May Serve as a Nutriment." *American Journal of Physiology* 41, no. 2 (Aug. 1): 258–65.

Hill-Briggs, F., N. E. Adler, S. A. Berkowitz, et al. 2021. "Social Determinants of Health and Diabetes: A Scientific Review." *Diabetes Care* 44, no. 1 (Jan.): 258–79.

Hillebrand, S. S., ed. 1974. *Is the Risk of Becoming Diabetic Affected by Sugar Consumption: Proceedings of the Eighth International Sugar Research Symposium.* Washington, DC: International Sugar Research Foundation.

Himsworth, H. P. 1949. "The Syndrome of Diabetes Mellitus and Its Causes." *The Lancet* 253, no. 6551 (Mar. 19): 465–73.

Himsworth, H. P. 1936. "Diabetes Mellitus: Its Differentiation into Insulin-Sensitive and Insulin-Insensitive Types." *The Lancet* 277, no. 5864 (Jan. 18): 127–30.

Himsworth, H. P. 1935. "Diet and the Incidence of Diabetes Mellitus." *Clinical Science,* no. 2: 117–48.

Himsworth, H. P. 1931. "Recent Advances in the Treatment of Diabetes." *The Lancet* 218, no. 5644 (Oct. 31): 978–79.

Himsworth, H. P., and E. M. Marshall. 1935. "The Diet of Diabetics Prior to the Onset of the Disease." *Clinical Science,* no. 2: 95–115.

Hoeg, J. M., and A. N. Klimov. 1993. "Cholesterol and Atherosclerosis: 'The New Is the Old Rediscovered.'" *American Journal of Cardiology* 7, no. 14 (Nov.): 1071–72.

Hoffman, W. 1979. "Meet Monsieur Cholesterol." University of Minnesota *Update* (Winter). Online: https://mbbnet.ahc.umn.edu/hoff/hoff_ak.html.

Holmes, F. L. 1986. "Claude Bernard, the 'Milieu Intérieur,' and Regulatory Physiology." *History and Philosophy of the Life Sciences* 8, no. 1: 3–25.

Holst, J. J., W. Holland, J. Gromada, et al. 2017. "Insulin and Glucagon: Partners for Life." *Endocrinology* 158, no. 4 (Apr.): 696–701.

Hooper, L., N. Martin, O. F. Jimoh, et al. 2020. "Reduction in Saturated Fat Intake for Cardiovascular Disease." *Cochrane Database of Systematic Reviews* 8, no. CD011737. Online: https://doi.org/10.1002/14651858.CD011737.pub3.

Houssay, B. A. 1952. "The Discovery of Pancreatic Diabetes, the Role of Oscar Minkowski." *Diabetes* 1, no. 2 (Mar.–Apr.): 112–16.

Howard, B. V., J. E. Manson, M. L. Stefanick, et al. 2006. "Low-Fat Dietary Pattern and Weight Change over 7 Years: The Women's Health Initiative Dietary Modification Trial." *JAMA* 295, no. 1 (Jan. 4): 39–49.

Howard, B. V., L. Van Horn, J. Hsia, et al. 2006. "Low-Fat Dietary Pattern and Risk of Cardiovascular Disease: The Women's Health Initiative Randomized Controlled Dietary Modification Trial." *JAMA* 295, no. 6 (Feb. 8): 655–66.

Hulley, S. B., J. M. Walsh, and T. B. Newman. 1992. "Health Policy on Blood Cholesterol. Time to Change Directions." *Circulation* 86, no. 3 (Sept.): 1026–29.

Hulley, S. B., W. S. Wilson, M. I. Burrows, and M. Z. Nichaman. 1972. "Lipid and Lipoprotein Responses of Hypertriglyceridaemic Outpatients to a Low-Carbohydrate Modification of the A.H.A. Fat-Controlled Diet." *The Lancet* 300, no. 7777 (Sept. 16): 551–55.

Hutton, S. K. Undated. *Health Conditions and Disease Incidence Among the Eskimos of Labrador.* London: Wessex Press.

Insull, W., Jr., T. Oiso, and K. Tsuchiya. 1968. "Diet and Nutritional Status of Japanese." *American Journal of Clinical Nutrition* 21, no. 7 (July): 53–77.

Institute of Medicine (IOM). 2005. *Dietary Reference Intakes for Energy, Carbohydrate,*

Fiber, Fat, Fatty Acids, Cholesterol, Protein, and Amino Acids. Washington, DC: National Academies Press. Online: https://nap.nationalacademies.org/catalog/10490/dietary-reference-intakes-for-energy-carbohydrate-fiber-fat-fatty-acids-cholesterol-protein-and-amino-acids.

International Sugar Research Foundation (ISRF). 1976. Memo to members, April 30. "Developments in Brief: ISRF Support of Health Research and International Symposia." Washington, DC. Internal document, Sugar Association, Inc., Records of the Great Western Sugar Company, Colorado Agricultural Archive, Colorado State University.

International Sugar Research Foundation (ISRF). 1975. "Planning the Research Effort." Bethesda, MD. Internal document, International Sugar Research Foundation, Inc., Records of the Great Western Sugar Company. Colorado Agricultural Archive, Colorado State University.

Inter-Society Commission for Heart Disease Resources. 1970. "Report of Inter-Society Commission for Heart Disease Resources. Prevention of Cardiovascular Disease. Primary Prevention of the Atherosclerotic Diseases." *Circulation* 42, no. 6 (Dec.): A55–A95.

Jacobs, D., H. Blackburn, M. Higgins, et al. 1992. "Report of the Conference on Low Blood Cholesterol: Mortality Associations." *Circulation* 86, no. 3 (Sept.): 1046–60.

Janes, R. G., and M. Prosser. 1947. "Influence of High Fat Diets on Alloxan Diabetes." *American Journal of Physiology* 151, no. 2 (Dec.): 581–87.

Jennings, K. 2021. "This $2 Billion Digital Health Startup Aims to Reverse Type 2 Diabetes." *Forbes* (Apr. 19). Online: https://www.forbes.com/sites/katiejennings/2021/04/19/this-2-billion-digital-health-startup-aims-to-reverse-type-2-diabetes/?sh=7108873b7044.

Johnson, H. W., and E. H. Rynearson. 1951. "A Diabetic Patient on a High Fat Diet for Twenty-Nine Years Without Complications." *Proceedings of the Staff Meetings of the Mayo Clinic* 26, no. 18 (Aug. 29): 329–31.

Johnson, R. E., F. Sargent II, and R. Passmore. 1958. "Normal Variations in Total Ketone Bodies in Serum and Urine of Healthy Young Men." *Quarterly Journal of Experimental Physiology and Cognate Medical Sciences* 43, no. 4 (Oct.): 339–44.

Jones, K. 2000. "Insulin Coma Therapy in Schizophrenia." *Journal of the Royal Society of Medicine* 93, no. 3 (Mar.): 147–49.

Jörgens, V. 2021. "The Discovery of Insulin in 1914: Georg Zülzer, from Berlin, and Camille Retuer, the Forgotten Chemist from Luxembourg." *Diabetes Metabolism* 47, no. 4 (July): 101180.

Joseph, J. J., P. Deedwania, T. Acharya, et al. 2022. "Comprehensive Management of Cardiovascular Risk Factors for Adults with Type 2 Diabetes: A Scientific Statement from the American Heart Association." *Circulation* 145, no. 9 (Mar.): e722–59.

Joslin, E. P. 1959. *Diabetic Manual for the Patient.* 10th edition. Philadelphia: Lea & Febiger.

Joslin, E. P. 1956. "Diabetes for the Diabetics: Ninth Banting Memorial Lecture of the British Diabetic Association." *Diabetes* 5, no. 2 (Mar.–Apr.): 137–45.

Joslin, E. P. 1950. "A Half-Century's Experience in Diabetes Mellitus." *British Medical Journal* 1, no. 4662 (May 13): 1095–98.

Joslin, E. P. 1948. *Diabetic Manual for the Doctor and Patient.* 8th edition. Philadelphia: Lea & Febiger.

Joslin, E. P. 1946. "Diabetes: Past, Present and Future." *Proceedings of the American Diabetes Association: Annual Meeting* 6: 161–69.

Joslin, E. P. 1933. "Fat and the Diabetic." *The New England Journal of Medicine* 209, no. 11 (Sept. 14): 519–28.

Joslin, E. P. 1930. "Arteriosclerosis in Diabetes." *Annals of Internal Medicine* 4, no. 1 (July): 54–55.

Joslin, E. P. 1929a. *A Diabetic Manual for the Mutual Use of Doctor and Patient.* 4th edition. Philadelphia: Lea & Febiger.

Joslin, E. P. 1929b. "Abolishing Diabetic Coma." *The Journal of the American Medical Association* 93, no. 1 (July 6): 33.

Joslin, E. P. 1928a. "Ideals in the Treatment of Diabetes and Methods for Their Realization." *The New England Journal of Medicine* 198, no. 8 (Apr. 12): 379–82.

Joslin, E. P. 1928b. "The Ten-Year Diabetic. What He Is. What He Should Be. How to Make Him So." *The American Journal of the Medical Sciences* 175, no. 4 (Apr.): 472–79.

Joslin, E. P. 1928c. *The Treatment of Diabetes Mellitus.* 4th edition. Philadelphia: Lea & Febiger.

Joslin, E. P. 1925. "The Reduction of the Death Rate from Diabetes, Particularly from Diabetic Coma, in Massachusetts." *Boston Medical and Surgical Journal* 193, no. 16 (Oct. 15): 707–12.

Joslin, E. P. 1923a. "The Routine Treatment of Diabetes with Insulin." *The Journal of the American Medical Association* 80, no. 22 (June 2): 1581–83.

Joslin, E. P. 1923b. *The Treatment of Diabetes Mellitus, with Observations Based upon Three Thousand Cases.* 3rd edition. Philadelphia: Lea & Febiger.

Joslin, E. P. 1921. "The Prevention of Diabetes Mellitus." *The Journal of the American Medical Association* 76, no. 2 (Jan. 8): 79–84.

Joslin, E. P. 1917a. "The Blood Lipoids in Diabetes." *The Journal of the American Medical Association* 79, no. 5 (Aug. 4): 375–78.

Joslin, E. P. 1917b. *The Treatment of Diabetes Mellitus with Observations upon the Disease Based upon Thirteen Hundred Cases.* Philadelphia: Lea & Febiger.

Joslin, E. P. 1916. *The Treatment of Diabetes Mellitus with Observations upon the Disease Based upon One Thousand Cases.* Philadelphia: Lea & Febiger.

Joslin, E. P. 1915. "Present-Day Treatment and Prognosis in Diabetes." *The American Journal of the Medical Sciences* 150, no. 4 (Oct.): 485–96.

Joslin, E. P. 1905. "The Improvement in the Treatment of Diabetes Mellitus." *Boston Medical and Surgical Journal* 153, no. 1 (July 6): 11–15.

Joslin, E. P., L. I. Dublin, and H. H. Marks. 1934. "Studies in Diabetes Mellitus. II: Its Incidence and the Factors Underlying Its Variations." *The American Journal of the Medical Sciences* 187, no. 4 (Apr.): 433–57.

Joslin, E. P., L. I. Dublin, and H. H. Marks. 1933. "Studies in Diabetes Mellitus. I. Characteristics and Trends of Diabetes Mortality Throughout the World." *The American Journal of the Medical Sciences* 180, no. 6 (Dec.): 753–73.

Joslin, E. P., H. F. Root, P. White, and A. Marble. 1940. *The Treatment of Diabetes Mellitus.* 7th edition. Philadelphia: Lea & Febiger.

Joslin, E. P., H. F. Root, P. White, A. Marble, and C. C. Bailey. 1946. *The Treatment of Diabetes Mellitus.* 8th edition. Philadelphia: Lea & Febiger.

Joslin, E. P., H. F. Root, P. White, A. Marble, and C. C. Bailey. 1937. *The Treatment of Diabetes Mellitus.* 6th edition. Philadelphia: Lea & Febiger.

Jovanovic, L., M. Druzin, and C. M. Peterson. 1981. "Effect of Euglycemia on the Outcome of Pregnancy in Insulin-Dependent Diabetic Women as Compared with Normal Control Subjects." *The American Journal of Medicine* 71, no. 6 (Dec.): 921–27.

Judson, H. F. 1979. *The Eighth Day of Creation: Makers of the Revolution in Biology.* New York: Simon and Schuster.

Kahneman, D. 2011. *Thinking, Fast and Slow.* New York: Farrar, Straus and Giroux.

Karam, J. H., G. M. Grodsky, and P. H. Forsham. 1963. "Excessive Insulin Response to Glucose in Obese Subjects as Measured by Immunochemical Assay." *Diabetes* 12 (May–June): 197–204.

Karolinska Institute. 1977. Press release: The 1977 Nobel Prize in Physiology or Medicine. Online at http://nobelprize.org/nobel_prizes/medicine/laureates/1977/press .html.

Karter, A. J., S. Nundy, M. M. Parker, H. H. Moffet, and E. S. Huang. 2014. "Incidence of Remission in Adults with Type 2 Diabetes: The Diabetes and Aging Study." *Diabetes Care* 37, no. 12 (Dec.): 3188–95.

Katz, L. N., and R. Pick. 1961. "Experimental Atherosclerosis as Observed in the Chicken." *Journal of Atherosclerosis Research* 1, no. 2 (Mar. 4): 93–100.

Kaufman, R. R. 2005. *Diabesity: The Obesity-Diabetes Epidemic That Threatens America— And What We Must Do to Stop It.* New York: Bantam Dell.

Keen, H., and R. K. Knight. 1962. "Self-Sampling for Blood-Sugar." *The Lancet* 279, no. 7238 (May 19): 1037–40.

Kellock, B. 1985. *The Fibre Man: The Life-Story of Dr. Denis Burkitt.* Herts, U.K.: Lion Publishing.

Kelsay, J. L. 1978. "A Review of Research on Effects of Fiber Intake on Man." *American Journal of Clinical Nutrition* 31, no. 1 (Jan.): 142–59.

Kempner, W., R. Lohmann Peschel, and C. Schlayer. 1958. "Effect of Rice Diet on Diabetes Mellitus Associated with Vascular Disease." *Postgraduate Medicine* 24, no. 4 (Oct.): 359–71.

Keys, A. 1994. "The Inception and Pilot Surveys." In *The Seven Countries Study: A Scientific Adventure in Cardiovascular Disease Epidemiology,* ed. D. Kromhout, A. Menotti, and H. Blackburn. Utrecht: Brouwer, 15–26.

Keys, A. 1975. "Coronary Heart Disease—The Global Picture." *Atherosclerosis* 22, no. 2 (Sept.–Oct.): 149–92.

Keys, A. 1957. "Diet and the Epidemiology of Coronary Heart Disease." *The Journal of the American Medical Association* 164, no. 17 (Aug. 24): 1912–19.

Keys, A. 1953a. "Atherosclerosis: A Problem in Newer Public Health." *Journal of the Mount Sinai Hospital, New York* 20, no. 2 (July–Aug.): 118–39.

Keys, A. 1953b. "Prediction and Possible Prevention of Coronary Disease." *American Journal of Public Health and the Nation's Health* 43, no. 11 (Nov.): 1399–407.

Keys, A. 1952. "Human Atherosclerosis and the Diet." *Circulation* 5, no. 1 (Jan.): 115–18.

Keys, A. 1951. "Cholesterol, 'Giant Molecules,' and Atherosclerosis." *The Journal of the American Medical Association* 147, no. 16 (Dec. 15): 1514–19.

Keys, A., J. Brozek, A. Henschel, O. Mickelsen, and H. L. Taylor. 1950. *The Biology of Human Starvation.* 2 vols. Minneapolis: University of Minnesota Press.

Keys, A., O. Mickelsen, E. von O. Miller, and C. B. Chapman. 1950. "The Relation in Man Between Cholesterol Levels in the Diet and in the Blood." *Science* 112, no. 2889 (July 21): 79–81.

Kimball, C. P., and J. R. Murlin. 1923. "Aqueous Extracts of Pancreas. III. Some Precipitation Reactions of Insulin." *Journal of Biological Chemistry* 58, no. 1 (Nov.): 337–46.

Kimmelstiel, P., and C. Wilson. 1936. "Intercapillary Lesions in the Glomeruli of the Kidney." *American Journal of Pathology* 12, no. 1 (Jan.): 83–97.

Kinsell, L. W., G. Michaels, L. DeWind, J. Partridge, and L. Boling. 1953. "Serum Lipids in Normal and Abnormal Subjects, Observations on Controlled Experiments." *California Medicine* 78, no. 1 (Jan.): 5–10.

Kinsell, L. W., D. B. Stone, M. Behrman, L. B. Flinn, G. J. Hamwi, et al. 1967. "Principles of Nutrition for the Patient with Diabetes Mellitus." *Diabetes* 16, no. 10 (Oct.): 738.

Kirtley, W. R., S. O. Waife, and F. B. Peck. 1953. "Effect of Glucagon in Stable and Unstable Diabetic Patients." *Proceedings of the Society of Experimental Biology and Medicine* 83, no. 2 (June): 387–89.

Knowler, W. C., J. P. Crandall, J.-L. Chiasson, and D. M. Nathan. 2018. "Prevention of Type 2 Diabetes." In *Diabetes in America*, 3rd edition. Bethesda, MD: National Institutes of Health, 38-1-38-21.

Koenig, R. J., C. M. Peterson, C. Kilo, A. Cerami, and J. R. Williamson. 1976. "Hemoglobin A1c as an Indicator of the Degree of Glucose Intolerance in Diabetes." *Diabetes* 25, no. 3 (Mar.): 230–32.

Kolata, G. 2012. "Diabetes Study Ends Early with a Surprising Result." *The New York Times* (Oct. 19). Online: http://www.nytimes.com/2012/10/20/health/in-study-weight-loss-did-not-prevent-heart-attacks-in-diabetics.html.

Kolata, G. 2008. "Diabetes Study Partially Halted After Deaths." *The New York Times* (Feb. 7). Online: http://www.nytimes.com/2008/02/07/health/07diabetes.html.

Kolata, G. 1987a. "Diabetics Should Lose Weight, Avoid Diet Fads." *Science* 235, no. 4785 (Jan. 9): 163–64.

Kolata, G. 1987b. "High-Carb Diets Questioned." *Science* 235, no. 4785 (Jan. 9): 164.

Kolata, G. 1985. "Heart Panel's Conclusions Questioned." *Science* 227, no. 4682 (Jan. 4): 40–41.

Kolata, G. 1982. "Heart Study Produces a Surprise Result." *Science* 218, no. 4567 (Oct. 1): 31–32.

Koop, C. E. 1988. "Message from the Surgeon General." In *The Surgeon General's Report on Nutrition and Health,* no. 88-50210. Washington, DC: U.S. Government Printing Office.

Korányi, A. 1963. "Prophylaxis and Treatment of the Coronary Syndrome." *Therapia Hungarica* 12: 17–20.

Korp, W., and E. Zweymüller. 1972. "50 Years of Insulin Treatment at the Vienna Hospital for Children: The Fate of Diabetic Children from the First Insulin Era." In *Diabetes: Its Medical and Cultural History,* ed. D. von Engelhardt. Berlin: Springer-Verlag, 437–50.

Krall, L. P., and A. P. Joslin. 1971. "General Plan of Treatment and Diet Regulation." In *Joslin's Diabetes Mellitus,* 11th edition, ed. A. P. Marble, R. F. Bradley, and L. P. Krall. Philadelphia: Lea & Febiger, 255–86.

Krebs, H. A. 1971. "How the Whole Becomes More Than the Sum of the Parts." *Perspectives in Biological Medicine* 14, no. 3 (Spring): 448–57.

Krebs, H. A. 1967. "The Making of a Scientist." *Nature* 215, no. 5109 (Sept. 30): 1441–45.

Kruszynska, Y. T., P. D. Home, I. Hanning, and K. G. Alberti. 1987. "Basal and 24-h C-Peptide and Insulin Secretion Rate in Normal Men." *Diabetologia* 30, no. 1 (Jan.): 16–21.

Ladd, W. S., and W. W. Palmer. 1923. "The Use of Fat in Diabetes Mellitus and the Carbohydrate-Fat Ratio." *The American Journal of the Medical Sciences* 166, no. 2 (Aug.): 157–69.

Landé, K. E., and W. M. Sperry. 1936. "Human Atherosclerosis in Relation to the Cholesterol Content of the Blood Serum." *Archives of Pathology* 22 (Sept.): 301–12.

Landsberg, L. 1986. "Diet, Obesity and Hypertension: An Hypothesis Involving Insulin, the Sympathetic Nervous System, and Adaptive Thermogenesis." *Quarterly Journal of Medicine* 61, no. 236 (Dec.): 1081–90.

Lane, M. A. 1907. "The Cytological Characters of the Areas of Langerhans." *American Journal of Anatomy* 7, no. 3 (Nov. 10): 409–22.

Lasker, R. D. 1993. "The Diabetes Control and Complications Trial: Implications for Policy and Practice." *The New England Journal of Medicine* 329, no. 14 (Sept. 30): 1035–36.

Laurell, S. 1956. "Plasma Free Fatty Acids in Diabetic Acidosis and Starvation." *Scandinavian Journal of Clinical and Laboratory Investigation* 8, no. 1: 81–82.

Laurence, W. L. 1960. "Diet and Arteries: Heart Association Unit Suggests Some Cut in Cholesterol Intake." *The New York Times* (Dec. 18): E7.

Laurence, W. L. 1957. "Drug for the Treatment of Diabetes Tested and Found of Great Importance." *The New York Times* (Feb. 24): E9.

Lawrence, J. 2012. *Diabetes, Insulin and the Life of R. D. Lawrence*. London: Royal Society of Medicine Press.

Lawrence, J. M., J. Divers, S. Isom, et al. 2021. "Trends in Prevalence of Type 1 and Type 2 Diabetes in Children and Adolescents in the US, 2001–2017." *JAMA* 326, no. 8 (Aug. 24–31): 717–27.

Lawrence, R. D. 1933. "An Address on Diabetes, with Special Reference to High Carbohydrate Diets." *British Medical Journal* 2, no. 3793 (Sept. 16): 517–21.

Leary, T. 1935. "Atherosclerosis, the Important Form of Arteriosclerosis, a Metabolic Disease." *The Journal of the American Medical Association* 105, no. 7 (Aug. 17): 475–81.

Leclercq, F. S. 1922. "Further Experiments with High Fat Diets in Diabetes." *Journal of Metabolic Research* 2, no. 2 (July): 39–55.

Lee, Y. H., M.-Y. Wang, X.-X. Yu, and R. H. Unger. 2016. "Glucagon Is the Key Factor in the Development of Diabetes." *Diabetologia* 59, no. 7 (July): 1372–75.

Lees, R. S., and D. E. Wilson. 1971. "The Treatment of Hyperlipidemia." *The New England Journal of Medicine* 284, no. 4 (Jan. 28): 186–95.

Leger, R. R. 1970. "Safety of Upjohn's Oral Antidiabetic Drug Doubted in Study: Firm Disputes Findings." *The Wall Street Journal* (May 21): 6.

Lennerz, B. S., A. Barton, R. K. Bernstein, et al. 2018. "Management of Type 1 Diabetes with a Very Low-Carbohydrate Diet." *Pediatrics* 141, no. 6 (June): e20173349.

Leren, P. 1966. *The Effect of Plasma Cholesterol Lowering Diet in Male Survivors of Myocardial Infarction. A Controlled Clinical Trial.* Stockholm: Scandinavian University Books.

Levine, R. 1967. "Insulin—The Biography of a Small Protein." *The New England Journal of Medicine* 277, no. 20 (Nov. 16): 1059–64.

Levine, R., and I. B. Fritz. 1956. "The Relation of Insulin to Liver Metabolism." *Diabetes* 5, no. 3 (May–June): 209–22.

Lichtenstein, A. 1938. "Free Diet in Children with Diabetes." *Journal of Pediatrics* 12, no. 2 (Feb.): 183–87.

Lindsay, S., and I. L. Chaikoff. 1963. "Naturally Occurring Arteriosclerosis in Animals: Comparison with Experimentally Induced Lesions." In *Atherosclerosis and Its Origins*, ed. M. Sandler and G. H. Bourne. New York: Academic Press, 350–438.

Look AHEAD Research Group. 2013. "Cardiovascular Effects of Intensive Lifestyle Intervention in Type 2 Diabetes." *The New England Journal of Medicine* 369, no. 2 (July 11): 145–54.

Lorenz, R. 1993. "Reply to 'Effects of Low Insulin and Low Carbohydrate on Frequency and Severity of Hypoglycemia.'" *The American Journal of Medicine* 92, no. 3 (Mar.): 340.

Lowy, C. 1998. "Home Glucose Monitoring, Who Started It?" *BMJ* 1, no. 7142 (May 9): 1467.

Luchsinger, J. A., C. Ryan, and L. J. Launer. 2018. "Diabetes and Cognitive Impairment." In *Diabetes in America*, 3rd edition. Bethesda, MD: National Institutes of Health, 24-1-24-19.

Lusk, G. 1928. *The Elements of the Science of Nutrition*. Philadelphia: W. B. Saunders.

Lusk, G. 1909. "Metabolism in Diabetes." *Archives of Internal Medicine* 3, no. 1 (Feb.): 1–22.

MacKay, E. M. 1943. "The Significance of Ketosis." *Journal of Clinical Endocrinology and Metabolism* 3, no. 2 (Feb.): 101–10.

Madison, L. L. and R. H. Unger. 1958. "The Physiologic Significance of the Secretion of Endogenous Insulin into the Portal Circulation. I. Comparison of the Effects of Glucagon-Free Insulin Administered via the Portal Vein and via a Peripheral Vein on the Magnitude of Hypoglycemia and Peripheral Glucose Utilization." *The Journal of Clinical Investigation* 37, no. 5 (May): 631–39.

Malone, J. I., J. M. Hellrung, E. W. Malphus, et al. 1976. "Good Diabetic Control—A Study in Mass Delusion." *Journal of Pediatrics* 88, no. 6 (June): 943–47.

Man, E. B., and J. P. Peters. 1935. "Serum Lipoids in Diabetes." *The Journal of Clinical Investigation* 14, no. 5 (Sept.): 579–94.

Mann, F. C., and T. B. Magath. 1922. "Studies in the Physiology of the Liver. II. The Effect of the Removal of the Liver on the Blood Sugar Level." *Archives of Internal Medicine* 30, no. 1 (July): 73–84.

Mann, F. C., and T. B. Magath. 1921. "Studies in the Physiology of the Liver. II. The Liver as a Regulator of the Glucose Concentration of the Blood." *American Journal of Physiology* 55, no. 2 (Mar. 1): 285–86.

Mann, G. V. 1977. "Diet-Heart: End of an Era." *The New England Journal of Medicine* 297, no. 12 (Sept. 22): 644–50.

Maratos-Flier, E., and J. S. Flier. 2005. "Obesity." In *Joslin's Diabetes Mellitus,* 14th edition, ed. C. R. Kahn, G. C. Weir, G. L. King, A. M. Jacobson, A. C. Moses, and R. J. Smith. New York: Lippincott, Williams & Wilkins, 533–45.

Marble, A. P., R. F. Bradley, and L. P. Krall, eds. 1971. *Joslin's Diabetes Mellitus*. 11th edition. Philadelphia: Lea & Febiger.

Marks, H. M. 1997. *The Progress of Experiment: Science and Therapeutic Reform in the United States, 1900–1990.* Cambridge, U.K.: Cambridge University Press.

Marks, H. P., and F. G. Young. 1939. "Observations on the Metabolism of Dogs Made Permanently Diabetic by Treatment with Anterior Pituitary Extract." *Journal of Endocrinology* 1, no. 1 (Jan.): 470–510.

Marriott, W. M. 1924. "The Food Requirements of Malnourished Infants, with a Note on the Use of Insulin." *The Journal of the American Medical Association* 83, no. 8 (Aug. 23): 600–603.

Marsh, P. L., L. H. Newburgh, and L. E. Holly. 1922. "The Nitrogen Requirement for Maintenance in Diabetes Mellitus." *Archives of Internal Medicine* 29, no. 1 (Jan.): 97–130.

Mattson, F. H., and S. M. Grundy. 1985. "Comparison of Effects of Dietary Saturated, Monounsaturated, and Polyunsaturated Fatty Acids on Plasma Lipids and Lipoproteins in Man." *Journal of Lipid Research* 26, no. 2 (Feb.): 194–202.

Mayer, J. 1955. "Regulation of Energy Intake and the Body Weight: The Glucostatic Theory and the Lipostatic Hypothesis." *Annals of the New York Academy of Sciences* 63, no. 1 (July 15): 15–43.

Mayer, J., and J. Dwyer. 1977. "Nutrition." *The Washington Post* (May 12): 93.

Mayer-Davis, E. J., L. M. Laffel, and J. B. Buse. 2018. "Management of Type 1 Diabetes with a Very Low-Carbohydrate Diet: A Word of Caution." *Pediatrics* 142, no. 2 (Aug.): e20181536B.

Mazaika, P. K., S. A. Weinzimer, N. Mauras, et al. 2016. "Variations in Brain Volume and Growth in Young Children with Type 1 Diabetes." *Diabetes* 65, no. 2 (Feb.): 476–85.

Mazur, A. 2011. "Why Were 'Starvation Diets' Promoted for Diabetes in the Pre-Insulin Period?" *Nutrition Journal* 10, no. 23 (Mar. 11): 1–9.

McElheny, V. K. 1977. "Coast Concern Plans Bacteria Use for Brain Hormone and Insulin." *The New York Times* (Dec. 2): 77, 85.

McGandy, R. B., and J. Mayer. 1973. "Atherosclerotic Disease, Diabetes, and Hypertension: Background Considerations." In *U.S. Nutrition Policies in the Seventies*, ed. J. Mayer. San Francisco: W. H. Freeman, 37–43.

McGarry, J. D. 2002. "Dysregulation of Fatty Acid Metabolism in the Etiology of Type 2 Diabetes." *Diabetes* 51, no. 1 (Jan.): 7–18.

McGarry, J. D. 1992. "What If Minkowski Had Been Ageusic? An Alternative Angle on Diabetes." *Science* 258, no. 5083 (Oct. 30): 766–70.

McGarry, J. D. 1979. "New Perspectives in the Regulation of Ketogenesis." *Diabetes* 28, no. 5 (May): 517–23.

McGarry, J. D., C. Robles-Valdes, and D. W. Foster. 1976. "Glucagon and Ketogenesis." *Metabolism* 25, no. 11, suppl. 1 (Nov.): 1387–91.

McGill, H. C., Jr., J. P. Strong, R. L. Holman, and N. T. Werthessen. 1960. "Arterial Lesions in the Kenya Baboon." *Circulation Research* 8, no. 5 (May): 670–79.

McLean Baird, I., and A. N. Howard, eds. 1969. *Obesity: Medical and Scientific Aspects.* London: E. & S. Livingstone.

Medvei, V. C. 1982. *A History of Endocrinology.* London: MTP Press.

Meinert, C. L. 2015. *The Trials and Tribulations of the University Group Diabetes Program: The UGDP.* Baltimore: Kelmscott Bookshop.

Mendosa, R. Undated a. "History of Blood Glucose Meters: 'Meter Memories'— Transcripts—Part III." Online: http://www.diabetes-book.com/meters-iii/.

Mendosa, R. Undated b. "History of Blood Glucose Meters: 'Meter Memories'— Transcripts—Part IV." Online: http://www.diabetes-book.com/meters-iv/.

Merton, R. K. 1973. *The Sociology of Science: Theoretical and Empirical Investigations.* Chicago: University of Chicago Press.

Metz, R. 1932. "Insulin in Malnutrition. Further Observations." *Annals of Internal Medicine* 6, no. 6 (Dec.): 743–50.

Micks, R. H. 1943. "The Diet in Diabetes." *British Medical Journal* 1, no. 4297 (May 15): 598–600.

Miettinen, M., O. Turpeinen, M. J. Karvonen, R. Elosuo, and E. Paavilainen. 1972. "Effect of Cholesterol-Lowering Diet on Mortality from Coronary Heart-Disease and Other Causes. A Twelve-Year Clinical Trial in Men and Women." *The Lancet* 300, no. 7882 (Oct. 21): 835–38.

Millard, E. B., and H. F. Root. 1948. "Degenerative Vascular Lesions and Diabetes Mellitus." *American Journal of Digestive Diseases* 15, no. 2 (Feb.): 41–51.

Mills, C. A. 1930. "Diabetes Mellitus, Sugar Consumption in Its Etiology." *Archives of Internal Medicine* 46, no. 4 (Oct.): 582–84.

Minkowski, O. 1989. "Historical Development of the Theory of Pancreatic Diabetes, 1929." *Diabetes* 38, no. 1 (Jan.): 1–6.

Mintz, D. H., J. S. Skyler, and R. A. Chez. 1978. "Diabetes Mellitus and Pregnancy." *Diabetes Care* 1, no. 1 (Jan.–Feb.): 49–63.

Mitchell, H. S. 1930. "Nutrition Survey in Labrador and Northern Newfoundland." *Journal of the American Dietetics Association* 6, no. 1 (June): 29–35.

Monod, J. 1965. "Nobel Lecture: From Enzymatic Adaption to Allosteric Transitions." Online: http://www.nobelprize.org/nobel_prizes/medicine/laureates/1965/monod-lecture.html.

Moore, T. J. 1989. *Heart Failure: A Critical Inquiry into American Medicine and the Revolution in Heart Care.* New York: Random House.

Moore, W. W. 1983. *Fighting for Life: The Story of the American Heart Association 1911–1975.* Dallas, TX: American Heart Association.

Müller, T. D., B. Finan, C. Clemmensen, R. D. DiMarchi, and M. H. Tschöp. 2017. "The New Biology and Pharmacology of Glucagon." *Physiological Reviews* 97, no. 2 (Apr.): 721–66.

Multiple Risk Factor Intervention Trial (MRFIT) Research Group. 1982. "Multiple Risk Factor Intervention Trial. Risk Factor Changes and Mortality Results." *JAMA* 248, no. 12 (Sept. 24): 1465–77.

Nahum, L. H., and H. E. Himwich. 1932. "Insulin and Appetite. I. A Method for Increasing Weight in Thin Patients." *The American Journal of the Medical Sciences* 183, no. 5 (May): 608–13.

Nasar, S. 1998. *A Beautiful Mind.* New York: Simon & Schuster.

National Analysts, Inc. 1974. "Attitudes Toward Sugar: A Study Conducted for the Sugar Association and the International Sugar Research Foundation." Records of the Great Western Sugar Company. Colorado Agricultural Archive, Colorado State University.

National Cholesterol Education Program (NCEP). 2002. "Third Report of the National Cholesterol Education Program (NCEP) Expert Panel on Detection, Evaluation, and Treatment of High Blood Cholesterol in Adults. (Adult Treatment Panel III) Final Report." *Circulation* 106, no. 25 (Dec. 17): 3143–421.

National Commission on Diabetes. 1995. *Report of the National Commission on Diabetes to the Congress of the United States. Volume 1. The Long-Range Plan to Combat Diabetes.* DHEW pub. no. (NIH) 76-1018. Washington, DC: Department of Health, Education, and Welfare.

National Institute of Diabetes and Digestive and Kidney Diseases (NIDDK). 2018. *Diabetes in America.* 3rd edition. Bethesda, MD: National Institutes of Health.

National Institute of Diabetes and Digestive and Kidney Diseases (NIDDK). 2016. "What Is Diabetes?" Online: https://www.niddk.nih.gov/health-information/diabetes /overview/what-is-diabetes.

National Research Council (NRC), Committee on Diet and Health, Food and Nutrition Board, Commission on Life Sciences. 1989. *Diet and Health: Implications for Reducing Chronic Disease Risk.* Washington, DC: National Academy Press.

Naunyn, B. 1924. "Preface to Karl Petrén's 'Diabetes-Studier.' " *Journal of Metabolic Research* 5 (Jan.–Mar.): 1–5.

Newburgh, L. H. 1950. "Obesity." In *Clinical Nutrition,* ed. N. Joliffe, F. F. Tisdall, and P. R. Cannon. New York: Paul B. Hoeber, 689–743.

Newburgh, L. H. 1931. "The Cause of Obesity." *The Journal of the American Medical Association* 97, no. 3 (Dec. 5): 1659–63.

Newburgh, L. H., and M. W. Johnston. 1930a. "The Nature of Obesity." *The Journal of Clinical Investigation* 8, no. 2 (Feb.): 197–213.

Newburgh, L. H., and M. W. Johnston. 1930b. "Endogenous Obesity—A Misconception." *Annals of Internal Medicine* 3, no. 8 (Feb.): 815–25.

Newburgh, L. H., and P. L. Marsh. 1923. "Further Observations on the Use of a High Fat Diet in the Treatment of Diabetes Mellitus." *Archives of Internal Medicine* 31, no. 4 (Apr.): 455–90.

Newburgh, L. H., and P. L. Marsh. 1921. "The Use of a High Fat Diet in the Treatment of Diabetes Mellitus. Second Paper: Blood Sugar." *Archives of Internal Medicine* 27, no. 6 (June): 699–704.

Newburgh, L. H., and P. L. Marsh. 1920. "The Use of a High Fat Diet in the Treatment of Diabetes Mellitus. First Paper." *Archives of Internal Medicine* 26, no. 6 (Dec.): 647–62.

Newgard, C. B. 2018. "John Denis McGarry, PhD: A Remembrance of a Master Metabolic Physiologist." *Diabetes Care* 41, no. 7 (July): 1330–36.

Nichol, E. S. 1932. "Insulin Fattening: Late Results in Sixty-Three Cases." *Southern Medical Journal* 25, no. 4 (Apr.): 405–10.

Nuttall, F. Q., and D. J. Brunzell. 1979. "Principles of Nutrition and Dietary Recommendations for Individuals with Diabetes Mellitus: 1979." *Diabetes* 28, no. 11 (Nov.): 1027–30.

O'Connor, A. 2018. "A Foe for Type 1 Diabetes." *The New York Times* (May 15): D4.

Olefsky, J. M., J. W. Farquhar, and G. M. Reaven. 1974. "Reappraisal of the Role of Insulin in Hypertriglyceridemia." *The American Journal of Medicine* 57, no. 4 (Oct.): 551–60.

Olson, R. E. 1998. "Discovery of the Lipoproteins, Their Role in Fat Transport and Their Significance as Risk Factors." *Journal of Nutrition* 128, no. 2, suppl. (Feb.): 439S–43S.

Opie, E. L. 1901a. "On the Relation of Chronic Interstitial Pancreatitis to the Islands of Langerhans and to Diabetes Mellitus." *Journal of Experimental Medicine* 5, no. 4 (Jan. 15): 397–428.

Opie, E. L. 1901b. "The Relation of Diabetes Mellitus to Lesions of the Pancreas. Hyaline Degeneration of the Islands of Langerhans." *Journal of Experimental Medicine* 5, no. 5 (Mar. 25): 527–40.

Osler, W. 1893. *The Principles and Practice of Medicine.* New York: D. Appleton.

Osler, W., and T. McCrae. 1923. *The Principles and Practice of Medicine.* 9th edition. New York: D. Appleton.

Osler, W., and T. McCrae. 1914. *The Principles and Practice of Medicine.* 8th edition. New York: D. Appleton.

Osmundsen, J. A. 1961. "New Views Given on Fats in Diet; Foods Rich in Starches and Sugars Appear to Raise Level of Triglycerides." *The New York Times* (May 4): 33, 39.

Page, I. H., F. J. Stare, A. C. Corcoran, H. Pollack, and C. F. Wilkinson, Jr. 1957. "Atherosclerosis and the Fat Content of the Diet." *Circulation* 16, no. 2 (Aug.): 163–78.

Pavy, F. W. 1862. *Researches on the Nature and Treatment of Diabetes.* London: John Churchill.

Peacock, I., and R. Tattersall. 1982. "Methods of Self Monitoring of Diabetic Control." *Clinics in Endocrinology and Metabolism* 11, no. 2 (July): 485–501.

Pellegrino, E. D. 1979. "The Sociocultural Impact of Twentieth-Century Therapeutics." In *The Therapeutic Revolution, Essays in the Social History of American Medicine,* ed. M. J. Vogel and C. E. Rosenberg. Philadelphia: University of Pennsylvania Press.

Peppa, M., J. Uribarri, and H. Vlassara. 2003. "Glucose, Advanced Glycation End Products, and Diabetes Complications: What Is New and What Works." *Clinical Diabetes* 21, no. 4 (Oct. 1): 186–87.

Petersen, A., R. Winkler, and S. A. O'Brien. 2023. "The $76 Billion Diet Industry Asks: What to Do About Ozempic?" *The Wall Street Journal* (April 10) Online: https://www.wsj.com/articles/ozempic-wegovy-mounjaro-weight-loss-industry-89419ecb.

Petersen, M. C., and G. I. Shulman. 2018. "Mechanisms of Insulin Action and Insulin Resistance." *Physiological Reviews* 98, no. 4 (Oct. 1): 2133–223.

Peterson, C. M., S. E. Forhan, and R. L. Jones. 1980. "Self-Management: An Approach to Patients with Insulin-Dependent Diabetes Mellitus." *Diabetes Care* 3, no. 1 (Jan.–Feb.): 82–87.

Peterson, C. M., R. L. Jones, A. Dupuis, et al. 1979. "Feasibility of Improved Blood Glucose Control in Patients with Insulin-Dependent Diabetes Mellitus." *Diabetes Care* 2, no. 4 (July–Aug.): 329–35.

Petrén, K., G. Blix, H. Malmoros, M. Odin, and E. Person. 1924. "Studies on Diabetes." *Journal of Metabolic Research* 5 (Jan.–Mar.): 7–82.

Phinney, S. D., B. R. Bistrian, W. J. Evans, E. Gervino, and G. L. Blackburn. 1983. "The Human Metabolic Response to Chronic Ketosis Without Caloric Restriction: Preservation of Submaximal Exercise Capability with Reduced Carbohydrate Oxidation." *Metabolism* 32, no. 8 (Aug.): 769–76.

Phinney, S. D., B. R. Bistrian, R. R. Wolfe, and G. L. Blackburn. 1983. "The Human Metabolic Response to Chronic Ketosis Without Caloric Restriction: Physical and Biochemical Adaptation." *Metabolism* 32, no. 8 (Aug.): 757–68.

Phinney, S. D., E. S. Horton, E. A. Sims, et al. 1980. "Capacity for Moderate Exercise in Obese Subjects After Adaptation to a Hypocaloric Ketogenic Diet." *The Journal of Clinical Investigation* 66, no. 5 (Nov.): 1152–61.

Pickup, J. C., H. Keen, J. A. Parsons, and K. G. Alberti. 1978. "Continuous Subcutaneous Insulin Infusion: An Approach to Achieving Normoglycaemia." *British Medical Journal* 1, no. 6107 (Jan. 28): 204–7.

Pi-Sunyer, F. X. 1994. "The Fattening of America." *JAMA* 272, no. 3 (July 20): 238.

Pitfield, R. L. 1923. "On the Use of Insulin in Infantile Inanition." *New York Medical Journal and Medical Record* 118, no. 4 (Aug. 15): 217–18.

Plath, S. 2005. *The Bell Jar*. New York: Harper Perennial.

Popper, K. R. 1979. *Objective Knowledge, An Evolutionary Approach*. Revised edition. Oxford: Clarendon Press.

Popper, K. R. 1963. *Conjectures and Refutations*. London: Routledge and Kegan Paul.

Porcellati, F., S. Di Mauro, A. Mazzieri, et al. 2021. "Glucagon as a Therapeutic Approach to Severe Hypoglycemia: After 100 Years, Is It Still the Antidote of Insulin?" *Biomolecules* 11, no. 9 (Aug. 27): 1281–99.

Prentice, R. L., B. Caan, R. T. Chlebowski, et al. 2006. "Low-Fat Dietary Pattern and Risk of Invasive Breast Cancer: The Women's Health Initiative Randomized Controlled Dietary Modification Trial." *JAMA* 295, no. 6 (Feb. 8): 629–42.

Presley, J. W. 1991. "A History of Diabetes Mellitus in the United States, 1880–1990." PhD dissertation. University of Texas at Austin.

Price, W. A. 1939. *Nutrition and Physical Degeneration: A Comparison of Primitive and Modern Diets and Their Effects*. New York: Paul B. Hoeber.

Prillaman, M. 2023. "The 'Breakthrough' Obesity Drugs that Have Stunned Researchers." *Nature* 613, no. 7942 (Jan. 5): 16–18.

Proctor, R. N. 1999. *The Nazi War on Cancer*. Princeton, NJ: Princeton University Press.

Rabinowitch, I. M. 1944. "Prevention of Premature Arteriosclerosis in Diabetes Mellitus." *Canadian Medical Association Journal* 51, no. 4 (Oct.): 300–306.

Rabinowitch, I. M. 1930. "Experiences with a High Carbohydrate–Low Calorie Diet for the Treatment of Diabetes Mellitus." *Canadian Medical Association Journal* 23, no. 4 (Oct.): 489–98.

Rabinowitch, I. M. 1928. "Unusual Fat Metabolism in a Case of Diabetes Mellitus." *The American Journal of the Medical Sciences* 176, no. 4 (Oct.): 489–91.

Rahbar, S. 1968. "An Abnormal Hemoglobin in Red Cells of Diabetics." *Clinica Chimica Acta* 22, no. 2 (Oct.): 296–98.

Rahbar, S., O. Blumenfeld, and H. M. Ranney. 1969. "Studies of an Unusual Hemoglobin in Patients with Diabetes Mellitus." *Biochemical and Biophysical Research Communications* 36, no. 5 (Aug. 22): 838–43.

Rasmussen, H. 1968. "Organization and Control of Endocrine Systems." In *Textbook of Endocrinology*, 4th edition, ed. R. H. Williams. Philadelphia: W. B. Saunders, 1–26.

Rawshani, A., N. Sattar, S. Franzén, et al. 2018. "Excess mortality and cardiovascular disease in young adults with type 1 diabetes in relation to age at onset: A nationwide, register-based cohort study." *The Lancet* 392, no. 10146 (Aug. 11): 477–86.

Reaven, G. M. 1988. "Banting Lecture 1988. Role of Insulin Resistance in Human Disease." *Diabetes* 37, no. 12 (Dec.): 1595–607.

Reaven, G. M. 1986. "Effect of Dietary Carbohydrate on the Metabolism of Patients with Non-Insulin Dependent Diabetes Mellitus." *Nutrition Reviews* 44, no. 2 (Feb.): 65–73.

Reaven, G. M. 1980. "How High the Carbohydrate?" *Diabetologia* 19, no. 5 (Nov.): 409–13.

Reaven, G. M., A. Calciano, R. Cody, C. Lucas, and R. Miller. 1963. "Carbohydrate Intolerance and Hyperlipemia in Patients with Myocardial Infarction Without Known Diabetes Mellitus." *Journal of Clinical Endocrinology and Metabolism* 23 (Oct.): 1013–23.

Reaven, G. M., A. M. Coulston, and R. A. Marcus. 1979. "Nutritional Management of Diabetes." *Medical Clinics of North America* 63, no. 5 (Sept.): 927–43.

Reaven, G. M., R. L. Lerner, M. P. Stern, and J. W. Farquhar. 1967. "Role of Insulin in Endogenous Hypertriglyceridemia." *The Journal of Clinical Investigation* 46, no. 11 (Nov.): 1756–67.

Reichard, G. A., Jr., O. E. Owen, A. C. Haff, P. Paul, and W. M. Bortz. 1974. "Ketone-Body Production and Oxidation in Fasting Obese Humans." *The Journal of Clinical Investigation* 53, no. 2 (Feb.): 508–15.

Relman, A. S. 1988. "Assessment and Accountability: The Third Revolution in Medical Care." *The New England Journal of Medicine* 319, no. 18 (Nov. 3): 1220–22.

Renold, A. E. 1970. "Insulin Biosynthesis and Secretion—A Still Unsettled Topic." *The New England Journal of Medicine* 282, no. 4 (Jan. 22): 173–82.

Renold, A. E., and G. F. Cahill, Jr., eds. 1965. *Handbook of Physiology. Section 5. Adipose Tissue.* Washington, DC: American Physiological Society.

Research Committee. 1965. "Low-Fat Diet in Myocardial Infarction, a Controlled Trial." *The Lancet* 286, no. 7411 (Sept. 11): 501–4.

Research Committee to the Medical Research Council. 1968. "Controlled Trial of Soya-Bean Oil in Myocardial Infarction." *The Lancet* 292, no. 7570 (Sept. 28): 693–99.

Review Panel of the National Heart Institute. 1969. "Mass Field Trials of the Diet-Heart Question: Their Significance, Timeliness, Feasibility and Applicability. Report of the Diet-Heart Review Panel of the National Heart Institute." American Heart Association Monograph no. 28.

Richardson, H. B., E. H. Mason, and G. F. Soderstrom. 1933. "Clinical Calorimetry. XXXIII. The Effect of Fasting in Diabetes as Compared with a Diet Designed to Replace the Foodstuffs Oxidized During a Fast." *Journal of Biological Chemistry* 57, no. 2 (Sept.): 587–611.

Richter, C. P. 1936. "Increased Salt Appetite in Adrenalectomized Rats." *American Journal of Physiology* 115, no. 1 (Feb.): 155–61.

Richter, C. P., and J. F. Eckert. 1939. "Mineral Appetite of Parathyroidectomized Rats." *The American Journal of the Medical Sciences* 198 (July): 9–16.

Richter, C. P., and E. C. H. Schmidt. 1941. "Increased Fat and Decreased Carbohydrate Appetite of Pancreatectomized Rats." *Endocrinology* 28, no. 2 (Feb.): 179–92.

Richter, C. P., E. C. H. Schmidt, and P. D. Malone. 1945. "Further Observations on the Self-Regulatory Dietary Selections of Rats Made Diabetic by Pancreatectomy." *Bulletin of the Johns Hopkins Hospital* 76: 192–219.

Ricketts, H. T. 1970. "Editorial Statement." *Diabetes* 19, suppl. 2: 747–49.

Riddle, M. C., W. T. Ambrosius, D. J. Brillon, et al. 2010. "Epidemiologic Relationships Between A1C and All-Cause Mortality During a Median 3.4-Year Follow-Up of Glycemic Treatment in the ACCORD Trial." *Diabetes Care* 33, no. 5 (May): 983–90.

Riddle, M. C., W. T. Cefalu, P. H. Evans, et al. 2021. "Consensus Report: Definition and Interpretation of Remission in Type 2 Diabetes." *Diabetes Care* 44, no. 10 (Aug. 30): 2438–44.

Riddle, M. C., and W. H. Herman. 2018. "The Cost of Diabetes Care—An Elephant in the Room." *Diabetes Care* 41, no. 5 (May): 929–32.

Rittenberg, D., and R. Schoenheimer. 1937. "Deuterium as an Indicator in the Study of Intermediary Metabolism: XI. Further Studies on the Biological Uptake of Deuterium into Organic Substances, with Special Reference to Fat and Cholesterol Formation." *Journal of Biological Chemistry* 121, no. 1 (Oct.): 235–53.

Rollo, J. 1798. *Cases of the Diabetes Mellitus; with the Results of the Trials of Certain Acids, and Other Substances, in the Cure of the Lues Venerea.* 2nd edition. London: T. Gillet.

Rollo, J. 1797a. *An Account of Two Cases of the Diabetes Mellitus: With Remarks, as They Arose During the Progress of the Cure; to Which Are Added, a General View of the Nature of the Disease and Its Appropriate Treatment.* Vol. 1. London: T. Gillet.

Rollo, J. 1797b. *An Account of Two Cases of the Diabetes Mellitus: With Remarks, as They*

Arose During the Progress of the Cure; to Which Are Added, a General View of the Nature of the Disease and Its Appropriate Treatment. Vol. 2. London: T. Gillet.

Root, H. F. 1950. "Diabetes Without Vascular Disorder After 25 Years." *British Medical Journal* 2, no. 4683 (Oct. 7): 840.

Root, H. F., R. H. Sinden, and R. Zanca. 1950. "Factors in the Rate of Development of Vascular Lesions in the Kidneys, Retinae and Peripheral Vessels of the Youthful Diabetic." *American Journal of Digestive Diseases* 17, no. 6 (June): 179–86.

Rosenquist, K. J., and C. S. Fox. 2018. "Mortality Trends in Type 2 Diabetes." In *Diabetes in America*, 3rd edition. Bethesda, MD: National Institutes of Health, 36-1-36-14.

Rosenzweig, J. L. 1994. "Principles of Insulin Therapy." In *Joslin's Diabetes Mellitus*, 13th edition, ed. C. R. Kahn and G. C. Weir. Philadelphia: Lea & Febiger, 460–88.

Roth, J., J. S. Volek, M. Jacobson, et al. 2004. "Paradigm Shifts in Obesity Research and Treatment: Roundtable Discussion." *Obesity Research* 12, suppl. 2 (Nov.): 145S–48S.

Rowe, A. H., and O. H. Garrison. 1932. "Lipodystrophy: Atrophy and Tumefaction of Subcutaneous Tissue Due to Insulin Injections." *The Journal of the American Medical Association* 99, no. 1 (July 2): 16–18.

Russ, E. M., H. A. Eder, and D. P. Barr. 1951. "Protein-Lipid Relationships in Human Plasma. I. In Normal Individuals." *The American Journal of Medicine* 11, no. 4 (Oct.): 468–79.

Ryan, J. R., M. C. Balodimos, B. I. Chazan, et al. 1970. "Quarter Century Victory Medal for Diabetes: A Follow-Up of Patients One to 20 Years Later." *Metabolism* 19, no. 7 (July): 493–501.

Rynearson, E. H. 1963. "Do Glands Affect Weight?" In *Your Weight and How to Control It,* ed. M. Fishbein. Garden City, NY: Doubleday, 69–78.

Rynearson, E. H., and C. F. Gastineau. 1949. *Obesity . . .* Springfield, IL: Charles C. Thomas.

Sackett, D. L., W. M. Rosenberg, J. A. Gray, et al. 1996. "Evidence Based Medicine: What It Is and What It Isn't." *BMJ* 1, no. 7023 (Jan. 12): 71–72.

Sacks, F. M., A. H. Lichtenstein, J. H. Y. Wu, et al. 2017. "Dietary Fats and Cardiovascular Disease: A Presidential Advisory from the American Heart Association." *Circulation* 136, no. 3 (July 18): e1–e23.

Sagan, C. 2011. *The Demon-Haunted World: Science as a Candle in the Dark.* New York: Random House.

Samaha, F. F., N. Iqbal, P. Seshadri, et al. 2003. "A Low-Carbohydrate as Compared with a Low-Fat Diet in Severe Obesity." *The New England Journal of Medicine* 348, no. 21 (May 22): 2074–81.

Sansum, W. D., N. R. Blatherwick, and R. Bowden. 1926. "The Use of High Carbohydrate Diets in the Treatment of Diabetes Mellitus." *The Journal of the American Medical Association* 86, no. 3 (Jan. 16): 178–81.

Saundby, R. 1908. "Diabetes Mellitus." In *A System of Medicine by Many Writers,* vol. 3, ed. C. Allbutt and H. D. Rolleston. London: Macmillan, 167–211.

Saundby, R. 1900. "An Address on the Modern Treatment of Diabetes Mellitus." *The Lancet* 155, no. 4003 (May 19): 1420–26.

Sawyer, L., and E. A. M. Gale. 2009. "Diet, Delusion and Diabetes." *Diabetologia* 52, no. 1 (Jan.): 1–7.

Schade, D. S., J. V. Santiago, J. S. Skyler, and R. A. Rizza. 1983. *Intensive Insulin Therapy.* Princeton: Excerpta Medica.

Schaffer, R. 2017. "As Obesity Rate Rises, 'Double Diabetes' Looms Large." *Endocrine Today* (Nov. 22). Online: https://www.healio.com/news/endocrinology/20171114/as-obesity-rate-rises-double-diabetes-looms-large.

Schatzkin, A., E. Lanza, D. Corle, et al. 2000. "Lack of Effect of a Low-Fat, High-Fiber

Diet on the Recurrence of Colorectal Adenomas. Polyp Prevention Trial Study Group." *The New England Journal of Medicine* 342, no. 16 (Apr. 20): 1149–55.

Schoenheimer, R. 1961. *The Dynamic State of Body Constituents*. Cambridge, MA: Harvard University Press.

Schoenheimer, R., and D. Rittenberg. 1940. "The Study of Intermediary Metabolism of Animals with the Aid of Isotopes." *Physiological Reviews* 20, no. 2 (Apr.): 218–48.

Schulz, K. 2010. *Being Wrong: Adventures in the Margin of Error*. EPub edition. New York: Harper Collins.

Schumacher, H., and J. Schumacher. 1956. "Then and Now: 100 Years of Diabetes Mellitus." In *Diabetes: Its Medical and Cultural History*, ed. D. Von Engelhardt. Berlin: Springer-Verlag, 238–65.

Schwartz, T. B., and C. L. Meinert. 2004. "The UGDP Controversy: Thirty-Four Years of Contentious Ambiguity Laid to Rest." *Perspectives in Biology and Medicine* 47, no. 4 (Autumn): 564–74.

Schweitzer, A. 1957. "Preface." In A. Berglas, *Cancer: Nature, Cause and Cure*. Paris: Institute Pasteur, ix.

Select Committee on Nutrition and Human Needs of the U.S. Senate. 1977. *Cardiovascular Disease*. Vol. 2, pt. 1 of *Diet Related to Killer Diseases: Hearings Before the Select Committee on Nutrition and Human Needs of the United States Senate, Ninety-Fifth Congress*. Feb. 1 and 2, 1977. Washington, DC: U.S. Government Printing Office.

Select Committee on Nutrition and Human Needs of the U.S. Senate. 1973. *Sugar in Diet, Diabetes, and Heart Disease: Hearing Before the Select Committee on Nutrition and Human Needs of the United States Senate, Ninety-Third Congress*. Part 2. Apr. 30 and May 1 and 2, 1973. Washington, DC: U.S. Government Printing Office.

Selvin, E., M. W. Steffes, H. Zhu, K. Matsushita, L. Wagenknecht, et al. 2010. "Glycated Hemoglobin, Diabetes, and Cardiovascular Risk in Nondiabetic Adults." *The New England Journal of Medicine* 362, no. 9 (Mar. 4): 800–811.

Sharman, M. J., W. J. Kraemer, D. M. Love, et al. 2002. "A Ketogenic Diet Favorably Affects Serum Biomarkers for Cardiovascular Disease in Normal-Weight Men." *Journal of Nutrition* 13, no. 7 (July): 1879–85.

Shaw, B. B. 1914. *Misalliance, the Dark Lady of the Sonnets, and Fanny's First Play. With a Treatise on Parents and Children*. London: Constable.

Shaw Dunn, J., H. L. Sheehan, and N. G. B. Mcletchie. 1943. "Necrosis of Islets of Langerhans Produced Experimentally." *The Lancet* 241, no. 6242 (Apr. 17): 484–87.

Sherrill, J. W., and E. M. MacKay. 1939. "Deleterious Effects of Experimental Protamine Insulin Shock." *Archives of Internal Medicine* 64, no. 5 (Nov.): 907–12.

Shurkin, J. N. 1992. "Terman's Kids: The Groundbreaking Study of How the Gifted Grow Up." Boston: Little, Brown.

Silink, M. 1982. "Home Blood Glucose Monitoring in Childhood Diabetes Mellitus." In *Diabetes Management in the '80s*, ed. C. M. Peterson. New York: Praeger Publishers, 176–83.

Silver, S., and J. Bauer. 1931. "Obesity, Constitutional or Endocrine?" *The American Journal of the Medical Sciences* 181, no. 6 (June): 769–77.

Silverhus, K. 2022. "Low-Carbohydrate and Very Low-Carbohydrate Eating Patterns in Adults with Diabetes: A Guide for Health Care Providers." American Diabetes Association.

Silverman, M. 1957. "Good News for Diabetics." *The Saturday Evening Post* (Aug. 24): 39–50.

Singer, P. 1993. "'Vindication' for a Diabetes Expert." *The New York Times* (July 18): 13.

Singer, P. 1988. "Diabetic Doctor Offers a New Treatment." *The New York Times* (Apr. 3): 12.

Singh, I. 1955. "Low-Fat Diet and Therapeutic Doses of Insulin in Diabetes Mellitus." *The Lancet* 268, no. 6861 (Feb. 26): 422–25.

Sinha, A., C. Formica, C. Tsalamandris, et al. 1996. "Effects of Insulin on Body Composition in Patients with Insulin-Dependent and Non-Insulin-Dependent Diabetes." *Diabetic Medicine* 13, no. 1 (Jan.): 40–46.

Skyler, J. S. 1978. "Nutritional Management of Diabetes Mellitus." In *Diabetes, Obesity, and Vascular Disease: Metabolic and Molecular Interrelationships, Part 2*, ed. H. M. Katzen and R. J. Majler. New York: Hemisphere Publishing, 645–98.

Skyler, J. S., R. Bergenstal, R. O. Bonow, et al. 2009. "Intensive Glycemic Control and the Prevention of Cardiovascular Events: Implications of the ACCORD, ADVANCE, and VA Diabetes Trials: A Position Statement of the American Diabetes Association and a Scientific Statement of the American College of Cardiology Foundation and the American Heart Association." *Circulation* 119, no. 2 (Jan. 20): 351–57.

Smith, A. H. 1889. *Diabetes. Mellitus and Insipidus*. Detroit, MI: George S. Davis.

Sokoloff, L. 1973. "Metabolism of Ketone Bodies by the Brain." *Annual Review of Medicine* 24, no. 1: 271–80.

Sondike, S. B., N. Copperman, and M. S. Jacobson. 2003. "Effects of a Low-Carbohydrate Diet on Weight Loss and Cardiovascular Risk Factor in Overweight Adolescents." *Journal of Pediatrics* 142, no. 3 (Mar.): 253–58.

Sönksen, P. H., S. L. Judd, and C. Lowy. 1978. "Home Monitoring of Blood-Glucose. Method for Improving Diabetic Control." *The Lancet* 311, no. 8067 (Apr. 8): 729–32.

Soskin, S. 1941. "The Blood Sugar: Its Origin, Regulation and Utilization." *Physiological Reviews* 21, no. 1 (Jan.): 140–93.

Stare, F. J., ed. 1975. "Sugar in the Diet of Man." In *World Review of Nutrition and Dietetics, volume 22*, ed. G. H. Bourne. Basel: S. Karger, 237–326.

Starling, E. H. 1923. "The Harveian Oration on the Wisdom of the Body." *The Lancet* 2, no. 3277 (Oct. 20): 685–90.

Starling, E. H. 1905. "The Croonian Lectures on the Chemical Correlation of the Functions of the Body I." *The Lancet* 166, no. 4275 (Aug. 5): 339–41.

Steinberg, D. 2004. "Thematic Review Series: The Pathogenesis of Atherosclerosis. An Interpretive History of the Cholesterol Controversy: Part I." *Journal of Lipid Research* 45, no. 9 (Sept.): 1583–93.

Steinberg, D. 1997. "Low Density Lipoprotein Oxidation and Its Pathobiological Significance." *Journal of Biological Chemistry* 272, no. 34 (Aug. 22): 20963–66.

Stitt, A. W., R. Bucala, and H. Vlassara. 1997. "Atherogenesis and Advanced Glycation: Promotion, Progression, and Prevention." *Annals of the New York Academy of Sciences* 811 (Apr. 15): 115–27; discussion 127–29.

Stoll, C. 1989. *The Cuckoo's Egg: Tracking a Spy Through the Maze of Computer Espionage*. New York: Doubleday.

Stolte, K. 1933. "Freie Diät Beim Diabetes." *Medizinische Klinik* 29 (Feb. 24): 288–89.

Stolte, K. 1931. "Freie Diät Beim Diabetes." *Medizinische Klinik* 27 (June 5): 831–38.

Stone, D. B., and W. E. Connor. 1963. "The Prolonged Effects of a Low Cholesterol, High Carbohydrate Diet upon the Serum Lipids in Diabetic Patients." *Diabetes* 12, no. 2 (Mar.–Apr.): 127–32.

Stout, R. W. 1982. *Hormones and Atherosclerosis*. Lancaster, U.K.: MTP Press.

Stout, R. W. 1968. "Insulin-Stimulated Lipogenesis in Arterial Tissue in Relation to Diabetes and Atheroma." *The Lancet* 292, no. 7570 (Sept. 28): 702–3.

Stout, R. W., E. L. Bierman, and R. Ross. 1975. "Effect of Insulin on the Proliferation of Cultured Primate Arterial Smooth Muscle Cells." *Circulation Research* 36, no. 2 (Feb.): 319–27.

Strom, A., and R. A. Jensen. 1951. "Mortality from Circulatory Diseases in Norway 1940–1945." *The Lancet* 257, no. 6647 (Jan. 20): 126–29.

Stunkard, A. 1973. "The Obese: Backgrounds and Programs." In *Nutrition Policies in the Seventies*, ed. J. Mayer. San Francisco: W. H. Freeman, 29–36.

Stunkard, A., and M. McLaren-Hume. 1959. "The Results of Treatment for Obesity: A Review of the Literature and Report of a Series." *Archives of Internal Medicine* 103, no. 1 (Jan.): 79–85.

Sun, M. 1980. "Insulin Wars: New Advances May Throw Market into Turbulence." *Science* 210, no. 4475 (Dec. 12): 1225–28.

Tamborlane, W. V., R. S. Sherwin, M. Genel, and P. Felig. 1979. "Reduction to Normal of Plasma Glucose in Juvenile Diabetes by Subcutaneous Administration of Insulin with a Portable Pump." *The New England Journal of Medicine* 300, no. 11 (Mar. 15): 573–78.

Tanner, T. H. 1870. *The Practice of Medicine*. 5th American edition. Philadelphia: Lindsay & Blakiston.

Tatem J. W., Jr. 1976. "President's Report." In board of directors meeting. Oct. 14, Internal document, Scottsdale, Ariz. Sugar Association, Inc., Records of the Great Western Sugar Company, Colorado Agricultural Archive, Colorado State University.

Tattersall, R. 2009. *Diabetes: The Biography*. Oxford, U.K.: Oxford University Press.

Tattersall, R. B. 1981. "Psychiatric Aspects of Diabetes—A Physician's View." *British Journal of Psychiatry* 139, no. 6 (Dec.): 485–93.

Tattersall, R., and P. Clarke. 1982. "Management of the Insulin-Dependent Diabetic: The English Experience." In *Diabetes Management in the '80s*, ed. C. M. Peterson. New York: Praeger Publishers, 208–16.

Tattersall, R., S. Walford, I. Peacock, E. Gale, and S. Allison. 1980. "A Critical Evaluation of Methods of Monitoring Diabetic Control." *Diabetes Care* 3, no. 1 (Jan.–Feb.): 150–54.

Taubes, G. 2020. *The Case for Keto*. New York: Alfred A. Knopf.

Taubes, G. 2016. *The Case Against Sugar*. New York: Alfred A. Knopf.

Taubes, G. 2008. "Diabetes. Paradoxical Effects of Tightly Controlled Blood Sugar." *Science* 322, no. 5900 (Oct. 17): 365–67.

Taubes, G. 2007. *Good Calories, Bad Calories*. New York: Knopf.

Taubes, G. 2002. "What If It's All Been a Big Fat Lie?" *The New York Times Magazine* (July 7): 22–27, 34, 45–47.

Taubes, G. 2001. "The Soft Science of Dietary Fat." *Science* 291, no. 5513 (Mar. 30): 2536–45.

Taubes, G. 1998a. "The Political Science of Salt." *Science* 281, no. 5379 (Aug. 14): 898–907.

Taubes, G. 1998b. "As Obesity Rates Rise, Experts Struggle to Explain Why." *Science* 280, no. 5368. (May 29): 1367–68.

Taubes, G. 1996. "Looking for the Evidence in Medicine." *Science* 272, no. 5258 (Apr. 5): 22–24.

Taubes, G. 1995. "Epidemiology Faces Its Limits." *Science* 269, no. 5221 (July 14): 164–69.

Teicholz, N. 2014. *The Big Fat Surprise*. New York: Simon & Schuster.

Thomas, W. A. 1927. "Health of a Carnivorous Race: A Study of the Eskimo." *The Journal of the American Medical Association* 88, no. 20 (May 14): 1559–60.

Thomson, A. P. 1924. "The Clinical Use of Insulin." *British Medical Journal* 1, no. 3298 (Mar. 15): 457–60.

Tisdall, F. F., A. Brown, T. G. H. Drake, and M. G. Cody. 1925. "Insulin in the Treatment of Malnourished Infants." *American Journal of Diseases of Children* 30, no. 1 (July): 101–18.

Tolstoi, E. 1950. "The Free Diet for Diabetic Patients." *American Journal of Nursing* 50, no. 10 (Oct.): 652–54.

Tolstoi, E. 1943. "Newer Concepts in the Treatment of Diabetes Mellitus with Protamine Insulin." *American Journal of Digestive Diseases* 10, no. 7 (July): 247–53.

Tompkins, W. A. 1977. *Continuing Quest: Dr. William David Sansum's Crusade Against Diabetes*. Santa Barbara, CA: Sansum Medical Research Foundation.

Toor, M., A. Katchalsky, J. Agmon, and D. Allalouf. 1960. "Atherosclerosis and Related Factors in Immigrants to Israel." *Circulation* 22, no. 2 (Aug.): 265–79.

Toverud, K. U. 1932. "The Result of 8 Years' Insulin Treatment of Diabetes Mellitus in Children." *Acta Paediatrica* 12, no. 4 (Mar.): 193–202.

Trowell, H. 1981. "Hypertension, Obesity, Diabetes Mellitus and Coronary Heart Disease." In *Western Diseases: Their Emergence and Prevention*, ed. H. C. Trowell and D. P. Burkitt. London: Edward Arnold, 1–32.

Trowell, H. C. 1978. "The Development of the Concept of Dietary Fiber in Human Nutrition." *American Journal of Clinical Nutrition* 31, 10 suppl. (Oct.): S3–S11.

Trowell, H. C. 1975a. "Obesity in the Western World." *Plant Foods for Man* 1, nos. 3–4: 157–68.

Trowell, H. C. 1975b. "Dietary-Fiber Hypothesis of the Etiology of Diabetes Mellitus." *Diabetes* 24, no. 8 (Aug.): 762–65.

Trowell, H. C. 1973. "Dietary Fibre, Ischaemic Heart Disease and Diabetes Mellitus." *Proceedings of the Nutrition Society* 32, no. 3 (Dec.): 151–57.

Trowell, H. C. 1972. "Ischemic Heart Disease and Dietary Fiber." *American Journal of Clinical Nutrition* 25, no. 9 (Sept.): 926–32.

Trowell, H. C. 1971. Letter to Peter Cleave. July 28, 1971, Wellcome Library Collection. #95714761.

Trowell, H. C. 1960. *Non-Infective Disease in Africa*. London: Edward Arnold.

Trowell, H. C., and S. A. Singh. 1956. "A Case of Coronary Heart Disease in an African." *East African Medical Journal* 33 (Oct.): 391–94.

Turkington, R. W. 1981. "Diabetes." *JAMA* 246, no. 17 (Oct. 23–30): 1965.

Turner, R. C., C. A. Cull, V. Frighi, and R. R. Holman. 1999. "Glycemic Control with Diet, Sulfonylurea, Metformin, or Insulin in Patients with Type 2 Diabetes Mellitus. Progressive Requirement for Multiple Therapies (UKPDS 49)." *JAMA* 281, no. 21 (June 2): 2005–12.

U.K. Prospective Diabetes Study Group (UKPDS). 1998. "Intensive Blood-Glucose Control with Sulphonylureas or Insulin Compared with Conventional Treatment and Risk of Complications in Patients with Type 2 Diabetes (UKPDS 33)." *The Lancet* 352, no. 9131 (Sept. 12): 837–53.

Ulijaszek, S. J. 2017. *Models of Obesity, from Ecology to Complexity in Science and Policy*. Cambridge, U.K.: Cambridge University Press.

Ungar, H., A. Laufer, and Z. Ben-Ishay. 1963. "Atherosclerosis and Myocardial Infarction in Various Jewish Groups in Israel." In *The Genetics of Migrant and Isolate Populations*, ed. E. Goldschmidt. New York: Williams & Wilkins, 120–26.

Unger, R. H. 1982. "Meticulous Control of Diabetes: Benefits, Risks, and Precautions." *Diabetes* 31, no. 6, pt. 1 (June): 479–83.

Unger, R. H. 1976. "The Banting Memorial Lecture 1975. Diabetes and the Alpha Cell." *Diabetes* 25, no. 2 (Feb.): 136–51.

Unger, R. H. 1966. "Glucoregulatory Hormones in Health and Disease. A Teleologic Model." *Diabetes* 15, no. 7 (July): 500–506.

Unger, R. H., A. M. Eisentraut, M. S. McCall, et al. 1959. "Glucagon Antibodies and Their Use for Immunoassay for Glucagon." *Proceedings of the Society of Experimental Biology and Medicine* 102, no. 3 (Dec.): 621–23.

Unwin, D., A. A. Khalid, J. Unwin, et al. 2020. "Insights from a General Practice Service Evaluation Supporting a Lower Carbohydrate Diet in Patients with Type 2 Diabetes Mellitus and Prediabetes: A Secondary Analysis of Routine Clinic Data Including HbA1c, Weight and Prescribing over 6 Years." *British Medical Journal Nutrition, Prevention and Health* 3, no. 2 (Nov.): 285–94.

U.S. Department of Agriculture (USDA) and U.S. Department of Health, Education, and Welfare (HEW). 1985. "Nutrition and Your Health: Dietary Guidelines for Americans." 2nd edition. *Home and Garden Bulletin* no. 232. Washington, DC.

U.S. Department of Agriculture (USDA) and U.S. Department of Health, Education, and Welfare (HEW). 1980. "Nutrition and Your Health: Dietary Guidelines for Americans." *Home and Garden Bulletin,* no. 228.

U.S. Department of Health, Education, and Welfare (HEW). 1988. *The Surgeon General's Report on Nutrition and Health,* no. 88-50210. Washington, DC: U.S. Government Printing Office.

U.S. Department of Health, Education, and Welfare (HEW). 1979. *Healthy People: The Surgeon General's Report on Health Promotion and Disease Prevention.* U.S. Dept of Health, Education, and Welfare pub. no. (NIH) 79-55071. Washington, DC: U.S. Government Printing Office.

U.S. Department of Health, Education, and Welfare (HEW). 1971. *Arteriosclerosis: A Report by the National Heart and Lung Institute Task Force on Arteriosclerosis.* 2 vols. U.S. Dept of Health, Education, and Welfare pub. no. (NIH) 72-137 and 72-219. Washington, DC: National Institutes of Health.

Valenstein, E. 1986. *Great and Desperate Cures: The Rise and Decline of Psychosurgery and Other Radical Treatments for Mental Illness.* New York: Basic Books.

Van Gaal, L. F. 1998. "Dietary Treatment of Obesity." In *Handbook of Obesity,* ed. G. A. Bray, C. Bouchard, and W. P. T. James. New York: Marcel Dekker, 875–90.

Van Itallie, T. B., and T. H. Nufert. 2003. "Ketones: Metabolism's Ugly Duckling." *Nutrition Reviews* 61, no. 10 (Oct.): 327–41.

Vernon, M. C., J. Mavropoulos, M. Transue, W. S. Yancy, Jr., and E. C. Westman. 2003. "Clinical Experience of a Carbohydrate-Restricted Diet: Effect on Diabetes Mellitus." *Metabolic Syndrome and Related Disorders* 1, no. 3 (Sept.): 233–37.

Veterans Health Administration (VHA). 2011. "Close to 25 Percent of VA Patients Have Diabetes." Online: http://www.va.gov/health/NewsFeatures/20111115a.asp.

Virchow, R. 1847. "Ueber die Standpunkte in der Wissenschftlichen Medicin." *Archiv für Pathologische Anatomie und Physiologie und für Klinische Medicin* 1: 1–19.

Virta Health. 2022. "Virta Health Highlights Lasting, Transformative Health Improvements in 5-Year Diabetes Reversal Study." (June 5). Online: https://www.virtahealth.com/blog/virta-sustainable-health-improvements-5-year-diabetes-reversal-study.

Visscher, M. B. 1965. "Frank Charles Mann, 1887–1962, a Biographical Memoir." Washington, DC: National Academy of Sciences.

Volek, J. S., A. L. Gómez, and W. J. Kraemer. 2000. "Fasting Lipoprotein and Postprandial Triacylglycerol Responses to a Low-Carbohydrate Diet Supplemented with N-3 Fatty Acids." *Journal of the American College of Nutrition* 19, no. 3 (June): 383–91.

Volek, J. S., A. L. Gómez, D. M. Love, et al. 2001. "Effects of a High-Fat Diet on Postabsorptive and Postprandial Testosterone Responses to a Fat-Rich Meal." *Metabolism* 50, no. 11 (Nov.): 1351–55.

Volek, J. S., M. J. Sharman, D. M. Love, et al. 2002. "Body Composition and Hormonal Responses to a Carbohydrate-Restricted Diet." *Metabolism* 51, no. 7 (July): 864–70.

Von Elm, E., D. G. Altman, M. Egger, et al. 2007. "The Strengthening the Reporting of Observational Studies in Epidemiology (STROBE) Statement: Guidelines for Reporting Observational Studies." *Journal of Clinical Epidemiology* 61, no. 4 (Apr.): 344–49.

von Noorden, C. 1912. *New Aspects of Diabetes, Pathology and Treatment.* New York: E. B. Treat.

von Noorden, C. 1905. *Clinical Treatises on the Pathology and Therapy of Disorders of Metabolism and Nutrition, Part VII, Diabetes Mellitus*. Trans. F. Buchanon and I. W. Hall. New York: E. B. Treat.

Wagener, H. P., T. J. Story Dry, and R. M. Wilder. 1934. "Retinitis in Diabetes." *The New England Journal of Medicine* 211, no. 25 (Dec. 20): 1131–37.

Walford, S., E. A. Gale, S. P. Allison, and R. B. Tattersall. 1978. "Self-Monitoring of Blood-Glucose. Improvement of Diabetic Control." *The Lancet* 311, no. 8067 (Apr. 8): 732–35.

Wallis, C. 1984. "Hold the Eggs and Butter." *Time* (Mar. 26): 56–63.

Wascovich, P. 2020. "In Memoriam: Dr. Roger H. Unger, Visionary Endocrinologist and Preeminent Authority on Diabetes." *Center Times Plus* (Sept. 1). Online: https://www.utsouthwestern.edu/ctplus/stories/2020/unger-obit.html.

Waymouth Reid, E. 1896. "A Method for the Estimation of Sugar in Blood." *Journal of Physiology* 20, nos. 4–5 (Oct.): 316–21.

Wennberg, J. E. 1984. "Dealing with Medical Practice Variations: A Proposal for Action." *Health Affairs* 3, no. 2 (Summer): 6–32.

Wertheimer, E. 1965. "Introduction—A Perspective." In *Handbook of Physiology, Section 5: Adipose Tissue*, ed. A. E. Renold and G. F. Cahill. Washington, DC: American Physiology Society, 5–11.

Wertheimer, E., and B. Shapiro. 1948. "The Physiology of Adipose Tissue." *Physiological Reviews* 28, no. 4 (Oct.): 451–64.

Werther, G. A., P. A. Jenkins, R. C. Turner, and J. D. Baum. 1980. "Twenty-Four-Hour Metabolic Profiles in Diabetic Children Receiving Insulin Injections Once or Twice Daily." *British Medical Journal* 2, no. 6237 (Aug. 9): 414–18.

West, F. H., E. D. Bond, J. T. Shurley, and C. D. Meyers. 1955. "Insulin Coma Therapy in Schizophrenia, a Fourteen-Year Follow-Up Study." *American Journal of Psychiatry* 111, no. 8 (Feb.): 583–89.

West, K. M. 1978. *Epidemiology of Diabetes and Its Vascular Lesions*. New York: Elsevier.

West, K. M. 1974. "Diabetes in American Indians and Other Native Populations of the New World." *Diabetes* 23, no. 10 (Oct.): 841–55.

West, K. M. 1973. "Diet Therapy of Diabetes: An Analysis of Failure." *Annals of Internal Medicine* 79, no. 3 (Sept.): 425–34.

Westman, E. C., and M. C. Vernon. 2008. "Has Carbohydrate-Restriction Been Forgotten as a Treatment for Diabetes Mellitus? A Perspective on the ACCORD Study Design." *Nutrition and Metabolism* 5 (Apr.): 10.

White, P. 1971. "Pregnancy in Diabetes." In *Joslin's Diabetes Mellitus*, 11th edition, ed. A. P. Marble, R. F. Bradley, and L. P. Krall. Philadelphia: Lea & Febiger, 581–98.

White, P., and E. P. Joslin. 1959. "The Etiology and Prevention of Diabetes." In *The Treatment of Diabetes Mellitus*, 10th edition, ed. E. P. Joslin, H. F. Root, P. White, and A. Marble. Philadelphia: Lea & Febiger, 47–98.

White, P. L. 1962. "Calories Don't Count." *JAMA* 179, no. 10 (Mar. 10): 828.

Whitehead, A. N. 1938. *Modes of Thought*. New York: The Free Press.

Wilder, R. M. 1958. "Recollections and Reflections on Education, Diabetes, Other Metabolic Diseases, and Nutrition in the Mayo Clinic and Associated Hospitals, 1919–1950." *Perspectives in Biology and Medicine* 1, no. 3 (Spring): 237–77.

Wilder, R. M. 1946. "Twenty-Five Years of the Insulin Era." *Proceedings of the American Diabetes Association: Annual Meeting* 6: 109–16.

Wilder, R. M. 1940. *Clinical Diabetes Mellitus and Hyperinsulinism*. Philadelphia: W. B. Saunders.

Wilder, R. M. 1924. "'Optimal' Diets for Diabetic Patients." *The Journal of the American Medical Association* 83, no. 10 (Sept. 6): 733–37.

Wilder, R. M., and D. L. Wilbur. 1938. "Diseases of Metabolism and Nutrition: Review of Certain Recent Contributions." *Archives of Internal Medicine* 61, no. 2 (Feb.): 297–365.

Wilder, R. M., and D. L. Wilbur. 1936. "Diseases of Metabolism and Nutrition: Review of Certain Recent Contributions." *Archives of Internal Medicine* 57, no. 2 (Feb.): 422–71.

Wilder, R. M., and D. L. Wilbur. 1935. "Diseases of Metabolism and Nutrition: Review of Certain Recent Contributions." *Archives of Internal Medicine* 54, no. 2 (Feb.): 304–43.

Wilding, J. P. H., R. L. Batterham, M. Davies, et al. 2022. "Weight Regain and Cardiometabolic Effects After Withdrawal of Semaglutide: The STEP1 Trial Extension." *Diabetes, Obesity & Metabolism* 24, no. 8 (Aug.): 1553-64.

Williams, J. R. 1922. "A Clinical Study of the Effects of Insulin on Severe Diabetes." *Journal of Metabolic Research* 2, nos. 5–6 (Nov.–Dec.): 729–51.

Williams, J. R. 1921. "An Evaluation of the Allen Method of Treatment of Diabetes Mellitus." *The American Journal of the Medical Sciences* 162, no. 1 (July): 62–72.

Williams, R. H., ed. 1968. *Textbook of Endocrinology*. 4th edition. Philadelphia: W. B. Saunders.

Wilson, D. E., and R. S. Lees. 1972. "Reciprocal Changes in the Concentrations of Very Low and Low Density Lipoproteins in Man." *The Journal of Clinical Investigation* 51, no. 5 (May): 1051–57.

Wilson, G. W. 1963. "Overweight and Underweight: The Psychosomatic Aspects." In *Your Weight and How to Control It*, ed. M. Fishbein. Garden City, NY: Doubleday, 133–26.

Wilson, N. L., ed. 1969. *Obesity*. Philadelphia: F. A. Davis.

Wing, R. R., D. Reboussin, and C. E. Lewis for the Look Ahead Research Group. 2013. "Intensive Lifestyle Intervention in Type 2 Diabetes." *The New England Journal of Medicine* 369, no. 24 (Dec. 12): 2358–59.

Winternitz, W. W. 1976. "Control of Blood Glucose in Diabetes." *The New England Journal of Medicine* 295, no. 9 (Aug. 26): 509–10.

Woloson, W. A. 2002. *Refined Tastes: Sugar, Confectionary, and Consumers in Nineteenth-Century America*. Baltimore: Johns Hopkins University Press.

Wood, F. C., and E. L. Bierman. 1972. "New Concepts in Diabetic Dietetics." *Nutrition Today* 7, no. 3 (May–June): 4–12.

Woodyatt, R. T. 1934. "Round Table Conference on Diabetes Mellitus: Dietary Trends." *Bulletin of the New York Academy of Medicine* 10, no. 6 (June): 335–46.

Woodyatt, R. T. 1921a. "Objects and Method of Diet Adjustment in Diabetes." *Transactions of the Association of American Physicians* 36: 269–92.

Woodyatt, R. T. 1921b. "Objects and Method of Diet Adjustment in Diabetes." *Archives of Internal Medicine* 28, no. 2 (Aug.): 125–41.

Woodyatt, R. T. 1916. "Acidosis in Diabetes." *The Journal of the American Medical Association* 66, no. 25 (June 17): 1910–13.

Woodyatt, R. T. 1909. "Prepared Foods and Diabetic Articles." *Illinois Medical Journal* 16, no. 6 (Dec.): 666–74.

World Health Organization (WHO). 2021. "Diabetes." Online: https://www.who.int/news-room/fact-sheets/detail/diabetes.

World Health Organization (WHO). 2006. "The World Health Organization Notes the Women's Health Initiative Diet Modification Trial, but Reaffirms That the Fat Content of Your Diet Does Matter." Online: http://web.archive.org/web/20160328120906/www.who.int/nmh/media/Response_statement_16_feb_06F.pdf.

World Health Organization (WHO). 1999. "Definition, diagnosis and classification of diabetes mellitus and its complications: Report of a WHO Consultation. Part 1: Diagnosis and classification of diabetes mellitus." Geneva: World Health Organization. Online: http:// whqlibdot.who.int/hq/1999/WHO_NCD_NCS_99.2.pdf.

Wortis, J. 1936. "On the Response of Schizophrenic Subjects to Hypoglycemic Insulin Shock." *Journal of Nervous and Mental Disease* 85, no. 5 (Nov.): 497–506.

Wrenshall, G. A., G. Hetenyi, and W. R. Feasby. 1962. *The Story of Insulin. Forty Years of Success Against Diabetes.* Bloomington: Indiana University Press.

Yalow, R. S., and S. A. Berson. 1960a. "Comparison of Plasma Insulin Levels Following Administration of Tolbutamide and Glucose." *Diabetes* 9 (Sept.–Oct.): 355–62.

Yalow, R. S., and S. A. Berson. 1960b. "Plasma Insulin Concentrations in Nondiabetic and Early Diabetic Subjects. Determinations by a New Sensitive Immuno-Assay Technic." *Diabetes* 9 (July–Aug.): 254–60.

Yalow, R. S., and S. A. Berson. 1960c. "Immunoassay of Endogenous Plasma Insulin in Man." *The Journal of Clinical Investigation* 39, no. 7 (July): 1157–75.

Yalow, R. S., S. M. Glick, J. Roth, and S. A. Berson. 1965. "Plasma Insulin and Growth Hormone Levels in Obesity and Diabetes." *Annals of the New York Academy of Sciences* 131, no. 1 (Oct. 8): 357–73.

Yancy, W. S., Jr., M. Foy, A. M. Chalecki, M. C. Vernon, and E. C. Westman. 2005. "A Low-Carbohydrate, Ketogenic Diet to Treat Type 2 Diabetes." *Nutrition and Metabolism* 1, no. 2 (Dec.): 34.

Yancy, W. S., Jr., M. K. Olsen, J. R. Guyton, R. P. Bakst, and E. C. Westman. 2004. "A Low-Carbohydrate, Ketogenic Diet Versus a Low-Fat Diet to Treat Obesity and Hyperlipidemia." *Annals of Internal Medicine* 140, no. 10 (May 18): 769–77.

Yancy, W. S., Jr., M. C. Vernon, and E. C. Westman. 2003. "A Pilot Trial of a Low-Carbohydrate, Ketogenic Diet in Patients with Type 2 Diabetes." *Metabolic Syndrome and Related Disorders* 1, no. 3 (Sept.): 239–43.

Yao, Y., T. Suo, R. Andersson, et al. 2017. "Dietary Fiber for the Prevention of Recurrent Colorectal Adenomas and Carcinomas." *Cochrane Database of Systematic Reviews* 1, no. 1 (Jan. 8): CD003430.

Yerushalmy, J. 1966. "On Inferring Causality from Observed Associations." In *Controversy in Internal Medicine*, ed. F. J. Ingelfinger, A. S. Relman, and M. Finland. Philadelphia: W. B. Saunders, 659–68.

Yerushalmy, J., and H. E. Hilleboe. 1957. "Fat in the Diet and Mortality from Heart Disease: A Methodologic Note." *New York State Journal of Medicine* 57, no. 14 (July 15): 2343–54.

Young, C. M. 1976. "Dietary Treatment of Obesity." In *Obesity in Perspective*, ed. G. A. Bray. DHEW pub. no. (NIH) 76-852, Washington, DC: U.S. Government Printing Office, 361–66.

Young, F. G. 1937. "Permanent Experimental Diabetes Produced by Pituitary (Anterior Lobe) Injections." *The Lancet* 230, no. 5946 (Aug. 14): 372–74.

Yudkin, J. 1988. "Sucrose, Coronary Heart Disease, Diabetes, and Obesity: Do Hormones Provide a Link?" *American Heart Journal* 115, no. 2 (Feb.): 493–98.

Yudkin, J. 1972. *Pure, White, and Deadly.* London: Davis-Poynter.

Yudkin, J. 1964. "Dietary Fat and Dietary Sugar in Relation to Ischaemic Heart-Disease and Diabetes." *The Lancet* 284, no. 7349 (July 4): 4–5.

Yudkin, J. 1963. "Nutrition and Palatability with Special Reference to Obesity, Myocardial Infarction, and Other Diseases of Civilization." *The Lancet* 281, no. 7295 (June 22): 1335–38.

Yudkin, J. 1957. "Diet and Coronary Thrombosis Hypothesis and Fact." *The Lancet* 273, no. 6987 (July 27): 155–62.

Zinman, B., J. S. Skyler, M. C. Riddle, and E. Ferrannini. 2017. "Diabetes Research and Care Through the Ages." *Diabetes Care* 40, no. 10 (Oct.): 1302–13.

Index

clinical trials on glycemic control of
2000s and, 347–51
continuous glucose monitors and,
395–96
control through diet, vs. insulin and
drug therapy, 5–9
cost of medical care for, 11, 389
curative vs. preventive medicine and,
393–94
DCCT study and, 306–16, 329, 347
defining, and blood glucose vs. insulin
levels, 59–60, 343
diet and, not truly understood, 8–9
dietary recommendations, and
discovery of insulin, 5–6
discovery of insulin and, 47, 49–50,
56–57
discovery of insulin and switch to high-
carbohydrate diet, 40–43, 88–91
discovery of pancreas's role in, 46–49
drug therapy vs. dietary restriction and,
402–4
early diagnosis, and urinary sugar, 4
early history of, 3–5, 27–41, 43, 45–49,
56, 62
endocrinology studies and, 50–55
errors in clinical judgment in research
on, 22–23
evidence-based medicine and, 390–94
failure of conventional dietary
guidelines, 14–15, 18
fat/cholesterol hypothesis disease and,
171, 173–206, 248–60, 266–67
fear of dietary fat and, 61–63
fear of ketoacidosis from high-fat diet
and, 76–77, 83, 87
first patients treated with insulin,
92–101
fundamental assumptions being
tested, 13
glucagon as counterbalance to insulin,
323–29
glycated proteins, or AGEs and, 272–74
heart disease and, 165–71, 205–6,
248–50
heart disease, carbohydrates, and
triglycerides and, 260–62

hemoglobin A1C, 269–71, 284–85
high blood insulin levels in type 2, 344
high blood sugar, high insulin levels,
and heart disease and, 268–79
homeostasis and wholistic thinking
and, 52–55, 317–23, 342–43
hypoglycemia avoided by carbohydrates
in diet, 98–103
insulin and diet and, 101–3
insulin and weight gain and, 128–35,
138–64, 344
insulin-delivery system advances and,
305–6
insulin resistance and, 57, 155–56, 250,
263–65, 275–76, 318
insulin sensitivity and, 155–56, 310–11
insulin shock, avoiding with
carbohydrates, 97–98
insulin's metabolic responses and, 56–57
insulin's role in fuel partitioning and,
57–59
insulin therapy, and carbohydrates in
diet and, 88–91
insulin therapy, and hypoglycemia,
314–18, 343–44
insulin therapy vs. healthy pancreatic
secretions, 276–77
Joslin and fear of dietary fat, 62–69, 72,
77–78, 83, 88–91, 180–81
lipoproteins (HDL, LDL) and, 254–58
liver and, 45–46, 56, 319–22, 325–29
low-carbohydrate, high-fat diet, and
glycemic control, 250–51, 287–88,
302, 329–41, 379–87, 396–402, 404
low-carbohydrate, high-fat diet for,
4–5, 62
low-carbohydrate diet, and weight loss
and, 357–79
low-carbohydrate diet trials and, 353–57
low-carbohydrate diet for type 1,
379–87, 398–402, 404
measurement of insulin in blood and,
156–57
medical practice vs. medical science
and, 19–22, 389–90, 402
metabolic syndrome and, 265–69, 287,
302